Human-Machine Interaction and IoT Applications for a Smarter World

Human-Machine Interaction and IoT Applications for a Smarter World

Edited by
Nishu Gupta
Srinivasa Kiran Gottapu
Rakesh Nayak
Anil Kumar Gupta
Mohammad Derawi
Jayden Khakurel

CRC Press
Taylor & Francis Group
Boca Raton London New York

CRC Press is an imprint of the
Taylor & Francis Group, an **informa** business

First edition published 2023
by CRC Press
6000 Broken Sound Parkway NW, Suite 300, Boca Raton, FL 33487-2742

and by CRC Press
4 Park Square, Milton Park, Abingdon, Oxon, OX14 4RN

CRC Press is an imprint of Taylor & Francis Group, LLC

ISBN: 9781032215228 (hbk)
ISBN: 9781032215235 (pbk)
ISBN: 9781003268796 (ebk)

DOI: 10.1201/9781003268796

Typeset in Times
by KnowledgeWorks Global Ltd.

Dr. Nishu Gupta dedicates this book to his

Family & Friends

Dr. Srinivasa Kiran Gottapu dedicates this book to his

**Parents (Jagannadham Gottapu & Krishna Veni Gottapu)
and Wife (Charitha Gottapu)**

Prof. Rakesh Nayak dedicates this book to his

Late Brother Mr. Rajesh Nayak

Dr. Anil Kumar Gupta dedicates this book to his

Parents, Family & Friends

Prof. Mohammad Derawi dedicates this book to his

Parents

Dr. Jayden Khakurel dedicates this book to his

Family

Contents

PART I   Introduction to Human-Machine Interaction (HMI) and Internet of Things (IoT)

PART II   HMI in IoT-Based Distributed Commercial Systems

PART III   Machine Interaction and Applications in the Healthcare Industry

PART IV   HMI and IoT in Cloud Setup

PART V   Sustainable Approaches Towards
Artificial Intelligence

Foreword

I am pleased to write this foreword because I feel this book, *Human-Machine Interaction and IoT Applications for a Smarter World*, deeply emphasizes the state-of-the-art technologies that comprise many research explorations in the field of artificial intelligence (AI) and that the book offers much valuable information about Internet of Things (IoT) applications through networking and automation.

Although the focus of this work is on human-machine interaction (HMI), it contains much more that will be of interest to those outside this field as well. In this book, I see that the editors and authors have conceptualized the intellectual foundations of embedded systems, elaborated its distinctive pedagogy in IoT applications, and studied its patterns and impact on the common man. This book will certainly help researchers and professional practitioners to develop a shared vision and understanding of interpretive discussion.

This book is a good initiative in a direction that addresses numerous issues related to IoT applications. I believe this book will provide a solid platform to various realms of the existing and upcoming technologies, especially in the field of AI, HMI, autonomous vehicles, and intelligent transportation. The authors can be confident that there will be many grateful readers who will have gained a broader perspective of the disciplines of machine interaction and its applications as a result of their efforts. I hope that this book will serve as a primer for industry and academia, professional developers, and upcoming researchers across the globe to learn, innovate, and realize the multifold capabilities of AI and IoT applications.

I wish good luck to the editors and contributors of this book.

Dr. V. Anil Kumar
Senior Principal Scientist
CSIR Fourth Paradigm Institute (CSIR-4PI),
Bengaluru, India

Foreword

I am delighted to write the foreword for this edited book on *Human-Machine Interaction and IoT Applications for a Smarter World*. This book highlights the importance that Internet of Things (IoT) and artificial intelligence (AI) technology can offer in taking care of various man-machine interaction-based applications through automation. It intends to demonstrate to its readers useful applications and architectures that cater to diversified technological requirements.

This book provides a window to the research and development in the field of man-machine interaction in a comprehensive way and explains the evolutions of contributing tools and techniques. The range of topics covered in this book is quite extensive, and every topic is discussed by experts in their own field. The advances and challenges are discussed with a focus on successes, failures and lessons learned, open issues, unmet challenges, and future directions.

I am confident that this book will provide an effective learning experience and a practical reference for researchers, professionals, and students who are interested in the integration of AI in IoT-based embedded technologies and its advances to the engineering field. The authors can be assured that because of their contributions, there will be many readers who will gain a better understanding of AI and its applications in various industrial sectors. I highly recommend this book to a variety of audiences, including academicians, industrial engineers, and researchers in the fields that use transportation systems, vehicular communication, Internet of Things, as well as to communication technology specialists and ad hoc communication networks students and scholars.

Dr. Ahmad Hoirul Basori
Professor
Department of Information Technology
Faculty of Computing and Information Technology in Rabigh
King Abdulaziz University
Rabigh, Saudi Arabia

Preface

Human-Machine Interaction (HMI) along with the Internet of Things (IoT) integrate the futuristic trends at the cutting edge of study and research by featuring enormous applications in a proficient, adaptable, and manageable way. This book covers the primary mainstays of the IoT world by thoroughly describing the present advancements, systems, and structures.

Written by international experts, this book intends to present its readers with day-to-day and upcoming trends concerning smart applications, interfaces, and paradigms by machine interaction and their subsequent prototypes. The primary purpose of this book is to make its readers understand in a very simple way the outcome of innovation in the field of interactive technologies. Readers will appreciate the diversified field of HMI, its interfaces, and their vast applications being catered to by the IoT. Such a book would create an awareness among general audiences about the fast-changing technological world, share findings with its stakeholders and prepare future researchers for making everything smart and easy. The methodology adopted in this book is to start with the basics and steadily build readers' confidence with the material. After reading this book, readers will have an elementary to intermediate level idea of the related technologies, their functioning, and their challenges. This would set a perfect platform for them to work upon the existing limitations and make their own contributions to the world of innovation and scientific research.

The book is a perfect blend of text as well as reference and suits almost all levels of technical education and research, as well as novices interested in an overview of applications of Cyber-Physical Systems, HMI, IoT, Industrial IoT technologies for the coming times, towards making almost everything smart, intelligent, and self-adaptive. The book is divided into different parts with multiple chapters. The contents of the book have been organized in a reader friendly manner. The book has a wide audience including university professors, graduate and PhD scholars, industry professionals, and researchers, particularly in the fields of computer communication, wireless communication, cyber-physical systems, machine learning, and sensor networks. The later chapters are committed to the use of a considerable number of ideas and principles in real case studies, so as to give helpful experiences that can be utilized as structure models and rules. It likewise introduces the meaning of classes of HMI-enabled IoT devices as hardware and software, and a test investigation identified with the creation and arrangement of an IoT test bed with heterogeneous gadgets.

This book contains contributions from all over the world, and we would like to thank all the authors for submitting their works. We extend our appreciation to the reviewers for their timely and focused review comments. We also gratefully acknowledge all the authors and publishers of the books quoted in the references.

Dr. Nishu Gupta
(NTNU, Gjøvik, Norway)

Dr. Srinivasa Kiran Gottapu
(University of North Texas, USA)

Prof. Rakesh Nayak
(O. P. Jindal University, Raigarh, India)

Dr. Anil Kumar Gupta
(C-DAC, Pune, India)

Prof. Mohammad Derawi
(NTNU, Gjøvik, Norway)

Dr. Jayden Khakurel
(University of Turku, Finland)

Acknowledgments

Dr. Nishu Gupta

I acknowledge the inspiration and blessings of my mother Smt. Rita Rani Gupta, father Prof. K. M. Gupta, and other family members. I am full of gratitude to my wife Smt. Anamika Gupta and son Master Ayaansh Gupta for the patience shown and encouragement given to complete this venture. I deeply acknowledge the blessings of my academic advisor and mentor Prof. Rajeev Tripathi. I am highly grateful to my PhD supervisor Assoc. Prof. Arun Prakash whose guidance has always encouraged me to do my best. He has been a major driving force towards this and many other such accomplishments. I wholeheartedly acknowledge the motivation given to me by friends Dr. Krishan Kumar (Assistant Professor, NIT Hamirpur, Himachal Pradesh, India), Ms. Isha Bharti (Capgemini Inc., USA), Mr. Gaurvendra Singh (Kotak Mahindra Bank, India), and Er. Jalaj Kumar Singh (KPMG Global Services), and other colleagues and friends for their support and motivation in several ways.

Dr. Srinivasa Kiran Gottapu

I strongly acknowledge the support and encouragement from Dr. Nishu Gupta in shaping this book. I would like to take this opportunity to thank my family members: father (Mr. Jagannadham Gottapu), mother (Smt. Krishna veni Gottapu), wife (Smt. Charitha Gottapu), brothers (Er. Siva Teja Gottapu and Er. Krishna Chaitanya Kota), sister (Keerthana Reddy Challa), and other family members for their love and patience.

Prof. Rakesh Nayak

I acknowledge the support and patience of my parents Mrs. Jyotsnamayee and Mr. Manabhanjan Nayak, my wife Ms. Tusarika Nayak, and my daughters Miss Srusti Nayak and Miss Nakshatra Nayak. I deeply acknowledge the blessings of my PhD supervisors Dr. C.V. Sastry and Prof. Jayaram Pradhan.

I acknowledge the blessings my late brother Mr. Rajesh Nayak, who always encouraged me to achieve greater heights in life. I miss you Dada.

Dr. Anil Kumar Gupta

I acknowledge the inspiration and blessings of my mother Late. Smt. Krishna Gupta, father Late. Shri Ram Kumar Gupta, and other family members. I am full of gratitude to my wife Smt. Preeti Aggarwal and son Akshat Gupta. I deeply acknowledge the help and support of Dr. Nishu Gupta, without whose support this would have never been possible.

Last but not the least, **we, the Editors** express our heartfelt gratitude to the Publisher and the team behind it for their continued support and cooperation in publishing this book.

Biography of Dr. Nishu Gupta

Dr. Nishu Gupta is a Senior Member, IEEE. He is a Postdoctoral Fellow in the Department of Electronic Systems, Faculty of Information Technology and Electrical Engineering, Norwegian University of Science and Technology (NTNU) in Gjøvik, Norway. He is a Research Fellow and Visiting Researcher in the Department of Informatics, University of Oviedo, Gijón, Spain under the research group on Systems for Multimedia and the Internet of Things (SMIOT). He is a Member of the Zero Trust Architecture working group of MeitY-C-DAC-STQC project under 'e-Governance Standards and Guidelines', Ministry of Electronics and Information Technology (MeitY), Government of India.

Dr. Nishu Gupta received his Ph.D. degree from the Department of Electronics and Communication Engineering, MNNIT Allahabad, Prayagraj, India which is an Institute of National Importance as declared by the Government of India. He is specialized in the field of Computer Communication and Networking. His major work is in the area of IoT-based enhanced safety applications in vehicular communication.

Dr. Gupta is the recipient of the Best Paper Presentation Award at the 4th International Conference on Computer and Communication Systems held at Nanyang Technological University, Singapore in 2019. He has published 5 patents and more than 42 research articles in IEEE Transactions, SCI and Scopus Indexed Journals. Dr. Gupta has supervised numerous theses at the Master's level and projects at the Bachelor level in his main line of work. He has authored and edited several books with international publishers like Taylor & Francis, Springer, Wiley, and Scrivener. Dr. Nishu is on the Editorial board of various internationally reputed journals and transactions. He serves as reviewer of more than twenty SCI indexed journals and transactions, IEEE Transactions on ITS, IEEE Access, IET Communications, etc. He was awarded twice for *Outstanding Contribution in Reviewing* by Elsevier.

Dr. Gupta has chaired several international conferences and played a key role in successfully organizing various international events. He was Conference Chair at the Second EAI/Springer International Conference on Cognitive Computing and Cyber Physical Systems (IC⁴S 2021), and Organizing Chair at the Second IOP International Conference on Computational Intelligence and Energy Advancements (ICCIEA-2021). He is an academic research collaborator with top academicians and researchers across the globe. He has served as Head of the ECE department; Chief Coordinator of Institute Innovation Cell, under MHRD-IIC, Govt. of India at his previous organization, besides holding many other key positions in the academic, administration, and research fields.

Before joining his current assignment, Dr. Gupta served as a faculty member in the Electronics and Communication Engineering Department, College of Engineering and Technology, SRM Institute of Science and Technology, Kattankulathur, Chennai, India. His research interests include Autonomous Vehicles, Edge Computing, Smart Wireless Systems, Internet of Things, Internet of Vehicles, Deep Learning, Ad Hoc Networks, Vehicular Communication, Driving Efficiency, Cognitive Computing, Human-Machine Interaction, Traffic Pattern Prediction, etc.

Biography of Dr. Srinivasa Kiran Gottapu

Dr. Srinivasa Kiran Gottapu received his MS degree in Electrical Engineering from the University of Wyoming, Laramie, WY, USA, in 2016. He received his PhD degree in Electrical Engineering from the University of North Texas, Denton, the United States, in 2019. He is a senior member of IEEE. Presently, he is an adjunct faculty member in the Department of Electrical Engineering, University of North Texas, USA. He is working as a Senior Cloud Engineer at HCL Technologies. His research interests include wireless communications and networks, cognitive radio networks, digital signal and image processing, information transmission and retrieval, and coding and information theory.

Biography of Prof. Rakesh Nayak

Prof. Rakesh Nayak is an author of two textbooks, is presently working as a Professor in the Department of Computer Science and Engineering at O. P. Jindal University, Raigarh, Chhatisgarh. He received his Master's degree in Computer Applications from Indira Gandhi National Open University in the year 2007 and MTech. (CSE) from Acharya Nagarjuna University in 2010. He received his PhD degree in Computer Science from Behrampur University in 2013. Prior to joining the computer science department in 2022, he has worked in various capacities in different engineering/MCA colleges. He has more than 21 years of teaching experience and has guided 11 MTech students. He has many publications in international journals to his credit.

Biography of Dr. Anil Gupta

Dr. Anil Gupta is a Senior Member, IEEE and Senior Member ACM. He has more than 24 years of industry experience. He received his Master's degree from IIT Roorkee, India, which is an Institute of National Importance as declared by the Govt. of India. Dr. Anil is working with CDAC Pune as Associate Director. His research interests are in HPC, SDN, IoT, 5G, data analytics NLP, computer vision, system software blockchains, and cybersecurity. He has more than 70 publications in various national/International conferences and journals. He has filed five patents in India, and multiple patents are in various stages of evaluation. He has guided more than 15 Master's dissertations. He has also guided more than 30 B Tech students for their projects.

Biography of Prof. Mohammad Derawi

Prof. Derawi is currently having a dual-career acting as the youngest professor in Norway and an extremely true innovator. He received his diplomas in Computer Science engineering from the Technical University of Denmark (DTU, Denmark) where he received both BSc (2007) and MSc (2009) degrees. In addition, he received the title as the youngest engineer of Denmark in 2009. Derawi has pursued his PhD in information security at the Norwegian Information Security Laboratory (NISLab), Gjøvik University (now NTNU, Norway). In the beginning of his PhD studies, he was a visiting researcher at the Center for Advanced Security Research Darmstadt (CASED), Germany. His PhD research interests included smart mobile technologies and also biometrics with specialization in behavioral biometric recognition in mobile devices. Derawi has been active in several European and national projects as well since 2009. Today he holds a professorship within electrical engineering and is specialized within information security, e-health, autonomous systems, biometric systems, wireless communications, IoT, digital fundamentals, and micro-controllers. He is also the head of the Smart Wireless Systems Research group at the Department of Electronic Systems (NTNU, Norway).

Prof. Derawi has next to his academic career also been working with truly innovative projects, which is also a part of his additional skills. He is a person who identifies the need within an industry, market segment or culture and spots opportunity in it. More importantly, he has the ability to identify needs that can be implemented in the market, to develop and refine solutions, take chances, push the envelope and create meaning.

Biography of Dr. Jayden Khakurel

Dr. Jayden Khakurel (DSc (tech), MBA) has ten years of experience in academia and industry. He has been working with numerous European Union (EU)-funded research projects in collaboration with leading research institutes across Europe, which emphasize payment, mobility, and healthcare. Moreover, he has been teaching software and innovation undergraduate and graduate-level courses in Finland and the United States.

He currently works with a conversational Artificial Intelligence (AI) using various presentation modalities (text, verbal, embodied with 3D avatar) to support child mental health across Finland as part of the Invest project (funded by the Academy of Finland). He has extensive expertise in user experience research for emerging healthcare technologies, and his current research can contribute to Human-Computer Interaction (HCI), public health, and child mental healthcare.

Part I

Introduction to Human-Machine Interaction (HMI) and Internet of Things (IoT)

1 Evolution of the Internet of Things and Fundamentals of Human-Machine Interaction

Anirban Chakraborty
University of North Texas,
Denton, TX, USA

CONTENTS

1.1 INTRODUCTION

The term "Internet of Things (IoT)" was first coined in the year 1999 by Kevin Ashton, cofounder and executive director of the Auto-ID Center at Massachusetts Institute of Technology. Ten years later, in a blog post [1], he explained his own rationale behind the phrase and his own visions of how it should be defined. According to him, the surrounding computers must know everything that is there to know about all the "Things" that are around us and use that knowledge to improve our lives. The roots of his visions date back to the invention of internet. Internet first came into existence in the 1960s

DOI: 10.1201/9781003268796-2

3

when the US Department of Defense funded a project on the creation of ARPANET, or the Advanced Research Projects Agency Network [2, 3]. In 1983, a new communication protocol called Transfer Control Protocol/Internet Protocol (TCP/IP) was developed that enabled communication between different computers on a different network and marked the genesis of internet [3]. Within a few years, the number of hosts or a computer system with a registered IP address increased. By 1989, there were around 80,000 hosts [4]. In the same year, Tim Berners-Lee, renowned British scientist, invented the World Wide Web (WWW) while working at CERN [5] to automate the information sharing between universities and institutions that are carrying out scientific study throughout the world. Finally, in 1993, CERN revealed the WWW project in the public domain and later released it with an open license to ensure its widespread reachability. Since then, we have been witnessing a radical transformation of the internet and today we seem unable to comprehend a device that is not connected to the internet.

An earlier application of the concepts involving IoT was first introduced by Mark Weiser [6], where he used the term "ubiquitous computing." He also proposed the phrase "a walk in the woods" to explain that machines must enter the human environment and adjust itself automatically rather than the opposite, consequently making the use of computers as refreshing as talking a walk in the woods. With the emergence of rapid technological growth, the relationship between humans, computers, and internet grew. This relationship was so intertwined that there arose a holistic need to further extend the concept of IoT in line with the demands of this century. The very first report on IoT published by the International Telecommunications Union (ITU) noted that:

> Machine-to-machine communications and person-to-computer communications will be extended to things, from everyday household objects to sensors monitoring the movement of the Golden Gate Bridge or detecting earth tremors. Everything from tyres to toothbrushes will fall within communications range, heralding the dawn of a new era, one in which today's internet (of data and people) gives way to tomorrow's Internet of Things. [7]

In 2008, the world witnessed a record. The number of devices connected to the internet surpassed the world population. The prevalent IP version 4 (IPv4) that assigned IP addresses to the devices connected to the internet exhausted, and IPv6 was introduced to resolve the issue. During the same time frame, open-source electronics such as Arduino (https://www.arduino.cc/) were also introduced, which paved the path to the current form of IoT.

Not only ordinary people but government organizations from several countries also started recognizing the importance of IoT. Notable among them is the Commission of the European Communities, which in 2009 published a report on IoT action plan for Europe and placed IoT at an elevated importance level among the intellectuals [8]. In fact, the rise of IoT has been so widespread that Gartner, Inc., forecasted in 2013 that by 2020 IoT will include 26 billion units [9].

1.2 DEFINITION OF IoT

Imagine a situation where there was a sudden plan for a feast on your birthday with your colleagues and you decide to prepare a roasted chicken for them. You decide to turn on your oven with a single click on your smartphone and let it preheat while returning from work. That way, when you reach home, your oven is fully ready, and you have one less thing to worry about. Now consider that as you and your colleagues approach your apartment, your smartphone detects your location and triggers the air conditioner to turn on and sets it to a predetermined temperature. Life becomes so easy when we can perform such basic but necessary activities with a simple click of the smartphone or automatically. More than a decade ago, this situation was beyond imagination, but now with internet and its widespread availability, the context does not seem to be a gimmick. In situations like this, IoT has provided us with the proper means where our devices interact with each other and take an active role in performing certain tasks automatically, without or with minimum human intervention.

IoT has been defined in various manners in the technological domain. According to Cisco: "The Internet of Things (IoT) is the next technology transition where devices will allow us to sense and control the physical world by making objects smarter and connecting them through an intelligent network. IoT is about connecting the unconnected" [10].

It is also defined as:

> The basic idea of this concept is the pervasive presence around us of a variety of things or objects—such as Radio-Frequency IDentification (RFID) tags, sensors, actuators, mobile phones, etc.—which, through unique addressing schemes, are able to interact with each other and cooperate with their neighbors to reach common goals. [11]

One of the common elements in both these definitions is interaction among devices. To achieve this interaction, internet is a basic requirement. When Kevin Ashton defined IoT, he clarified that in the first version of the internet, the data was created by human beings by typing, manual digitization of data, etc. He envisioned that with the large-scale implementation of IoT, the data will be created by the things. Considering this concept, a succinct definition of IoT has been put forward as "An open and comprehensive network of intelligent objects that have the capacity to auto-organize, share information, data and resources, reacting and acting in face of situations and changes in the environment" [12]. The integrated relationship between all the surrounding devices will thus make the devices smarter and improve the well-being of all individuals.

When these devices interact among themselves and exchange data, the question of reliability and security comes into the picture. To address these areas, various protocols, concepts, and technologies came into existence and form an integral part of IoT. For example, IoT includes the communication networks, sensory devices, context awareness, and remote service management. IoT aims to unify all these major components. While communicating across devices, they must comply with certain communication protocols. WLANs, WSN, Mobile networks, etc., form the pillars of the communication and networking aspects of IoT. Hence, by connecting multiple sensor-enabled physical devices within the network, an auxiliary level of digital intelligence is facilitated. This enables real-time cross-platform/device data sharing using multifarious secured and standardized communication protocols.

1.3   IoT PHASES

The basic premise of IoT is to connect the unconnected devices surrounding us and enable a harmonious synergy between them. By ensuring this, IoT enables communication between the devices, leading to innovative services that benefit the user.

IoT thus aims to bring values to our life. The entire process of a successful and sustainable IoT implementation for one's own business has been coherently put forward by Siemens. According to the article, a successful IoT implementation must include five different phases, which are described below:

1. **Strategy development:** Any business model must aim to align the business strategies, operations, and associated technologies starting from the very beginning. These measures must precede and be addressed before the start of IoT implementation. An initial holistic strategy development will help businesses to remodel their business approaches and leverage their assets to their full extent.

2. **Idea and prototype:** Continuous brainstorming is necessary to develop the quality of the service and the use case generated by the business. It is necessary to consider all the skills and capacities available to the business from internal and external sources to strengthen the business ideas.

3. **Connect, adapt, and integrate systems:** To thrive with the competition, the business must provide certain services that are unique and improve the quality of the consumers. This is achieved through data that is collected from all the things or devices that are connected via a network. This phase can, therefore, be subdivided into the following phases:

   a. **Connected devices (sensors/actuators):** Sensors capture the information from the connected devices, and this information is then digitized. Therefore, in this phase, devices with no sensors are outfitted with sensors that are compatible with the IoT framework. Actuators automate certain tasks. Actuators are designed in such a way that they are intelligent and can therefore perform certain tasks automatically based on the data that is collected by the sensors.

   b. **Sensor data acquisition/aggregation:** This stage primarily involves the data acquisition systems (DAS) and the internet gateway. DAS is integrated into the sensor network and the internet gateway comprises Wi-Fi and wired LANs. The data collected from the previous stage is then compressed and converted into an optimal form so that the optimized form can be properly analyzed in the later stages. In this stage, the aggregation of the data and its digitization are also carried out.

   c. **Communication:** In this phase, the data that is detected by the sensors and actuators are sent to the required destination. In many scenarios, the data collected is sent to the cloud. All the subsequent phases use this cloud-based data repository to perform their tasks.

4. **Analysis and visualization:** This phase is all about processing the raw data collected from the previous phases and extract meaningful information from them. At the beginning of this phase, the data is preprocessed by enhanced forms of machine learning and visual representations by the IT world. Once this preprocessing is complete, the data is then further analyzed by skilled data scientists and domain experts. Domain experts and data scientists are therefore crucial in establishing the efficacy of the business model. Domain experts are tasked with a full interpretation of the data that was gathered

to establish the distinction between the correlation between the data and the causalities. Once the job of domain experts is over, data scientists take control of their interpretation and develop the algorithms that are needed to implement the use case and extend services to the consumers.

5. **Action:** The expertise of the domain experts and data scientists will go into vain unless relevant measures are taken to ensure that the system is properly maintained. This requires routine checkup of the infrastructures, software bug fixes, and incorporation of additional features based on user feedback. It must be ensured that any development in any of the previous phases is properly incorporated into the system.

1.4 IoT FRAMEWORKS

Some characteristics of any IoT-based frameworks are:

1. **Connectivity:** Connectivity among different devices is the key to any IoT implementation. With no connectivity, no collective intelligence can be accumulated and the basic premise of IoT will fail.
2. **Sensing:** The sensor must be calibrated with high accuracy such that any change of the surrounding is being detected. The detection must be accurate and instantaneous. Any erroneous or untimely detection of any change of any magnitude will cause an unexpected outcome of the framework.
3. **Intelligence:** Once the data is gathered, intelligent analysis must be performed to provide a meaning to the raw data and put it to a good use for the society.
4. **Dynamic:** IoT must be prepared in collecting and interpreting data from dynamic systems. The framework, therefore, needs to be dynamic as well.
5. **Scalability:** IoT aims to bring as many devices as possible under its connectivity spectrum and creates a symbiotic relationship among each other. The number of such connected devices is bound to increase with time and with it, the data. Naturally, data management becomes very critical with time. In addition, sufficient means must be available to analyze and create meaningful interpretations of the data. The increase in complexity should in no way affect the functioning of the connected devices.
6. **Heterogeneity:** Usually, all the devices connected to an IoT framework are not of the same kind, and they usually communicate among themselves over different networks using different communication protocols. In such cases, the data collected is different as well. Proper means should exist to interpret various varieties of data that are collected from the framework.
7. **Security:** An IoT framework is completely and constantly reliant on network connectivity to facilitate communication between devices. Therefore, an IoT framework is susceptible to cyberattacks and data breaches. An essential feature that every IoT framework should incorporate is security from such unwanted events. Necessary measures should be taken to ensure that all the devices in the network, the data, and the cloud are properly secured as per the latest threat definitions.

1.5   IoT APPLICATIONS

IoT aims to improve the quality of our life by achieving a synergistic relationship between devices and human beings. These days, it is often said that the Fourth Industrial Revolution is because of the rise and potential of IoT [13–16]. An exhaustive documentation on the domains that have used IoT for their development can be found in Ref. [17]. The prime areas of application are smart cities, transportation, industry, agriculture, healthcare, smart homes, buildings, and environment. Depending on the area, the IoT model is remodeled. For example, the IoT model used for the automatic functioning and synchronization of the equipment and devices in an industrial building will differ from that of the IoT model used to perform the same in an ordinary household. Naturally, this means the security measures undertaken, the accuracy target, and the robustness of the IoT model will vary to a great extent depending on the scenario where the IoT framework is deployed. Therefore, it is often said that, in case of IoT, there is no "one-size-fits-all" solution (Hassan et al., 2018).

1.5.1   HEALTHCARE

IoT has proved immensely beneficial in healthcare. Consider that there is a patient admitted to a hospital and suddenly the patient undergoes some physical distress. The distress can be an abnormal rise in blood pressure, imminent cardiac attack, etc. One of the low-power biosensors attached to the patient's body measures the stress and signals the doctor's and nurse's pager for an immediate action. The medical personnel respond, and the patient received necessary treatment. There could have been some serious consequences if the hospital lacked this kind of infrastructure. Use of low-power wearable biosensors has been well documented [18–22]. These days, IoT frameworks are also used to build up the tools for elderly care. Starting from glucose level monitoring, cardiac rhythm assessment, sleep, and movement tracking, IoT has been benefiting the daily life of these individuals. Several non-healthcare-based industries have been exemplar in establishing a productive value of IoT in healthcare. For example, Apple Inc. has introduced a feature in their Apple watch that automatically sends a copy of the ECG report to the concerning doctor of an apple watch user if it finds irregular rhythms in it [23]. In this way, any slight abnormalities in the cardiac rhythm can be detected and timely intervention of healthcare personnel can help in recovery of the watch user. Not only that, but Apple Inc. has also introduced handwashing reminder [24] based on the location of the user, considering the recent COVID-19 pandemic. It also keeps a track of whether the hand is being washed as per the CDC recommended duration.

1.5.2   SMART HOMES AND BUILDINGS

Several home appliances are available that are aimed at improving the well-being and stay of the persons in the home. Appliances range from automatic room lightning, baby-monitoring devices, pet-monitoring devices, air-conditioning control, energy saving, fire control, and threat control management. Google Nest is an excellent example that integrates a lot of the features and available easily to the common people. Any smart home-based IoT applications focus primarily on energy efficiency and energy saving. By providing innovative home solutions, they provide a certain level of luxury, comfort, and security.

Not only homes, buildings these days are also fitted with sensors to detect the total number of residents at a time. These basically include motion sensors and video surveillance cameras. Motion sensors are primarily used for automatic control of the lights in the common areas of the building, thus saving energy when not required. Smart buildings often employ heating, ventilation, and air-conditioning system which are popularly known as HVAC system along with temperature sensors for the building management system (BMS). To facilitate integration of all IoT frameworks into a common framework, smart buildings usually employ a building autonomous system (BAS). When different sensors come and fit into the framework, heterogeneity is introduced. Heterogeneous systems require much more attention and are thus considered as one challenge faced by IoT. For a detailed study, refer [25–30].

1.5.3 SMART CITIES

In cities and urban development, IoT plays a crucial role in water and waste management by using dumpsters which are smart [31–33], lightning and energy control [34, 35], parking management, safety precaution, etc. IoT considers the city as a single entity and its monitoring is essential for the betterment of the citizens in the cities. In the United States, AT&T is playing an important part in implementing integrated smart city solutions [36]. The European Commission's Horizon started a new CITYkeys 2020 program that aimed at assessing and evaluating the indicators of smartness in various smart cities throughout Europe. Not only the developed nations, several other third-world countries like India have also undertaken a project on establishing hundred smart cities in 2014 [37–40].

1.5.4 TRANSPORTATION

Vehicles these days are fitted with several sensors and cameras to make the drive secured and easy like lane departure sensors, front and back traffic determination, blind spot determination, etc. The various sensors that are present in a car add up to increase the smartness of the vehicle. Things have escalated a lot from there. Consider an electric car. Besides the ubiquitous sensors in ordinary cars, electric cars use various kinds of additional sensors and take several steps forward toward automation. Self-driving is one of the most sought features in such cars. Navigation automatically finds the charge stations on the route and pet care inside the car has become the norm in such vehicles. Modern vehicles also monitor the driver's comfort during travel and keep track of all possible vehicular diagnostics. Maintaining a soothing temperature inside the car based on outside temperature has now become a basic need. Maps and navigation applications like Waze rely on user-contributed data to monitor traffic and prepare the driver for imminent situations on the road. It is popularly said that Tesla has been the pioneer in setting up the IoT standards in vehicular industries, starting from its very inception in 2003 [41].

1.6 INDUSTRY 4.0 AND ITS EVOLUTION

Industries dedicated to manufacturing are constantly developing to survive and thrive in a competitive environment. Industry 4.0, which comprises Industrial IoT (I-IoT) and Smart Manufacturing (SM), is at the core of these businesses. I-IoT represents the interconnected machines over the network in each industrial setup. There is a subtle difference between I-IoT and IoT. Where I-IoT pertains to industries, IoT connotes consumer items. I-IoT is built on the premise of IoT, but it involves gathering and speedily analyzing data received from connected machines in real time that are mission critical to the proper functioning of businesses. Industry 4.0 therefore aims to employ smart digital technologies, machine learning, artificial intelligence, and big data to establish a well-connected ecosystem for the industries that are trying to automate the manufacturing and supply chain management.

1.6.1 INDUSTRY 1.0 TO 4.0

1. **The First Industrial Revolution:** This is the period from late 1700s and early 1800s when manual and animal labors were prevalent. In some advanced cases, the use of machines like steam-powered engines was also used.
2. **The Second Industrial Revolution:** This started at the beginning of the 20th century and involved the use of steel and electricity to make the manufacturing process more efficient and boost productivity.
3. **The Third Industrial Revolution:** This period witnessed the use of electronic and eventually computer technologies in the factories, which led to the use of digital automated technologies over the analog and mechanical ones.
4. **The Fourth Industrial Revolution or Industrial 4.0:** This period takes control of the automated technologies of the previous era and integrates with the smartness of IoT. This resulted in real time assessing of the data, which ultimately sped up the manufacturing process. Industry 4.0 has resulted in constant communication and symbiosis among various departments of an industry, thus resulting in overall profitability of the business.

1.7 THE RISE OF HUMAN-MACHINE INTERFACES IN IoT

Human-machine interface (HMI) is a part and parcel of our everyday life. For example, when we visit an ATM booth, we use a smart device interface to perform our banking transactions. Even using a mobile phone or a computer to perform our operations for us is an example of HMI. HMI can therefore be deemed as a means that translates the human instruction into a machine-interpretable instruction. In the era of Industry 4.0, HMI plays a very critical role. In an industrial context, HMIs include all the smart terminals that let a worker to connect and interact with an industrial system for facilitating an industrial operation. Mostly they are the touchscreen displays, display monitors on machine tops, standalone terminals, etc. Industrial literatures sometimes refer to HMI as operator interface terminal (OIT) or man-machine interface (MMI). Therefore, in a nutshell, HMI in a manufacturing concern is an interface that aims to bring in a human intervention to the automation process which brings in a level of control, monitoring, and visualization between the humans, process, and the machines [42].

1.7.1 HMI Platforms

1. **Proprietary:** The software that runs on the smart device interface is developed by a company which is not the parent industry where it has been set up.
2. **Open:** The complete smart device setup that includes the hardware, and the software has been designed by the parent company where it has been set up.

HMIs are further classified into two different categories:

1. **Standard:** These HMIs are designed primarily for plants, buildings, and other indoor uses.
2. **Rugged:** These are used in extremely harsh and challenging environmental conditions.

1.7.2 Importance of HMI

1. **Risk:** It has been found that a lack of proper HMI infrastructure might lead to increased risk for the workers and the assets [43].
2. **Homogeneity:** Workers feel safer and more used to while using smartphones and other smart devices owing to the rising popularity of the said technologies within the consumers.
3. **Digital Revolution:** HMI has the potential to integrate the practices used in Industry 4.0 tightly and pave the path of a better industrial output.

1.7.3 Future of HMI

1. **Intuitive control panel:** Constant focus must be given on making the usability of the HMIs easy. If the usability is complex for an ordinary worker, then it will be very difficult to boost up the productivity. The user must be able to interpret the contents that is delivered in the smart devices for interpretation.
2. **On-machine applications:** The operator must be able to access important information easily from the machines, its performance metrics, operating condition, and current state. Modern HMIs must be able to host machine operation apps like maintenance of the machines, troubleshooting documents, etc. [44].
3. **I-IoT-connected HMI:** HMI and I-IoT must be tightly integrated to streamline all the industrial operations. They must be able to connect with a greater number of complicated machines, collect data from them, and present the data to the user intuitively that can easily be analyzed. It must also allow the communication between devices from anywhere, not just the terminal.
4. **Mobile HMI:** People are mobile these days. We prefer doing our tasks on the go. In future, HMIs must be designed in such a manner that the users can access all the pertaining data regarding the industrial workflows right from their smartphones or tablets or even smartwatches. This way manufactures will monitor their operation keenly.

1.8 SUMMARY

IoT is shaping up to be the technology of the future. The potential benefits of IoT in various domains have been identified and several organizations from different countries have harnessed its potential to create several use cases and promote the well-being of their citizens. IoT is extensively used in healthcare, transportation, homes, etc., and several other places are being recognized where its benefits can be reaped. Despite being undoubtedly beneficial, IoT also poses several challenges. Regular device monitoring and maintenance is extremely necessary to ensure that the IoT framework functions flawlessly. In addition, data management and data security over the network have become an area of concern for cybersecurity experts. When all its components are properly configured, the IoT framework will lead to a promising future. The era of the Fourth Industrial Revolution is already here, and IoT is the mother of this era. With HMI, IoT aims to streamline the productivity of big industrial concerns as well. The functionality and efficacy of HMI-enabled IoT are rapidly rising, and we are about to set our path to a mobile IoT implementation that is reliable, smart, and intuitive.

REFERENCES

1. K. Ashton, "That 'Internet of Things' Thing | RFID Journal." [Online]. Available: https://www.rfidjournal.com/that-internet-of-things-thing. [Accessed: 20-Sep-2021]
2. E. Andrews, "Who Invented the Internet?" [Online]. Available: https://www.history.com/news/who-invented-the-internet. [Accessed: 15-Mar-2021]
3. "A Brief History of the Internet." [Online]. Available: https://www.usg.edu/galileo/skills/unit07/internet07_02.phtml. [Accessed: 15-Mar-2021]
4. "Hobbes' Internet Timeline—The Definitive ARPAnet & Internet history." [Online]. Available: https://www.zakon.org/robert/internet/timeline/#Growth. [Accessed: 08-Mar-2021]
5. "The birth of the Web | CERN." [Online]. Available: https://home.cern/science/computing/birth-web. [Accessed: 15-Mar-2021]
6. M. Weiser, "The Computer for the 21st Century," *Scientific American*, vol. 265, no. 3, pp. 94–105, 1991, doi: 10/c2ntdd. [Online]. Available: http://www.jstor.org/stable/24938718
7. "ITU Internet Reports 2005: The Internet of Things." [Online]. Available: https://www.itu.int/osg/spu/publications/internetofthings/. [Accessed: 08-Mar-2021]
8. "Internet of Things—An action plan for Europe." [Online]. Available: https://www.eesc.europa.eu/en/our-work/opinions-information-reports/opinions/internet-things-action-plan-europe. [Accessed: 08-Mar-2021]
9. "Forecast: The Internet of Things, Worldwide, 2013." [Online]. Available: https://www.gartner.com/en/documents/2625419/forecast-the-internet-of-things-worldwide-2013. [Accessed: 16-Mar-2021]
10. "Public Safety Blog Series—Connecting the Unconnected in Public Safety Response." [Online]. Available: https://blogs.cisco.com/government/connecting-the-unconnected-in-public-safety-response. [Accessed: 16-Mar-2021]
11. L. Atzori, A. Iera, and G. Morabito, "The Internet of Things: A Survey," *Computer Networks*, vol. 54, no. 15, pp. 2787–2805, 2010, doi: 10/fnq7ds. [Online]. Available: https://www.sciencedirect.com/science/article/pii/S1389128610001568
12. S. Madakam, R. Ramaswamy, and S. Tripathi, "Internet of Things (IoT): A Literature Review," *Journal of Computer and Communications*, vol. 3, no. 5, pp. 164–173, 2015, doi: 10/ggtms3. [Online]. Available: 10.4236/jcc.2015.35021

13. "Lead the Fourth Industrial Revolution with Industrial IoT (IIoT)." [Online]. Available: https://www.dxc.technology/manufacturing/insights/143344-lead_the_fourth_indus-trial_revolution_with_industrial_iot_iiot. [Accessed: 21-Mar-2021]
14. T. Maddox, "How IoT Will Drive the Fourth Industrial Revolution." [Online]. Available: https://www.zdnet.com/article/how-iot-will-drive-the-fourth-industrial-revolution/. [Accessed: 21-Mar-2021]
15. N. Negahban, "Kinetica BrandVoice: The Internet of Things Is Powering the Data-Driven Fourth Industrial Revolution." [Online]. Available: https://www.forbes.com/sites/kinetica/2019/05/31/the-internet-of-things-is-powering-the-data-driven-fourth-industrial-revolution/. [Accessed: 21-Mar-2021]
16. T. P. T. Premium, "The Fourth Industrial Revolution: Why You Need a Global IoT strat-egy." [Online]. Available: https://www.techrepublic.com/article/the-fourth-industrial-revolution-why-you-need-a-global-iot-strategy/. [Accessed: 21-Mar-2021]
17. "AIOTI WG Research and Partnerships | AIOTI." [Online]. Available: https://aioti.eu/research/. [Accessed: 21-Mar-2021]
18. N. Bui and M. Zorzi, *Health Care Applications: A Solution Based on the Internet of Things*, p. 131, 2011, doi: 10.1145/2093698.2093829.
19. A. Dohr, R. Modre-Osprian, M. Drobics, D. Hayn, and G. Schreier, "The Internet of Things for Ambient Assisted Living," in *2010 Seventh International Conference on Information Technology: New Generations*, 2010, pp. 804–809, doi: 10.1109/itng.2010.104. [Online]. Available: 10.1109/ITNG.2010.104
20. M. C. Domingo, "An Overview of the Internet of Things for People with Disabilities," *Journal of Network and Computer Application*, vol. 35, no. 2, pp. 584–596, 2012, doi: 10.1016/j.jnca.2011.10.015. [Online]. Available: https://www.sciencedirect.com/science/article/pii/S1084804511002025
21. C. Doukas and I. Maglogiannis, "Bringing IoT and Cloud Computing towards Pervasive Healthcare," in *2012 Sixth International Conference on Innovative Mobile and Internet Services in Ubiquitous Computing*, vol. 1, pp. 922–926, 2012, doi: 10.1109/imis.2012.26. [Online]. Available: 10.1109/IMIS.2012.26
22. G. Yang, *et al.*, "A Health-IoT Platform Based on the Integration of Intelligent Packaging, Unobtrusive Bio-Sensor, and Intelligent Medicine Box," *IEEE Transactions on Industrial Informatics*, vol. 10, no. 4, pp. 2180–2191, 2014, doi: 10.1109/tii.2014.2307795. [Online]. Available: 10.1109/TII.2014.2307795
23. "Healthcare—Apple Watch." [Online]. Available: https://www.apple.com/healthcare/apple-watch/. [Accessed: 21-Mar-2021]
24. "Wash Your Hands with Apple Watch Series 4 or Later." [Online]. Available: https://support.apple.com/en-us/HT211206. [Accessed: 21-Mar-2021]
25. M. Jia, A. Komeily, Y. Wang, and R. S. Srinivasan, "Adopting Internet of Things for the Development of Smart Buildings: A Review of Enabling Technologies and Applications," *Automation in Construction*, vol. 101, pp. 111–126, 2019, doi: 10.1016/j.autcon.2019.01.023. [Online]. Available: https://www.sciencedirect.com/science/article/pii/S0926580518307064
26. R. K. Kodali and S. Yerroju, "IoT Based Smart Emergency Response System for Fire Hazards," in *2017 3rd International Conference on Applied and Theoretical Computing and Communication Technology (iCATccT)*, 2017, pp. 194–199, doi: 10.1109/icatcct.2017.8389132. [Online]. Available: 10.1109/ICATCCT.2017.8389132
27. Z. Li, T. Wang, Z. Gong, and N. Li, "Forewarning Technology and Application for Monitoring Low Temperature Disaster in Solar Greenhouses Based on Internet of Things," *Transactions of the Chinese Society of Agricultural Engineering*, vol. 29, no. 4, pp. 229–236, 2013. [Online]. Available: https://www.ingentaconnect.com/content/tcsae/tcsae/2013/00000029/00000004/art00029
28. M. V. Moreno, M. A. Zamora, and A. F. Skarmeta, "User-Centric Smart Buildings for Energy Sustainable Smart Cities," *Transactions on Emerging Telecommunications*

Technologies, vol. 25, no. 1, pp. 41–55, 2014, doi: 10.1002/ett.2771. [Online]. Available: 10.1002/ett.2771

29. C.-S. Ryu, "IoT-Based Intelligent for Fire Emergency Response Systems," *International Journal of Smart Home*, vol. 9, no. 3, pp. 161–168, 2015, doi: 10.14257/ijsh.2015.9.3.15. [Online]. Available: 10.14257/ijsh.2015.9.3.15

30. C. Wei and Y. Li, "Design of Energy Consumption Monitoring and Energy-Saving Management System of Intelligent Building based on the Internet of Things," in *2011 International Conference on Electronics, Communications and Control (ICECC)*, pp. 3650–3652, 2011, doi: 10.1109/icecc.2011.6066758. [Online]. Available: 10.1109/ICECC.2011.6066758

31. I. Hong, S. Park, B. Lee, J. Lee, D. Jeong, and S. Park, "IoT-Based Smart Garbage System for Efficient Food Waste Management," *The Scientific World Journal*, vol. 2014, pp. 1–13, 2014, doi: 10.1155/2014/646953. [Online]. Available: 10.1155/2014/646953

32. S. Phithakkitnukoon, M. I. Wolf, D. Offenhuber, D. Lee, A. Biderman, and C. Ratti, "Tracking Trash," *IEEE Pervasive Computing*, vol. 12, no. 2, pp. 38–48, 2013, doi: 10.1109/mprv.2013.37. [Online]. Available: 10.1109/MPRV.2013.37

33. "Smartup Cities—Smart City Solutions Marketplace." [Online]. Available: https://www.smartupcities.com/. [Accessed: 22-Mar-2021]

34. M. Castro, A. J. Jara, A. F. G. Skarmeta, "Smart Lighting Solutions for Smart Cities," in *2013 27th International Conference on Advanced Information Networking and Applications Workshops*, pp. 1374–1379, 2013, doi: 10.1109/waina.2013.254. [Online]. Available: 10.1109/WAINA.2013.254

35. D. Kyriazis, T. Varvarigou, A. Rossi, D. White, and J. Cooper, "Sustainable Smart City IoT Applications: Heat and Electricity Management & Eco-conscious Cruise Control for Public Transportation," in *2013 IEEE 14th International Symposium on "A World of Wireless, Mobile and Multimedia Networks" (WoWMoM)*, pp. 1–5, 2013, doi: 10.1109/wowmom.2013.6583500. [Online]. Available: 10.1109/WoWMoM.2013.6583500

36. "AT&T Launches Smart Cities Framework with New Strategic Alliances, Spotlight Cities, and Integrated Vertical Solutions I AT&T." [Online]. Available: https://about.att.com/story/launches_smart_cities_framework.html. [Accessed: 22-Mar-2021]

37. "What Will It Take to Create Smart Cities in India? I Smart Cities Dive." [Online]. Available: https://www.smartcitiesdive.com/ex/sustainablecitiescollective/what-will-it-take-create-smart-cities-india/1036661/. [Accessed: 22-Mar-2021]

38. "6th Smart Cities India 2021 Expo, International Exhibitions and Conferences", 24–26 March 2021, Pragati Maidan, New Delhi. [Online]. Available: https://www.smartcities-india.com/default.aspx. [Accessed: 22-Mar-2021]

39. "India's 2014 Budget Promises Cash for 100 Smart Cities, Metros & Much More I Smart Cities Dive." [Online]. Available: https://www.smartcitiesdive.com/ex/sustaina-blecitiescollective/indias-2014-budget-promises-cash-100-smart-cities-metros-much-more/281501/. [Accessed: 22-Mar-2021]

40. "List of 100 Smart Cities Selected I National Centre for Promotion of Employment for Disabled People." [Online]. Available: https://www.ncpedp.org/list_of_100_smartcities. [Accessed: 22-Mar-2021]

41. "The Tesla IoT Case Study." [Online]. Available: https://study.com/academy/lesson/the-tesla-iot-case-study.html. [Accessed: 22-Mar-2021]

42. "Industrial IoT HMI in Advanced Control Visualization-AIS." [Online]. Available: http://www.aispro.com/technologies/human-machine-interface-technology. [Accessed: 23-Sep-2021]

43. E. Flaspöler, *et al.*, "The Human Machine Interface as an Emerging Risk," *EU-OSHA (European Agency for Safety and Health at Work). Luxemburgo*, 2009.

44. "The Future Human Machine Interfaces (HMI) in an IoT World I Tulip." [Online]. Available: https://tulip.co/blog/the-future-human-machine-interfaces-hmi-in-an-iot-world/. [Accessed: 23-Sep-2021]

2 Technologies and Infrastructure in Internet of Things (IoT)

Sanjay Hosur
Union College,
Schenectady, NY, USA

Vinay Ramakrishnaiah
Los Alamos National Laboratory,
Los Alamos, NM, USA

CONTENTS

2.1 INTRODUCTION

Internet of Things (IoT) applications have specific requirements such as long range, low data rate, low energy consumption, and cost-effectiveness. With explosive growth of IoT technologies, there is an increasing number of practical applications in many fields such as security, asset tracking, agriculture, smart metering, smart cities, and smart homes (Ratasuk 2015).

Today, the term IoT is used very broadly and defined in many ways based on the context it is used in. IoT describes the network of internet-connected devices/ physical objects/things that are embedded with technologies such as sensors and software and are capable of functioning by exchanging data with other devices. The International Telecommunication Union defines IoT as "a global infrastructure for the Information Society, enabling advanced services by interconnecting (physical and virtual) things based on existing and evolving, interoperable information and communication technologies." The term IoT has been around for about more than two decades now. The first use of the term IoT is attributed to the work of the Auto-ID Labs at the Massachusetts Institute of Technology (MIT) on networked radio-frequency identification (RFID) infrastructures (Atzori 2010).

The multiple dimensions of IoT have been summarized in Figure 2.1 (i-scoop 2020), which takes a broad view on the following six categories:

1. Device connection
2. Data sensing
3. Communication
4. Data analytics
5. Data value
6. Human value

The focus of this chapter is on category 3, which is communication. Communication can occur over short distances or over a long range to a very long range. Examples include Wi-Fi, LPWA network technologies such as LoRa or Narrow Band IoT (NB-IoT), cloud computing, fog computing, etc.

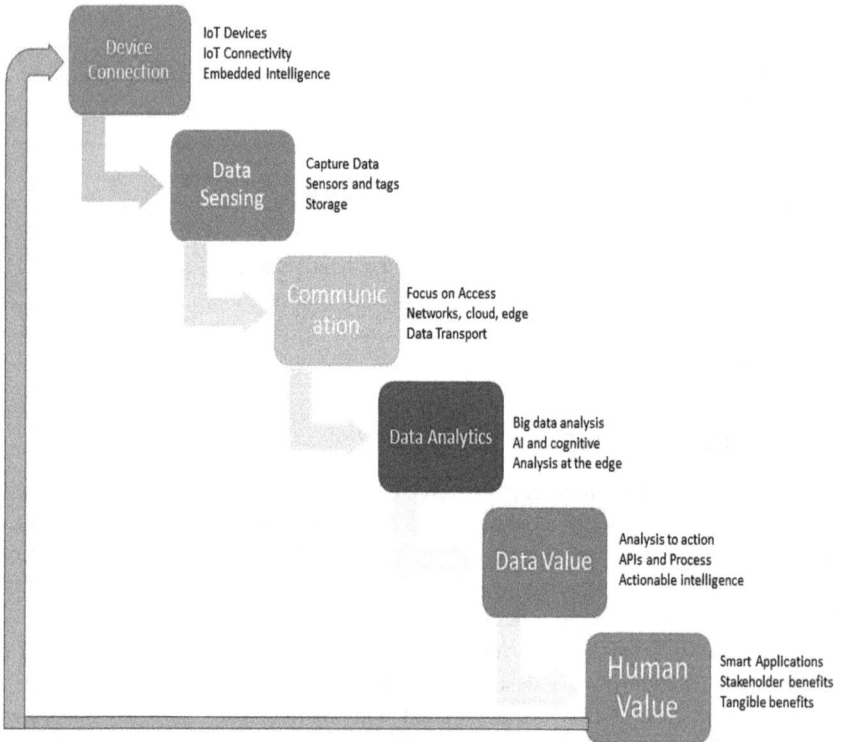

FIGURE 2.1   Broad categories defining the scope of Internet of Things.

In this chapter, we will briefly discuss the following two categories of technologies, which are widely used in IoT, and discuss the pros and cons of each technology:

- Fog computing
- Low-power wide-area network (LPWAN)

Finally, before concluding, a brief overview of newer IoT technologies—Zigbee Alliance and Connected Home over Internet Protocol (IP)—is also discussed.

2.2   FOG COMPUTING AND IoT

Fog computing, also called edge computing, is intended for distributed computing where numerous devices connect to a cloud. The word "fog" is used to suggest a cloud's periphery or edge. The idea behind fog computing is to do as much processing as possible using computing units co-located with the data-generating devices, so that processed rather than raw data is forwarded, and bandwidth requirements are reduced. Also, the processed data is most likely needed by the same set of devices that generated them, therefore, by processing locally rather than remotely, the latency between the input and the response is minimized. This idea is similar to using special-purpose hardware/co-processors, such as digital signal processing units for performing fast Fourier transforms, which have long been used to reduce latency and burden on the central processing units (CPU) (Bonomi 2012).

Some of the important characteristics that make fog computing a nontrivial extent of cloud computing are:

- Edge location, location awareness, and low latency
- Geographical distribution
- Large-scale sensor networks
- Support for mobility
- Real-time interactions
- Predominance of wireless access
- Heterogeneity
- Interoperability and federation
- Support for online analytics and interplay with the cloud

Recently, a lot of research has started focusing on fog computing integrated with IoT. The term fog computing was first introduced by Cisco and is defined as a highly virtualized platform that provides computing, storage, and networking services between end devices. On the other hand, traditional cloud computing data centers are typically (but not exclusively) located at the edge of the network (Bonomi 2012).

Fog computing acts as a bridge between the IoT devices and large-scale cloud computing and storage services. According to Cisco, fog computing is considered an extension of the cloud computing paradigm from the core of the network to the edge of the network. Fog computing will bring cloud networking, computing, and storage capabilities down to the edge of the network, which will address the real-time issue of IoT devices and provide secure and efficient IoT applications (Ketel 2017).

TABLE 2.1
Cloud and Fog Computing Comparison

Parameter	Cloud	Fog
Latency	High	low
Client server distance	Multiple hops	Single hop
Server node location	Anywhere in internet	Edge of local network
Deployment	Centralized	Distributed
Location awareness	No	Yes
Data attack	Lower possibility	Higher possibility
Security	Defined	Hard to define
Deployment	Centralized	Distributed
Storage	Scalable	Limited
Computing power	Higher	lower

Fog computing provides different services and applications with widely distributed deployments. The fog has the ability to provide efficient real-time communication between different IoT applications, such as connected vehicles, through proxy and access points positioned alongside highways and tracks. Fog computing is considered to be the best choice for applications with low latency requirements such as video streaming, gaming, augmented reality, etc. (Skarlat 2016).

Now, let us look at some of the major differences between fog computing and cloud computing (Atlam 2018). These are summarized in Table 2.1.

Next, Figure 2.2 shows the basic architecture of fog computing. Fog computing provides advantages that suit applications with low latency requirements such as emergency and healthcare-related services, video streaming, gaming, and augmented

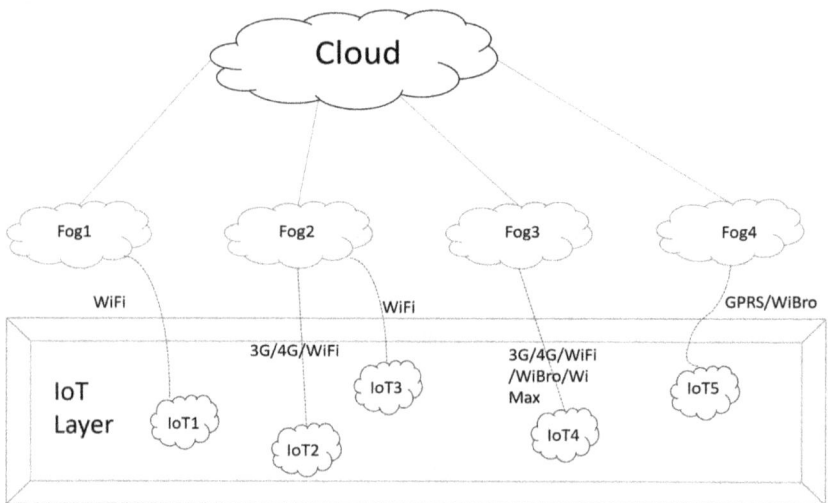

FIGURE 2.2   Basic architecture of fog computing.

reality, among others (Aazam and Huh 2016). For smart communication, fogs are going to play an important role. For example, the fog will be able to deliver high-quality streaming to mobile nodes, such as moving vehicles, through proxies and access points positioned accordingly, along highways and tracks.

Let us now understand the layered architecture of fog computing for practical implementation starting from the bottommost layer:

1. Physical and virtualization: In this layer, different types of nodes such as physical nodes, virtual nodes, and virtual sensor networks are present. These nodes are maintained and managed according to their types and service demands. Different types of sensors that are geographically distributed are used to sense the surroundings and send the collected data to upper layers via gateways for further processing (Liu 2017).
2. Monitoring: In this layer, all nodes are tracked and monitored for resource utilization, availability of sensors, fog nodes, and network elements. The performance and status of all applications and services deployed on the infrastructure are monitored by observing the tasks performed on each node at a given instance of time, predicting the resource and energy requirements (Aazam and Huh 2015).
3. Preprocessing: The preprocessing layer performs data management tasks. Collected data is analyzed and processed in this layer to extract meaningful information.
4. Temporary storage: The preprocessed data is stored temporarily in the temporary storage layer. When the data is transmitted to the cloud, it is generally deleted from the temporary storage media as it is no longer needed (Aazam and Huh 2015). This layer includes storage devices with appropriate storage space virtualization for data distribution, replication, and de-duplication.
5. Security: In this layer, the encryption or decryption is done along with some measures to protect data integrity and avoid data tampering.
6. Transport: In this layer, the preprocessed data is uploaded to the cloud to allow the cloud to extract and create more useful services.

The layered architecture of fog computing is shown in Figure 2.3 (Atlam 2018).

As with any new technology, there are numerous challenges associated with the dream of making the fog a reality. A few of the challenges faced are listed in Figure 2.4 (Atlam 2018) and are briefly discussed here.

1. Security: This section can be divided into the following subsections (Mukherjee 2017):
   a. Authentication: This is a resource-intensive process. Unfortunately, many fog computing devices will not have the memory, or the CPU capabilities required to carry out cryptographic operations.
   b. Rogue node detection: The presence of rogue nodes is a serious threat and it is a difficult task to identify them. A lot of research is being carried out in this direction to enable better detection.

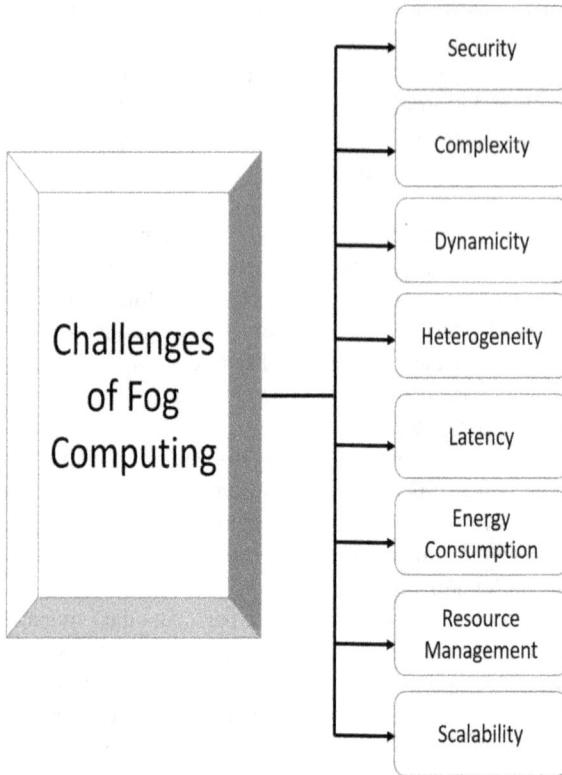

FIGURE 2.3   Layered architecture of fog computing.

    c. Privacy: Resource constrained IoT devices lack the ability to encrypt or decrypt generated data, which make them vulnerable to adversaries.

    d. Access control: IoT introduces new challenges in access control because we are dealing with a large number of devices with limited resources.

2. Scalability (Luan 2015): As the number of IoT devices increases day by day, so does the generated data, which poses a challenge in terms of resource requirements.

3. Resource management (Mouradian 2017): Sensible management of fog resources is required for efficient operation of the fog computing environment as it is difficult for such devices to match the resource capacity of traditional servers.

4. Energy consumption: Reducing energy consumption in fog computing is an important challenge that needs to be addressed as the computation is distributed and can be less energy-efficient than the centralized cloud model of computation (Ni 2017).

5. Latency: The devices at the fog edge generally have limited resources, which can add to the undesirable latency and this must be addressed when implementing IoT (Caiza 2020).

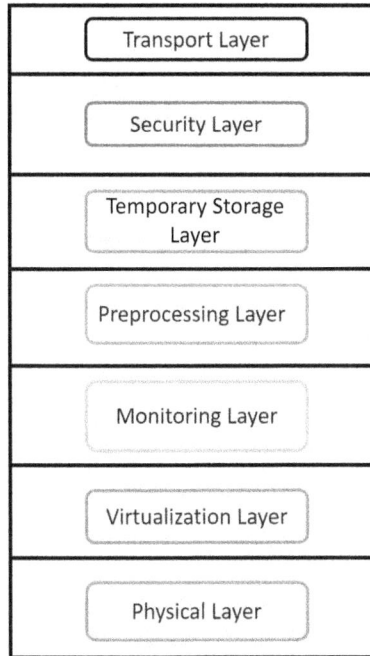

FIGURE 2.4   Challenges of fog computing.

6. Heterogeneity: The management and coordination of networks, heterogeneous IoT devices, and the selection of the appropriate resources is a big challenge as there are many IoT devices and sensors that are designed by different manufacturers with different capabilities (Yi 2015).

7. Dynamicity: Fog nodes will need automatic and intelligent reconfiguration of the topological structure and assigned resources as devices suffer from software and hardware aging, which will result in changing workflow behavior and device properties (Luan 2015).

8. Complexity: For IoT devices and sensors designed by different manufacturers, choosing the optimal set of components is becoming very complicated, especially with different software and hardware configurations and personal requirements (Luan 2015).

Fog computing is new technology and there are many research avenues available, of which, some of the open challenges are listed below (Atlam 2018):

1. Communication between the fog and the cloud: Selecting the appropriate communication between the fog and the cloud that ensures high performance and low latency of the fog nodes is a key challenge (Skarlat, Resource Provisioning for IoT Services in the Fog 2016).

2. Communication between fog servers: Each fog server will manage a pool of resources at different locations. Communication and collaboration between

fog servers is necessary to maintain service provision and content delivery among them (Luan 2015).

3. Fog computing deployment: Fog servers are deployed at different locations, so they need to adapt their services regarding management and maintenance costs.

4. Parallel computation algorithm: Optimization algorithms are typically time- and resource-consuming when applied on a large scale. Therefore, parallel approaches will be required to accelerate the optimization process (Yin 2017).

5. Security: Fog computing has different security requirements than the cloud due to its different characteristics, notable mobility, heterogeneity, and large-scale geo-distribution (Mukherjee 2017).

6. End-user privacy: Preserving the end user's privacy is a significant issue that faces fog computing as fog nodes are closer to end users, which allows them to collect more sensitive data (Mukherjee 2017).

2.3   LOW-POWER WIDE-AREA NETWORK (LPWAN)

Industry and research communities are focusing more on the low-power wide-area network (LPWAN) due to its low power, long range, and low-cost communication characteristics (Baharudin and Yan 2016; Guibene 2017; Vondrouš et al. 2016).

The three leading technologies in LPWAN are LoRa, NB-IoT, and Sigfox. First, we present a brief description of each technology, and the technical differences are highlighted in Table 2.2.

TABLE 2.2
Summary of Major LPWAN Technologies

Parameter	Sigfox	LoRa	NB-IoT
Frequency	Unlicensed ISM bands	Unlicensed ISM bands	Licensed LTE
Bandwidth	100 Hz	125 kHz and 250 kHz	200 kHz
Max data rate	100 bps	50 kbps	200 kbps
Bidirectional	Limited	Yes	Yes
Maximum messages per day	140 (uplink(UL)) 40 (downlink(DL))	Unlimited	Unlimited
Modulation	BPSK	CSS	QPSK
Maximum payload length (bytes)	12 (UL) 8 (DL)	243	1600
Range	10 km (urban) 40 km (rural)	5 km (urban) 20 km (rural)	1 km (urban) 10 km (rural)
Encryption	Not supported	AES	LTE
Interference immunity	Very high	Very high	Low
Adaptive data rate	No	Yes	No
Allow private network	No	Yes	No
Standardization	Sigfox company with ESTI	LoRa	3GPP

2.3.1 SIGFOX

Sigfox was founded by Ludovic Le Moan and Christophe Fourtet in 2010 with a vision to connect every object in our physical world to the digital universe. It is now present in more than 70 countries. Sigfox is the first player to build the largest IoT ecosystem in the world, from big manufacturers to hundreds of startups and device-makers on four continents (Sigfox website 2021).

According to Sigfox (Sigfox website 2021), the standard is based on the following principles:

1. The provision of a minimal channel for transferring small messages that would complement other communication protocols such as Wi-Fi, Bluetooth, Satellite, 3G, 4G, and 5G;
2. Setting up a backup channel in case the main communication links fail, or in cases where a network goes down after a natural disaster or a malicious act;
3. Increasing and guaranteeing the security of networks and exchanges to stabilize our digitalized economy;
4. Simplifying access to different networks to increase adoption;
5. Reducing energy consumption linked to telecommunications networks.

2.3.2 LoRa

LoRa (long range) (Alliance 2015) is a proprietary spread spectrum modulation technique by Semtech. It is a derivative of chirp spread spectrum (CSS). Semtech's LoRa devices and wireless radio-frequency technology are a long-range, low-power wireless platform that has become the de facto technology for IoT networks worldwide. LoRa devices and the open LoRaWAN® protocol enable smart IoT applications that solve some of the biggest challenges facing our planet; examples of which include energy management, natural resource reduction, pollution control, infrastructure efficiency, and disaster prevention.

LoRa wireless RF technology is a Semtech innovation, and the LoRa devices offer compelling features for IoT applications, including long range, low power consumption, and secure data transmission. The technology can be utilized by public, private, or hybrid networks and provides greater range than cellular networks. LoRa technology can easily plug into existing infrastructure and enables low-cost battery-operated IoT applications. Semtech's LoRa chipsets are incorporated into devices manufactured by a large ecosystem of IoT solution providers and connected to LoRaWAN-based networks around the globe. Simply stated, LoRa connects devices (or all things) to the cloud.

The LoRaWAN open specification is a low-power, wide-area networking (LPWAN) protocol based on LoRa technology. Designed to wirelessly connect battery-operated things to the internet in regional, national, or global networks, the LoRaWAN protocol leverages the unlicensed radio spectrum in the industrial, scientific, and medical (ISM) band. The specification defines the device-to-infrastructure of LoRa physical layer parameters and the LoRaWAN protocol and provides seamless interoperability between devices. While Semtech provides the LoRa radio chips, the LoRa Alliance®, a nonprofit association and the fastest growing technology alliance, drives

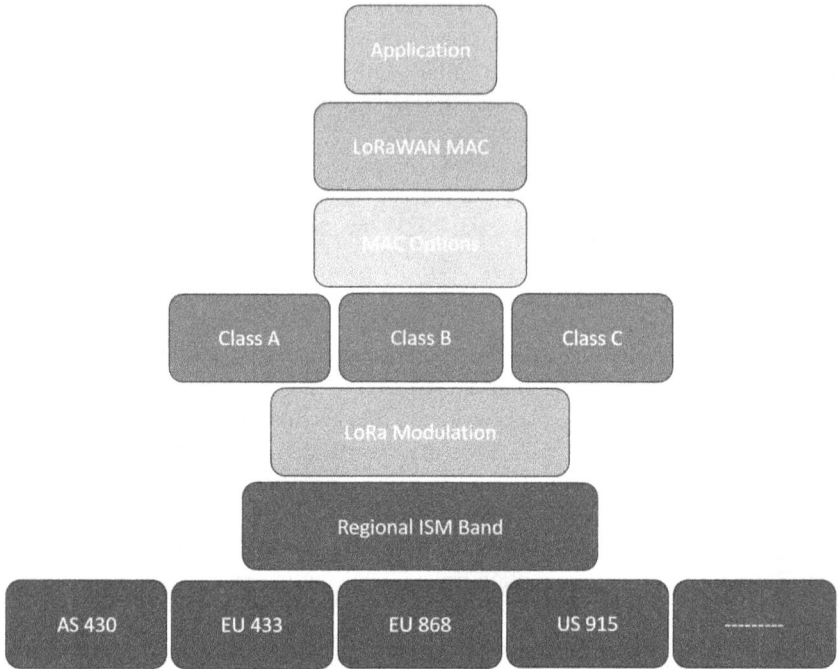

FIGURE 2.5 Building blocks of LoRa.

the standardization and global harmonization of the LoRaWAN protocol (Semtech 2021). Figure 2.5 illustrates the different blocks of LoRa, while Figure 2.6 shows the communication between end device and base station for LoRaWAN class A.

LoRaWAN divides the end devices into three classes based on different requirements of a wide range of IoT applications, e.g., latency requirements.

Class A devices allow bidirectional communications whereby each end device's uplink transmission is followed by two short downlink receive windows as shown in Figure 2.6. The transmission lot scheduled by the end device is based on its own communication needs with a small variation based on a random time basis.

Class B devices, in addition to the random receive windows of class A, open extra receive windows at scheduled times.

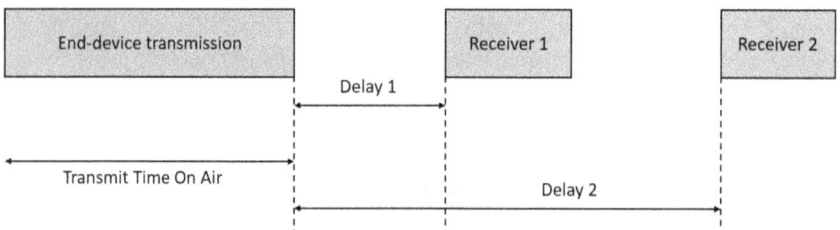

FIGURE 2.6 LoRaWAN class A bidirectional communication.

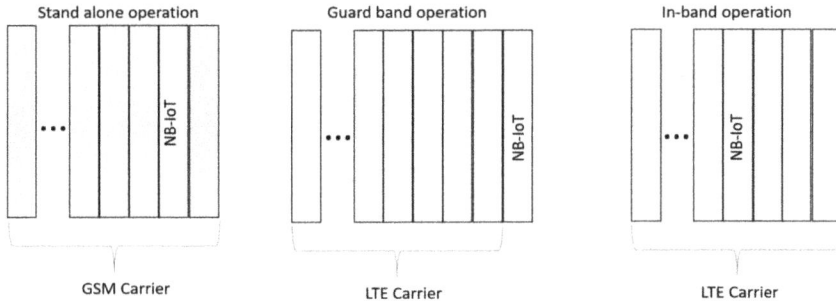

FIGURE 2.7 NB-IoT modes of operation.

Class C end devices almost continuously open receive windows and only close when transmitting at the expense of excessive energy consumption.

2.3.3 NB-IoT

NB-IoT is an IoT technology set up by 3GPP as a part of Release 13. Although it is integrated into the Long-Term Evolution(LTE) standard, it can be regarded as a new air interface (Schlienz and Raddino 2016). It is kept as simple as possible in order to reduce device costs and minimize battery consumption; thus, it removes many features of LTE, including handover, measurements to monitor the channel quality, carrier aggregation, and dual connectivity (Sinha 2017).

The NB-IoT modes of operation are as follows:

1. Stand-alone operation: A possible scenario is the utilization of the currently used GSM frequency band.
2. Guard band operation: Utilizing the unused resource blocks within an LTE carrier's guard band.
3. In-band operation: Utilizing resource blocks within an LTE carrier.

The modes are illustrated in Figure 2.7.

2.4 COMPARISON OF LPWAN TECHNOLOGIES WITH IOT FACTORS

The major IoT factors for which the above mentioned technologies are measured against are as follows (see Table 2.3):

1. QoS: Quality of Service manages network capabilities and resources to provide a reliable backbone to IoT connectivity (Link Labs 2016).
2. Payload length:
   a. NB-IoT offers the maximum payload length and allows transmission of data up to 1600 bytes.
   b. LoRaWAN allows sending a maximum of 243 bytes.
   c. Sigfox proposes the lowest payload length of 12 bytes.

TABLE 2.3

Comparison of LPWAN Technologies with IoT Factors

Factor	Sigfox	LoRa	NB-IoT
Scalability	Medium	Medium	High
Range	High	Medium	Medium
Coverage	High	Medium	Low
Deployment	Medium	High	Low
Cost efficiency	High	High	Low
Battery life	High	High	Low
Qos	Low	Low	High
Payload length	Low	Low	High
Latency performance	Low	Medium	High

3. Latency: For IoT applications with low latency connectivity, NB-IoT and LoRaWAN-Class C are better choices.
4. Scalability: NB-IoT has higher scalability compared to LoRaWAN and Sigfox. NB-IoT allows connectivity of more than 100K devices per base station compared to 50K per cell for Sigfox and LoRaWAN (Mikhaylov 2016).
5. Range: Sigfox has a range >40 km, while LoRa has <20 km and NB-IoT has <10 km (Sinha 2017).
6. Coverage: The coverage is a direct function of range; thus, the descending order of coverage is - Sigfox, LoRa, and NB-IoT.
7. Deployment: The technology standards were published in different years and below is a list showing the year in which each technology was introduced.
   a. NB-IoT—2016
   b. LoRa—2009 (Alliance 2015)
   c. Sigfox—2010
8. Cost efficiency: Sigfox and LoRa have an advantage over NB-IoT as the spectrum used in Sigfox and LoRa are free, while spectrum used in NB-IoT is expensive.
9. Battery life: The NB-IoT end devices consume additional energy due to synchronous communication and QoS handling, and its OFDM/FDMA access modes require more peak current (Oh and Shin 2016). This additional energy consumption reduces the NB-IoT end device lifetime compared to Sigfox and LoRaWAN. In general, Sigfox, LoRaWAN, and NB-IoT end devices are in sleep mode for most of the time as long as the application needs, which reduces energy consumption.

2.5   OTHER TECHNOLOGIES

2.5.1   ZIGBEE ALLIANCE (ZIGBEE ALLIANCE 2020)

The Zigbee Alliance (currently rebranded as Connectivity Standards Alliance (CSA)), established in 2002, is an organization of hundreds of companies creating, maintaining, and delivering open, global standards for the IoT. The vision of this alliance is to develop and promote the adoption of a new, royalty-free connectivity standard, simplifying development for manufacturers and increasing compatibility for customers and consumers. The Zigbee Alliance board of directors comprises of executives from Amazon, Apple, ASSA ABLOY, Comcast, Google, Huawei, IKEA, The Kroger Co., LEEDARSON, Legrand, Lutron Electronics, NXP Semiconductors, Resideo, Schneider Electric, Signify (formerly Philips Lighting), Silicon Labs, SmartThings, Somfy, STMicroelectronics, Texas Instruments, Tuya, and Wulian.

To ensure commercial success, the projects' commercial strategy group has more than 50 industry leaders, including Allegion, DSP Group, DSR Corporation, Latch, Legrand, MMB Networks, Nordic Semiconductor, NXP Semiconductors®, OSRAM, Qorvo, Schneider Electric, Signify (formerly Phillips Lighting), Silicon Labs, Somfy, and ubisys. The major objectives as listed on the website are as follows:

1. Clarify the commercial use cases that can readily be supported by the project's initial specification.
2. Define the new features required for additional commercial use cases.
3. Facilitate conversation and collaboration among members to strengthen the use and adoption of IP-based connectivity standards in the commercial market globally.
4. Advocate and encourage others to join and contribute to this market-leading effort.

2.5.2 CONNECTED HOME OVER IP

Project Connected Home over IP is a new working group within the Zigbee Alliance. This working group plans to develop and promote the adoption of a new, royalty-free connectivity standard to increase compatibility among smart home products, with security as a fundamental design tenet.

The goal of the Connected Home over IP project is to simplify the development for manufacturers and increase compatibility for consumers. The project is built around a shared belief that smart home devices should be secure, reliable, and seamless to use. By building upon IP, the project aims to enable communication across smart home devices, mobile apps, and cloud services and define a specific set of IP-based networking technologies for device certification.

As can be seen from the website and other sources, the Zigbee Alliance working on Project Connected Home over IP is still in infancy and more development will be carried out in future.

2.6 CONCLUSION

In this chapter, we saw some of the current leading technologies and infrastructure in IoT along with their pros and cons. Also, we studied the comparison between the different technologies like fog vs cloud computing and Sigfox vs LoRa vs NB-IoT. Based on these discussions, we can conclude that each technology has advantages in certain scenarios; thus, the most suitable technology to be used is dependent on the application. Some applications might also use a combination of these technologies as needed. Finally, we also highlighted some future directions with fog computing and overviewed newly developed technologies such as Zigbee and Connected Home over IP.

REFERENCES

Aazam, Mohammad and Huh, Eui-Nam. 2015. "Fog computing micro datacenter based dynamic resource estimation and pricing model for IoT." *2015 IEEE 29th International Conference on Advanced Information Networking and Applications*. IEEE. 687–694.

Aazam, Mohammad and Huh, Eui-Nam. 2016. "Fog computing: The cloud-IoT/IoE middleware paradigm." *IEEE Potentials* 35 (3): 40–44.

Alliance, LoRa. 2015. *A technical overview of LoRa® and LoRaWAN*. https://lora-alliance.org/wp-content/uploads/2020/11/what-is-lorawan.pdf, LoRa Alliance.

Atlam, Hany F, Walters, Robert J and Wills, Gary B. 2018. "Fog computing and the internet of things: A review." *Big Data and Cognitive Computing* 2: 10.

Atzori, Luigi, Iera, Antonio and Morabito, Giacomo. 2010. "The internet of things: A survey." *Computer Networks* 54: 2787–2805.

Baharudin, Ahmad Muzaffar and Yan, Wanglin. 2016. "Long-range wireless sensor networks for geo-location tracking: Design and evaluation." *2016 International Electronics Symposium (IES)*. IEEE. 76–80.

Bonomi, Flavio, Milito, Rodolfo, Zhu, Jiang and Addepalli, Sateesh. 2012. "Fog computing and its role in the internet of things." *Proceedings of the First Edition of the MCC Workshop on Mobile Cloud Computing*. 13–16.

Caiza, Gustavo, Saeteros, Morelva, Oñate, William and Garcia, Marcelo V. 2020. "Fog computing at industrial level, architecture, latency, energy, and security: A review." *Heliyon* 1–7. Volume 6

Guibene, Wael, Nowack, Johannes, Chalikias, Nikolaos, Fitzgibbon, Kevin, Kelly, Mark and Prendergast, David. 2017. "Evaluation of LPWAN technologies for smart cities: River monitoring use-case." *2017 IEEE Wireless Communications and Networking Conference Workshops (WCNCW*. San Francisco. 1–5.

2020. *i-scoop*. https://www.i-scoop.eu/internet-of-things-guide/internet-of-things/.

Ketel, Mohammed. 2017. "Fog-cloud services for iot." *Proceedings of the SouthEast Conference*. 262–264.

2016. *Link Labs*. Dec 13. Accessed 2021. https://www.link-labs.com/blog/quality-of-service-qos-lpwan-iot.

Liu, Yang, Fieldsend, Jonathan E and Min, Geyong. 2017. "A framework of fog computing: Architecture, challenges, and optimization." *IEEE Access* 25445–25454. Volume 5

Luan, Tom H, Longxiang, Gao, Li, Zhi, Xiang, Yang, Wei, Guiyi and Sun, Limin. 2015. "Fog computing: Focusing on mobile users at the edge." *arXiv preprint arXiv:1502.01815* 1–11.

Mekki, Kais, Bajic, Eddy, Chaxel, Frederic and Meyer, Fernand. 2019. "A comparative study of LPWAN technologies for large-scale IoT deployment." *ICT Express* 1–7. Volume 5

Mikhaylov, Konstantin, Petaejaejaervi, Juha and Haenninen, Tuomo. 2016. "Analysis of capacity and scalability of the LoRa low power wide area network technology." *European Wireless 2016; 22th European Wireless Conference*. VDE. 1–6.

Mouradian, Carla, Naboulsi, Diala, Yangui, Sami, Glitho, Roch H, Morrow, Monique J and Polakos, Paul A. 2017. "A comprehensive survey on fog computing: State-of-the-art and research challenges." *IEEE Communications Surveys & Tutorials* 416–464. Volume 20

Mukherjee, Mithun, Matam, Rakesh, Shu, Lei, Maglaras, Leandros, Ferrag, Mohamed Amine, Choudhury, Nikumani and Kumar, Vikas. 2017. "Security and privacy in fog computing: Challenges." *IEEE Access* 19293–19304. Volume 5

Ni, Jianbing, Zhang, Kuan, Lin, Xiaodong and Shen, Xuemin. 2017. "Securing fog computing for internet of things applications: Challenges and solutions." *IEEE Communications Surveys & Tutorials* 601–628. Volume 20

Oh, Sung-Min and Shin, JaeSheung. 2016. "An efficient small data transmission scheme in the 3GPP NB-IoT system." *IEEE Communications Letters* 660–663. Volume 21

Ratasuk, Rapeepat, Mangalvedhe, Nitin and Ghosh, Amitava. 2015. "Overview of LTE enhancements for cellular IoT." *2015 IEEE 26th Annual International Symposium on Personal, Indoor, and Mobile Radio Communications (PIMRC).* IEEE. 2293–2297.

J. Schlienz and D. Raddino. Narrowband Internet of Things Whitepaper. Accesses: Feb. 15, 2018. [Online]. Available: https://www.rohdeschwarz.com/appnote/1MA2662021.

Semtech. Accessed 2021. https://www.semtech.com/lora

2021. *Sigfox website.* Accessed 2021. https://www.sigfox.com/en/sigfox-story.

Sinha, Rashmi Sharan, Wei, Yiqiao and Hwang, Seung-Hoon. 2017. "A survey on LPWA technology: LoRa and NB-IoT." *Ict Express* 3 (1): 14–21.

Skarlat, Olena, Schulte, Stefan, Borkowski, Michael and Leitner, Philipp. 2016. "Resource provisioning for IoT services in the fog." *2016 IEEE 9th International Conference on Service-Oriented Computing and Applications (SOCA).* Macau: IEEE. 32–39.

—. 2016. "Resource provisioning for IoT services in the fog." *2016 IEEE 9th International Conference on Service-Oriented Computing and Applications (SOCA).* Macau, China: IEEE. 32–39.

Vondrouš O, Kocur Z, Hégr T, Slavíček O. 2016. "Performance evaluation of IoT mesh networking technology in ISM frequency band." *2016 17th International Conference on Mechatronics-Mechatronika (ME).* IEEE. 1–8.

Yi, Shanhe, Hao, Zijiang, Qin, Zhengrui and Li, Qun. 2015. "Fog computing: Platform and applications." *2015 Third IEEE Workshop on Hot Topics in Web Systems and Technologies (HotWeb)*, 73–78.

Yin, Bo, Shen, Wenlong, Cheng, Yu, Cai, Lin X and Li, Qing. 2017. "Distributed resource sharing in fog-assisted big data streaming." *2017 IEEE International Conference on Communications (ICC).* Chengdu, China: IEEE. 1–6.

2020. *Zigbee Alliance.* Nov 18. Accessed 2021 https://csa-iot.org/.

3 Applications of IoT through Machines' Integration to Provide Smart Environment

Manikandan Jagarajan and Ramkumar Jayaraman
Department of Computing Technologies
SRM Institute of Science and Technology,
Chennai, TN, India

Amrita Laskar
Advanced Consultant, Capgemini Engineering,
Gothenburg, Västra Götaland County, Sweden

CONTENTS

DOI: 10.1201/9781003268796-4

3.1 INTRODUCTION

The sustainability level of Internet of Things (IoT) is considered an important medium to communicate the world through IoT devices and make everything possible through that medium. Through IoT devices, we can achieve our daily activities in an easier way. Some of the IoT devices include smart phones, smart E-bikes, autonomous cars, smart temperature sensors, etc., which can be applied in various fields. As a result of this tremendous result, data generation will be in large volume from IoT devices and their supporting platforms. These data should be transmitted and stored through back-end storage centers such as cloud infrastructure [1]. Some special computing capabilities are needed to compute a large volume of data other than normal traditional big data analysis. IoT devices generate continuous streaming data that is in raw format. These data directly can't be handled for storage and other processes. These data should be processed and cleaned to extract the meaningful information and knowledge from the raw data. Next step is to handle the heterogeneity of data from different IoT devices that is dependent on the application domain e-health, smart vehicles, smart electricity grids, smart home management, etc. Impact of machine learning (ML) and deep learning (DL) will be more in job sector of the abovementioned units. Since DL and ML may be appropriate for many applications in this smart world environment, it leads to increase in demand for ML and DL products in huge numbers. Table 3.1 briefs the economic impact of IoT in different applications, and from that, we can infer that IoT applications through machine interactions are playing a key role in healthcare industry and manufacturing industry unit as represented in Table 3.1.

These approaches directly and indirectly have a strong impact on the growth of industries on economic things. The economic impact of DL and ML approaches is defined under knowledge work automation in McKinsey's report that says that complex problems, knowledgeable classification, and creative problem-solving can be achieved using computers [2]. We will talk about various ML and DL approaches in this segment, which can be simply applied to the burden of data processing, analysis, and timely decisions connecting to IoT devices.

3.2 CHARACTERISTICS OF IoT DATA

IoT data can be streaming continuous data or storage of big data. IoT data can be analyzed as streaming data, which should be processed and analyzed on time to take decisions [3] and batch processing data which can be stored and processed for future processes. Some of the examples of streamed data that are to be processed on time are patient's information to doctors, autonomous cars [4], etc., and for batch

TABLE 3.1
Demand for Machine Learning and Deep Learning Products

Industry	Percentage of Economic Impact of IoT	Scenarios
Healthcare	41	Continuous monitoring of health status of patients
Industry	33	Industry 4.0, manufacturing
		Industry, data analytics of production profit and loss, power management in smart industry.
Energy	7	Management of power production and consumption, oil and gas leakages in pipelines.
Transportation	6	Cargo monitoring, vehicle fuel level monitoring, vehicle tracking, route management.
Urban infrastructure	4	Traffic management, smart waste management, smart parking.
Agriculture	5	Smart farming, unmanned aerial vehicles such as agricultural drones, smart greenhouse, climatic condition analysis.

processing, data include weather report, monthly accident records for road safety, etc. Some of the characteristics of IoT data that need the help of DL and ML algorithms for processing and analyzing are as follows:

1. IoT continuous streaming data
2. IoT as big data and analytics

3.2.1 IoT CONTINUOUS STREAMING DATA

IoT streaming data are data that flows in huge volume, data generated or received in a very short duration of time, and should be clearly analyzed and processed to extract immediate information and make fast decisions. These kinds of analytics can be deployed and handled in high-performance computing systems as per many researchers' views and it can be parallel mode, that is, data parallelism and increased processing system. Many subtasks that are divided from the large task can be performed by data parallelism. Increased processing in the sense implements many processors to handle, which runs in pipeline mode. Methodologies such as ML and DL technologies shall be implemented in cloud/device level edges to handle the streaming data in an effective way. The sustainability of IoT data can be achieved through DL methodologies.

3.2.1.1 How DL Works Here

For instance, we will take one scenario to understand the purpose of DL/ML. Take road traffic of any city; normally, it will have separate lanes for each type of vehicles such as two wheelers, four wheelers, pedestrian, etc. for any emergency conditions; automatic traffic control should be there to give way to ambulance;

through DL image classification algorithms, we can identify the objects and alert the crowd by giving alarms. We can pass the patient data to nearby hospitals and doctors can take prior steps to treat the patient. At signals, traffic inspector can receive the ambulance data in prior and based on that automatic traffic management system can run.

3.2.2   IoT Data as Big Data and Analytics

Sensors, gadgets, online media, medical care software, temperature sensors, and other programming applications and advanced gadgets that consistently generate a large amount of organized, unstructured, or semi-organized data produce more data and its data volume is expanding. This huge amount of data leads to big data [5]. IoT data characteristics such as heterogeneity, noise, rapid growth, and variety are different from normal big data characteristics due to different sensors and objects involved during data collection and it can be further categorized as big data's volume, variety, and velocity. From IoT big data analytics, we can extract knowledge information, data patterns, extracting hidden information for data prediction and taking faster dynamic decisions. Machine learning methods are most widely used in data mining, data analytics from IoT big data. Through data analysis, if we extract 80 percent of the unseen patterns from healthcare, industry, and smart city sectors, it can benefit business industries and individuals too.

3.3   COMPUTING FRAMEWORK OF IoT DATA

The IoT data processing should be done in a computing framework; it can be of either cloud, edge, or fog computing structure. These frameworks usage for IoT data processing depend on the applications and its nature of work environment. The application can be immediate response in need of a batch processing application.

3.3.1   Cloud Computing

In cloud, data is sent to data centers for processing, where it is processed and analyzed before being made available. Since most of the processing runs at high speed, this architecture implies load balancing and high latency, suggesting that it is insufficient for processing IoT data. Processing of big data leads to the increase in the cloud server's CPU blocks because the volume of data is high [6]. Various cloud-based services are as follows:

1. Infrastructure as a service: It provides Internet-based virtualized computing resources.
2. Platform as a service (PaaS): This platform provides an environment or infrastructure, which can be used to place the abovementioned equipment on Internet for rent.
3. Software as a Service (SaaS): A distributed software model is represented in this chapter. In this model, all functional implementations will be launched by a service provider and users can access it through the Internet [7,8].

3.3.2 FOG COMPUTING/EDGE COMPUTING

Fog architecture makes use of end-device facilities (switches, routers, multiplexers, and so on) for computing, storage, and processing. Fog computing architecture is made up of physical and logical network components, software, and hardware that come together to form a wide network of interconnected devices. It is used to transport data from the data center to the server's edge. Fog is being implemented in many fields like e-health, smart traffic management, and military applications [9,10]. In edge computing, the processing of data is run at edge devices themselves or the nearby edge servers, which will be physically close to the end devices. Autonomous cars face recognition application with reduced response time [11] for several applications.

3.4 MACHINE INTERACTIONS IN SMART CITIES

Interactions between machines can be drawn from the multidimensional selection of the smart cities and scenarios; shortly this can be viewed as human-computer interaction. Dimensionalities include the features such as urban regions, rural regions, smart mobility, smart economy, smart people, citizen engagement, creativity, and innovation [12]. Different authors provide their own way of exploring the dimensions and attributes of the smart world [13,14]. For efficient smart world services, various advanced technologies such as unmanned aerial vehicles, robotics, cloud, fog, and edge computing infrastructures are needed along with reliable and robust communication protocols and networking infrastructure are also needed [15–20]. Network architecture protocols differ based on the smart city application domains. For applications that fall within a short range of communications use personal area network (PAN) class, which includes IEEE 802.15.4, IEEE 801.15.1, some of its applications are smart buildings, smart electricity management, etc. [21,22]. Longer range communications require LAN (IEEE 802.11), which includes applications like manufacturing industry unit, smart transport system. Wide area communications can use 4G, 5G, satellite services, and WAN, which in turn, IEEE 802.16, includes applications such as smart grids [23]. Some of the needed parameters along with protocols for machine interactions are bandwidth, security, delay tolerance, latency time, heterogeneity of devices [24,25].

3.5 MACHINE LEARNING AND DEEP LEARNING AS A FAMILY FOR IoT

Gaining knowledge from streaming information requires ongoing (or close to continuous) updates to a model. There are numerous significant perspectives to consider in this "learning stage," like improving latency delay, reducing delay response time [26]. In this section, we will discuss the various ML-based and advanced ML-based, that is, DL approaches to gather IoT information data, analyze, process it and other operational features [27].

Machine learning algorithms take input samples as training dataset and testing dataset. Generally, there are three types of learning: supervised, unsupervised, and

TABLE 3.2

Machine Learning Models

ML Methods/DL Methods	Data Processing Type	Outcome	Reference Work
K-nearest neighbor (KNN)	Classification	Tree-based search	[38]
Support vector machine (SVM)	Classification and regression	Detection of microcalcification clusters in digital mammogram	[39]
Multiple linear regression	Regression	Estimation of the market value of football players who play in the forward positions	[40]
Random forests	Classification and regression	Analysis of large genome-wide association (GWA) datasets	[41]
Daily consumption estimation network and intraday load forecasting network	Edge devices, control center in cloud	Smart meters (for different electronic gadgets)	[42]

reinforcement learning [28–30]. The training set in supervised learning consists of input vector samples and their corresponding appropriate labels, also called target vectors [31]. The training set does not need to be labeled in unsupervised learning. Reinforcement learning is concerned with determining the best action or series of actions to take in order to maximize payoff in a given scenario.

It's important to consider the three principles below to figure out which model is best for handling and dynamic on shrewd knowledge produced by things in the IoT [32–36]. The first is an application for the IoT. The knowledge qualities of the IoT are the second, and the information-driven vision of artificial intelligence (AI) calculations is the third [37]. These ideas can help you understand how ML models can be applied to IoT data in a smart way in Table 3.2.

3.5.1 MACHINE LEARNING APPROACHES

3.5.1.1 K-Nearest Neighbor

The aim of K-nearest neighbors (KNNs) algorithm is to sort out appearing unknown data points by examining the data points closest to K in the input or feature space's training set. Hence, to find out new data points from KNNs, parameters such as distance metric, Euclidian distance, Manhattan distance, and Minkowski distance are needed. If new unknown data point is arrived, then we can classify the data point based on value of K and the maximum voters from the nearest neighbor. If the value of K is low may be $k = 1$ or 2, then the data can be noisy. If the value of K is larger, then the classification will be smoother.

The distance functions can be given by the following formula:

$$\text{Euclidean Distance:} \sqrt{\sum_{i=1}^{k} \left(x_i - y_i \right)^2}$$

$$\text{Manhattan Distance: } \sum_{i=1}^{k} |x_i - y_i|$$

Minkowski distance: $\left(\sum_{i=1}^{k} \left(|x_i - y_i|\right)^q\right)^{1/q}$ —q can be real value. If $q = 1$, then it is equivalent to Manhattan distance and if $q = 2$, then it is equivalent to Euclidean distance.

The above three distance calculations are only valid for continuous variables. For categorical variable, Hamming distance will be used.

$$\text{DH} = \sum_{i=1}^{k} |x_i - y_i|, \text{ if } x = y \text{ then } D = 0$$

$$x \neq y \text{ then } D = 1$$

An object's classification is determined by a maximum vote among its neighbors, with the object being assigned to the class with the most members among its k closest neighbors (k is a positive integer, typically small). If the value of k is 1 ($k = 1$), the object is simply assigned to the class of its closest neighbor. KNN has the disadvantage of requiring the entire training set to be stored, rendering it unscalable for large data sets in Figure 3.1.

The above flow chart defines the steps involved in KNN algorithm for identifying the new data point in which category it falls. K value can be random number, most preferably K value will be 5.

3.5.1.2 Support Vector Machine (SVM)

SVM (support vector machine) is a supervised ML algorithm that can be used to solve classification and regression problems. SVMs are non-probabilistic binary classifiers that seek out the separating hyperplane that divides all of the training set's classes by the greatest possible margin. Then, depending on side of the hyperplane, which it falls, the unseen new data point with the predicted label is resolved [16]. We will discuss linear and non-linear function of the input variables in the below section. The hyperplane's normal vector is denoted by w, and the parameter influencing the hyperplane's offset from the origin along its normal vector is denoted by b. For each training point x_i, a vector ξ_i called a slack variable, which gives the distance in units of $|w|$ by which this training point violates the margin. A constrained optimization problem is used to describe binary linear classification, as shown below:

$$min_{(w,b,\xi)} f\left(w,b,\xi\right) = \frac{1}{2} w^T w + C \sum_{i=1}^{n} \xi_i \text{ with reference to } y_i \left(w^T x_i + b\right) - 1$$

$$+ \xi_i \geq 0 \ i = 1,2,\ldots,n$$

```
        ┌─────────────────────┐
        │  Find K Number of   │
        │     neighbors       │
        └─────────────────────┘
                  │
                  ▼
        ┌─────────────────────┐
        │ Find Euclidean Distance of │
        │ K- number of neighbors     │
        └─────────────────────┘
                  │
                  ▼
        ┌─────────────────────┐
        │ Choose K nearest Neighbors as │
        │ per the Euclidean distance    │
        │ calculation.                  │
        └─────────────────────┘
                  │
                  ▼
        ┌─────────────────────┐
        │ Count number of Data points in │
        │ each group in these K neighbors │
        └─────────────────────┘
                  │
                  ▼
        ┌─────────────────────┐
        │ Assign the new data points to │
        │ the group which is having the │
        │ number of maximum             │
        │ neighbors.                    │
        └─────────────────────┘
                  │
                  ▼
        ┌─────────────────────┐
        │  KNN model algorithm │
        │    end process       │
        └─────────────────────┘
```

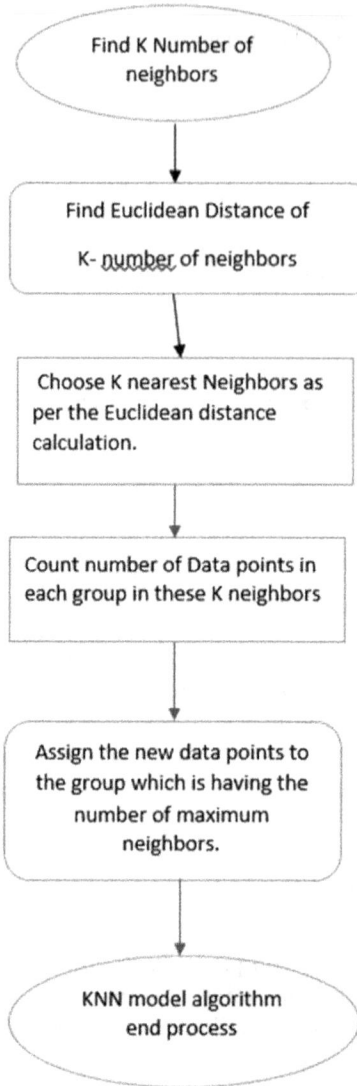

FIGURE 3.1   Represents the KNN algorithm.

This is L1 norm representation for penalty term; there are other norms that exist such as L2 norm for penalty that can be decided based on application. Cross-validation or Bayesian optimization can be used to choose hyperparameter C. SVM's can be of two types, first one is linear SVM and second one is non-linear SVM.

3.5.1.3   Linear SVM

It is used to separate data from datasets linearly. If two classes of data sets can be classified or separated by a line or using a line function, then it is termed linear SVM and the classifier is known as linear SVM classifier.

3.5.1.4 Non-Linear SVM

It is used to separate data in a hyperplane, which cannot be separated by using a straight line, that is, non-linearly separated data. Special functions are used to separate the datasets such as by converting two-dimensional data to three-dimensional data, circle function in two-dimensional space, etc. [43]. The above figure represents the classification of datasets in two-dimensional and three-dimensional space using SVMs.

Regression analysis is a group of ML methods for predicting one or more target variables' (x) values to predict the value of an outcome continuous variable (y). In a nutshell, the aim of a regression model is to create a mathematical equation that defines y as a function of x variables. On the basis of new values of predictor variables (x), the mathematical equation can then be used to predict the outcome (y).

3.5.1.5 Linear Regression

It is the most basic and widely used method for a continuous variable prediction. It denotes a relationship that is linear between the outcome and predictor variables. The aim of linear regression is to learn a function $f(x,y)$; the linear regression equation can be represented as

$$y = a0 + a*x + c,$$

where $a0$ represents intercept;

> a represents weight of regression inextricably linked to the predictor variable x,
> c represents residual error.

The linear regression coefficients are determined in such a way that the error in predicting the outcome value is minimized. The ordinary least squares method is used for computing beta coefficients. When there are multiple predictor variables, such as $x1$ and x, the regression equation can be written as $y = b0 + b1*x1 + b2*x2 + e2$. In some cases, there may be an association effect between some predictors, such as when the value of a predictor variable $x1$ is increased, the efficacy of the predictor $x2$ in explaining variance in the outcome variable is increased.

The equation for regression can be written as $y = a0 + a1*x1 + a2*x2 + c$. When there are more predictor variables, such as $x1$ and $x2$, there may be a correlation effect between some predictors in some situations, such as when the value of a predictor variable $x1$ increases, the predictor $x2$'s effectiveness in explaining variation in the outcome variable increases.

There are many methods for training the model, including ordinary least squares, root mean squared error (RMSE), Bayesian linear regression, and least-mean-squares (LMS). LMS is appealing because it is faster, scalable accurate to larger data sets, and uses the stochastic gradient descent (also called sequential gradient descent) technique to learn parameters online. The best model is the one with the least amount of prediction error.

The model prediction error is calculated by the RMSE. It's the average difference between the model's expected value and the observed known outcome values. It can be computed as

$RMSE = \sqrt{\left(\text{mean} \left(\text{observed} - \text{predicted} \right)^2 \right)}$. The model is better, if the RMSE is low.

3.5.1.6   Classification and Regression Trees (CART)

In classification and regression trees (CART), the input space is divided into axis-aligned cuboid regions RK, and each region is allocated a different regression or classification model for label predictions of data points that fall within that region. A binary tree that represents a sequential decision-making process can be used to define the process of the corresponding target label prediction provided an unknown, new input vector (data point) x.

3.6   DEEP LEARNING FOR IoT

Deep learning has become an important part of data analysis and analytics of IoT. One of the top three strategic technical innovations in 2017 has been identified as DL, along with IoT [44]. This is because of the lagging of traditional ML techniques to fulfill the needs of analytics of IoT. Excellent results in numerous fields such as image processing and recognition, natural language processing, retrieval of information, indoor translation, speech recognition, and so on have been shown by DL models and these services are the foundation for IoT applications [2]. Deep learning can be defined as the subset of ML which is influenced by the human brain structure that tries to take decisions and conclusions as humans.

3.6.1   RESTRICTED BOLTZMANN MACHINE (RBM)

It is a generative stochastic artificial neural network that can learn a probability distribution based on its inputs. It consists of two layers: visible and hidden layers. The input is received by the visible layer, while the hidden layer stores the latent variables. In restricted Boltzmann machine (RBM), visible layer's each neuron is connected to the hidden layer neurons and, in reverse, but within the same layer, no connection should be there between two neurons. Every neuron in both layers is attached with a bias. Backpropagation and gradient descent techniques are used in the training phase to optimize the network's connection weights. In DL approaches, also RBM can be applied. Deep belief networks (DBNs), in particular, can be created by "stacking" RBMs and fine-tuning the resulting deep network via gradient descent and backpropagation. The algorithm to train RBM, which is focused on optimizing the weight vector W is contrastive divergence algorithm in Figure 3.2.

3.6.2   DEEP BELIEF NETWORKS

The DBN is a generative form of DNN with many hidden layers and a visible layer. DBNs have the ability to extract hierarchical meaning from data. Any DBN layer is RBMs, and the previous RBM's secret layer serves as the current DBN layer's input layer. A DBN may learn to probabilistically recreate its inputs when trained on a set of instances with no supervision. Feature detectors are then used by the layers. A DBN can be further trained to do classification under supervision after completing this learning

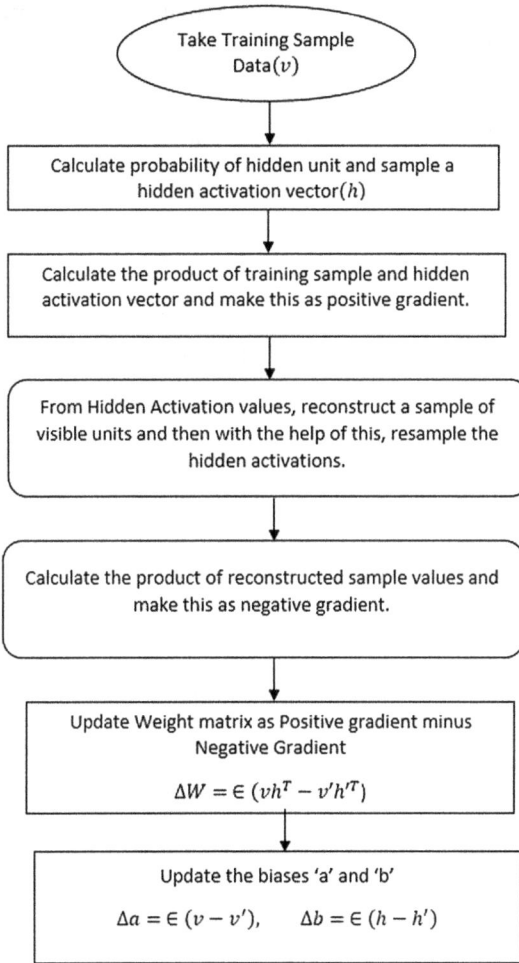

FIGURE 3.2   Represents the RBM algorithm.

stage. DBNs are made up of simple, unsupervised networks like RBMs or autoencoders, with the hidden layer of each sub-network serving as the visible layer for the next.

3.6.3   CONVOLUTION NEURAL NETWORK

A convolutional neural network (CNN) takes input as a two-dimensional image or a voice signal and uses a series of hidden layers to extract hierarchical characteristics. In CNN, the hidden layers consist of convolutional layers and a fully connected layer. The convolutional layer is made up of filters that will perform operations with the input data. The dimension size of the filters is smaller than the dimensions of the input. The output of the convolutional layers is the feature maps that come out by the product of the input and the filter. To reduce representation dimension, these feature maps are then sent through a series of pooling layers called max pooling. This is done to speed up computation and prevent overfitting. Another essential component of CNN is the rectified linear units (ReLU), which consists of neurons with a SoftMax activation function in the form of $f(y) = \max(0, y)$ [45]. A typical convolution operation is given by an equation; Figure 3.2 [46] shows the CNN architecture like Figures 3.3 and 3.4.

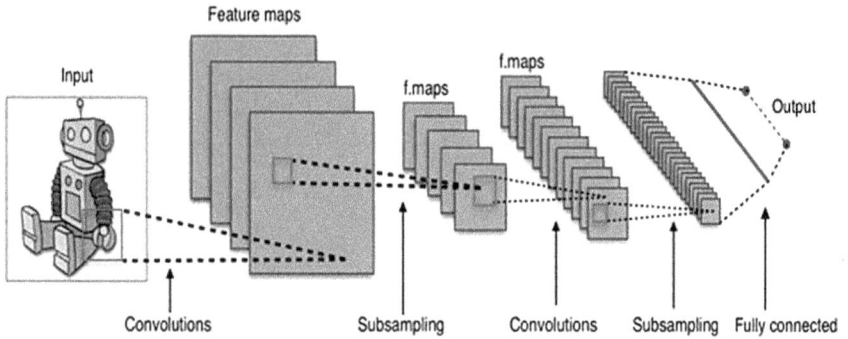

FIGURE 3.3   CNN architecture.

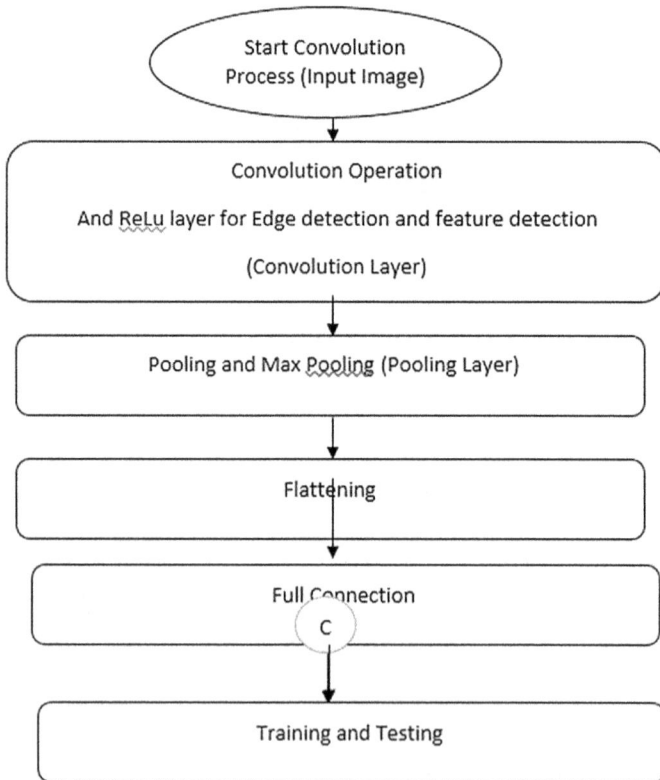

FIGURE 3.4   Represents the CNN algorithm.

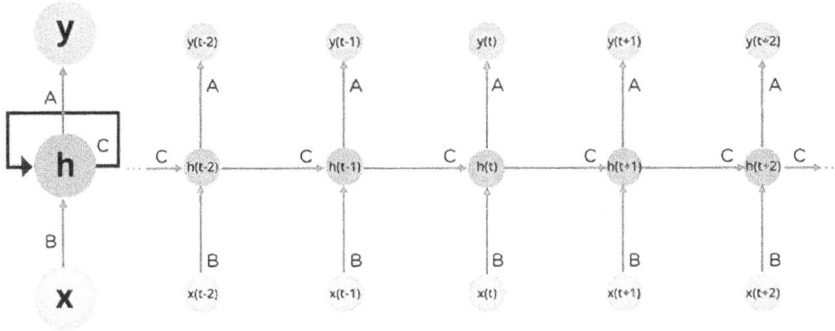

FIGURE 3.5   Architecture of RNN.

$$(f * g)(t) = \int_{-\infty}^{\infty} f(\top) g(t - \top) d\top$$

3.6.4   Recurrent Neural Network (RNN)

For modeling time series tasks, feed-forward neural network is not convenient. Recurrent neural network (RNN) was designed to solve those kinds of tasks. The RNN receives both the current and previous input samples as input. RNN's output at time step "t" is dependent on RNN's output at time step "t–1." Because of this, every neuron's output is given as input to the next following step. As a consequence, each RNN's neuron has an internal memory that stores data from the previous input. Backpropagation through time (BPTT), a version of the backpropagation algorithm, is used to train the network. RNN architecture is shown in Figure 3.5.

3.7   APPLICATIONS OF IoT IN SMART WORLD SCENARIO—A CASE STUDY

3.7.1   IoT in Smart Waste Management System

These days, waste management is a global concern. According to World Bank research, the world creates 2.01 billion tonnes of municipal solid garbage each year, with at least 33% of it not being managed in an environmentally sound manner, and it is growing in a fast manner. In trash collection and maintenance process, IoT plays a very important role by reducing the operational cost, resource maintenance, classification of wastes. We can use wastebins with intelligence using IoT sensors [47], which will collect data and transfer it to the server using Internet. Based on the data, collected future data can be predicted and placements of wastebins can be located where waste is segregating more. Machine learning plays an important role in IoT-based waste management through effective algorithm such as classification, regression, and clustering. With less human intervention and smart IoT technologies, waste dustbins can be categorized into solid wastes and liquid wastes, reusable wastes. These can be achieved using a single microcontroller with a ultrasound sensor used for checking the level of dustbins and transferring information using LoRa E32 technology [24]. The smart bin can be solar-powered waste management bin which will automatically compress the wastes in the space and easily manages the wastes so that instead of filling the wastebins once by humans can be equivalent to filling five times using IoT sensors, further data can be transferred to cloud for managing, processing.

3.7.2   IoT in Smart Agriculture

In terms of Agriculture sensors, unmanned drones, autonomous vehicles, and other types of IoT devices are used mostly in recent trends. Through IoT, farmers can monitor the soil moisture through remote access and use proper fertilizers for proper crops through sensors. Through data analytics from the food chain management collected data one can improve productivity. Jeetendra Shenoy et al. [25] proposed a solution to reduce the number of middle hops and agents between the farmer and the end consumer by using an IOT-based solution, which in turn reduces transportation costs and price predictability based on historical data analytics and current market conditions; the author [25] discussed polyhouse that will maintain data on humidity, soil pH, temperature, soil moisture and the data will be sent to cloud for analysis and process, then from the server, instructions will be sent to IoT controller board to control the water flow, to maintain the soil moisture. Even crop analysis also can be achieved through IoT data analysis.

3.8   APPROACHES OF DEEP LEARNING WITH USE CASES FOR IoT DATA ANALYSIS

Deep learning can be used for analyzing hidden data patterns from the data, because of this for data analytics, deep learning is considered as one of the most important tools which has grown higher. To integrate heterogeneous mixture of data sensed from various modalities, specific DL models have been proposed. RBMs have a lot of promise when it comes to feature extraction and classification. It's been applied and used in many IoT use cases such as intrusion detection [48], emergency detection [49], forecasting demand for electricity [50]. DBNs excel at extracting input features from a large set of input data. DBNs have been used in lot of IoT use cases such as efficient load balancing scheme [51], classification of emotions based on electroencephalography (EEG) [52] in Table 3.3.

3.9   CONCLUSION

This chapter deals with the processing of raw data that are coming from IoT devices that can be captured, computed, stored, processed either at edge, cloud, or fog computing level. In order to do data analysis and data analytics of IoT, we are in need of framework that can support data heterogeneity, handling larger volume of data. Data analysis and finding insights from data patterns that are done at cloud can't be suitable for IoT streaming data applications; owing to this, we should move for a technology that can suit the abovementioned problems. Deep learning, in combination with ML, has been found to be successful in analyzing difficult, data patterns that are more complex in nature produced by IoT. Maximum knowledge of the domain can be extracted

TABLE 3.3
Existing Works with Usecase Description

Paper	Description	Use case	Application
[48]	This paper proposes a restricted Boltzmann machine-based anomaly detection algorithm on healthcare robots for emergency detection.	Emergency detection	Healthcare industry
[49]	The use of restricted Boltzmann for intrusion detection is demonstrated in this analysis.	Intrusion detection	In all sector applications.
[50]	This paper proposes a method based on a restricted Boltzmann machine variant, for predicting electricity demand. The results show that the proposed method for forecasting power demand outperforms traditional benchmark approaches.	Power demand forecasting	Smart grid
[51]	A DBN algorithm was used for examining air conditioner use patterns in order to come up with a strategy in smart homes for optimized energy consumption in this research. The accuracy of air conditioners in office rooms is 97.69 percent.	Analysis of air conditioners' electricity consumption	Smart home
[52]	This research uses a deep belief network to achieve the equipped and best load balancing in IoT scenarios.	Efficient load balancing scheme	All sector applications
[53]	Deep belief networks are used in this paper to classify two emotional groups based on EEG. In contrast to other state-of-the-art approaches, the proposed approach produces more accurate results.	EEG-based emotion classification	Smart healthcare

from the data insights through ML and DL. The IoT data through multiple filters that can be used for business decisions, healthcare sector improvements, agricultural improvements, and almost in all sectors. Moreover, DL models generate prediction outcome more accurate. This chapter gives a detailed view on DL and ML approaches towards IoT data analysis and processing. Deep learning along with ML can be implemented either in edge infrastructure or fog environment to handle huge amounts of live streaming data that need very less response time. More research has to be carried out in this particular area to bring out the advantages of DL and ML approaches in edge infrastructure in the upcoming years, which will be having data in Zettabytes.

REFERENCES

1. Adi, E., Anwar, A., Baig, Z. Sherali, Z., "Machine learning and data analytics for the IoT". Neural Comput Appl, 32:16205–16233, 2020.
2. J. Manyika, M. Chui, J. Bughin, R. Dobbs, P. Bisson, and A. Marrs, "Disruptive technologies: Advances that will transform life, business, and the global economy". McKinsey Global Institute, San Francisco, CA, 180, 2013.
3. M. Mohammadi, A. Al-Fuqaha, S. Sorour and M. Guizani, "Deep learning for IoT big data and streaming analytics: A survey". IEEE Commun Surv Tutorials, 20(4): 2923–2960, Fourthquarter 2018.
4. M.S. Mahdavinejad, M. Rezvan, M. Barekatain, P. Adibi, P. Barnaghi, A.P. Sheth, "Machine learning for internet of things data analysis: a survey", Digital, 2018.

5. K. Kambatla, G. Kollias, V. Kumar, A. Grama, "Trends in big data analytics". J Parall Distrib Comput, ISSN 0743-7315, 74(7): 2561–2573, 2014.

6. M. Marjani, F. Nasaruddin, A. Gani, A. Karim, I.A. Targio Hashem, A. Siddiqa, "Big IoT data analytics: Architecture, opportunities, and open research challenges". IEEE Access, 5: 5247–5261, 2017.

7. A. Papageorgiou, M. Zahn, E. Kovacs, "Efficient auto-configuration of energy-related parameters in cloud-based IoT platforms". 2014 IEEE 3rd International Conference on Cloud Networking (CloudNet), 236–241, 2014.

8. L. Wang, R. Ranjan, "Processing distributed internet of things data in clouds". IEEE Cloud Comput, 2(1): 76–80, Jan.–Feb. 2015.

9. Y. Shi, G. Ding, H. Wang, H.E. Roman, S. Lu, "The fog computing service for health-care", Future Information and Communication Technologies for Ubiquitous HealthCare (Ubi-HealthTech), 2nd International Symposium on IEEE, 1–5, 2015.

10. F. Ramalho, A. Neto, K. Santos, N. Agoulmine, J.B. Filho, "Enhancing ehealth smart applications: A Fog-enabled approach". 17th International Conference on E-health Networking, Application & Services (HealthCom), IEEE, 323–328, 2015.

11. S. Raza, S. Wang, M. Ahmed, M.R. Anwar, "A survey on vehicular edge comput-ing: Architecture, applications, technical issues, and future directions". Wireless Communications and Mobile Computing, 2019.

12. J.R. Gil-Garcia, J. Zhang, G. Puron-Cid, "Conceptualizing smartness in government: an integrative and multi-dimensional view". Gov Inf Q, 33: 524–534, 2016.

13. H.V. Jagadish, B.C. Ooi, K.-L. Tan, C. Yu, R. Zhang, "iDistance: An adaptive bþ-tree based indexing method for nearest neighbor search". ACM Trans Database Syst (TODS), 30(2): 364–397, 2005.

14. A. McCallum, K. Nigam, "A comparison of event models for naive Bayes text classifi-cation". AAAI-98 Workshop on Learning for Text Categorization, 752, Citeseer, 41–48, 1998.

15. H. Zhang, "Exploring conditions for the optimality of naïve Bayes". Int J Pattern Recog Artif Intell, 19(2): 183–198, 2005.

16. C. Cortes, V. Vapnik, "Support-vector networks". Mach Learn, 20(3): 273–297, 1995.

17. B. Scholkopf, A.J. Smola, Learning with Kernels: Support Vector Machines, Regu-larization, Optimization, and beyond. MIT Press, 2001

18. I. Guyon, B. Boser, V. Vapnik, "Automatic capacity tuning of very large vc-dimension classifiers". in: Advances in Neural Information Processing Systems, 147–147, 1993.

19. C.-C. Chang, C.-J. Lin, "LIBSVM: A library for support vector machines". ACM Trans Intell Syst Technol, 2(3): 1–27, 2011.

20. N. Cristianini, J. Shawe-Taylor, An Introduction to Support Vector Machines and Other Kernel-based Learning Methods. Cambridge University Press, 2000.

21. K. Pearson, "Liii. on lines and planes of closest fit to systems of points in space". Lond Edinb Dublin Philos Mag J Sci, 2(11): 559–572.

22. H. Abdi, L.J. Williams, "Principal component analysis". Wiley Interdiscip Rev Comput Stat, 2(4): 433–459, 2010.

23. I. Jawhar, N. Mohamed, J. Al-Jaroodi, "Networking architectures and protocols for smart city systems". J Internet Serv Appl, 9: 26, 2018.

24. T.A. Khoa, C.H. Phuc, P.D. Lam, L.M.B. Nhu, N.M. Trong, N.T.H. Phuong, N. Van Dung, N. Tan-Y, H.N. Nguyen, D.N.M. Duc, "Waste management system using IoT-based machine learning in university". Wireless Commun Mobile Comput, 2020.

25. J. Shenoy and Y. Pingle, "IOT in agriculture". 2016 3rd International Conference on Computing for Sustainable Global Development (INDIACom), 2016, 1456–1458.

26. R. Bro, A.K. Smilde, "Principal component analysis". Anal Methods, 6(9): 2812–2831, 2014.

27. J. Ali, R. Khan, N. Ahmad, I. Maqsood, "Random forests and decision trees". IJCSI Int J Comput Sci Issues, 9(5):1694–0814, 2012.

28. D. Barber, Bayesian Reasoning and Machine Learning. Cambridge University Press, 2012.
29. C.M. Bishop, Pattern Recognition and Machine Learning. Springer, 2006.
30. K.P. Murphy, Machine Learning: A Probabilistic Perspective. MIT Press, 2012.
31. N.I. Gould, P.L. Toint, "A quadratic programming bibliography". Numer Anal Group Intern Rep, 1, 2000.
32. G.A. Seber, A.J. Lee, Linear Regression Analysis. vol. 936, John Wiley & Sons, 2012.
33. A. Coates, A.Y. Ng, "Learning feature representations with K-Means", in: G. Montavon, G.B. Orr, K.R. Müller (Eds.), Neural Networks: Tricks of the Trade. Lecture Notes in Computer Science, vol. 7700, Springer, Berlin, Heidelberg, 2012.
34. V. Jumutc, R. Langone, J.A. Suykens, "Regularized and sparse stochastic k-means for distributed large-scale clustering". Big Data (Big Data), 2015 IEEE International Conference on, IEEE, 2535–2540, 2015.
35. Y. Qin, Q.Z. Sheng, N.J. Falkner, S. Dustdar, H. Wang, A.V. Vasilakos, "When things matter: a survey on data-centric internet of things". J Netw Comput Appl, 64: 137–153, 2016.
36. M.A. Kafi, Y. Challal, D. Djenouri, M. Doudou, A. Bouabdallah, N. Badache, "A study of wireless sensor networks for urban traffic monitoring: applications and architectures". Proc Comput Sci, 19: 617–626, 2013.
37. J. Neter, M.H. Kutner, C.J. Nachtsheim, W. Wasserman, Applied Linear Statistical Models, 4, Irwin, Chicago, IL, 1996.
38. S. Yi, Z. Hao, Z. Qin, Q. Li, "Fog computing: Platform and applications," in Proc. 3rd IEEE Workshop Hot Topics Web Syst. Technol. (HotWeb), Washington, DC, USA, 73–78, 2015.
39. I. El-Naqa, Y. Yang et al., "A support vector machine approach for detection of microcalcifications". IEEE Trans Med Imaging, 21(12), 2002.
40. Y. Kologlu, H. Birinci, S.I. Kanalmaz, B. Ozyilmaz, "A Multiple Linear Regression Approach For Estimating the Market Value of Football Players in Forward Position", 2018.
41. B.A. Goldstein et al., "An Application of Random Forests to a genome-wide association dataset: Methodological considerations & new findings", 2010.
42. L. Li, K. Ota and M. Dong, "When weather matters: IoT-based electrical load forecasting for smart grid,". IEEE Commun Mag, 55(10): 46–51, Oct. 2017
43. https://towardsdatascience.com/support-vector-machine-introduction-to-machine-learning-algorithms-934a444fca47.
44. K. Panetta, "Gartner's top 10 strategic technology trends for 2017", 2016. [Online]. Available: http://www.gartner.com/smarterwithgartner/ gartners-top-10-technology-trends-2017.
45. T.J. Saleem, M.A. Chishti, "Deep learning for Internet of Things data analytics". Procedia Comput Sci, 163: 381–390, 2019.
46. https://en.wikipedia.org/wiki/File:Typical_cnn.png.
47. L. Breiman," Random forests". Mach Learn, 45 (1): 5–32, 2001.
48. T. Aldwairi, D. Perera, M.A. Novotny, "An evaluation of the performance of restricted Boltzmann machines as a model for anomaly network intrusion detection". Comput Netw, 144: 111–119, 2018.
49. H.-G. Kim, S.-H. Han, H.-J. Choi, Discriminative Restricted Boltzmann Machine for Emergency Detection on Healthcare Robot. IEEE, 2017.
50. E. Mocanu, E.M. Larsen, P.H. Nguyen, P. Pinson, M. Gibescu, "Demand Forecasting at Low Aggregation Levels using Factored Conditional Restricted Boltzmann Machine, 2016".
51. H.-Y. Kim, J.-M. Kim, "A load balancing scheme based on deep-learning in IoT". Cluster Comput, Volume 20 Issue 1: 873–878, 2016.
52. W.-L. Zheng, J.-Y. Zhu, Y. Peng, and B.-L. Lu. "EEG-Based Emotion Classification Using Deep Belief Networks." IEEE Conference on Multimedia and Expo (ICME), 2014.
53. W. Song, N. Feng, Y. Tian, S. Fong, and K. Cho, "A deep belief network for electricity utilisation feature analysis of air conditioners using a smart IoT platform". J Inform Process Syst, 14(2): 162–175, 2018.

Part II

HMI in IoT-Based Distributed Commercial Systems

4 Smart Energy Management on the Wearable Devices Based on Edge Computing

Aileni Raluca Maria
Politehnica University of Bucharest,
Bucharest, Romania

Suciu George
Politehnica University of Bucharest and
Beia Consult International,
Bucharest, Romania

*Poenaru Carmen, Anghel Madalina,
Mocanu Cristian, Subea Oana, and Orza Oana*
Beia Consult International,
Bucharest, Romania

CONTENTS

DOI: 10.1201/9781003268796-6

4.1 INTRODUCTION

For a long time, the role of the remote devices has been to assimilate, store, and send data to cloud systems. However, today edge computing aims to store and process data within the measurement devices. Among the benefits of edge computing are reducing data transfer costs to the Cloud, reducing the reaction time, carrying out an immediate analysis (for example, in the field), increased cybersecurity (by avoiding to send and store raw data in the cloud access to temporal data for real-time analysis the ability to work without even connecting to the Cloud).

The increased number of patients with chronic diseases and who are demanding constant monitoring has created a significant stimulus for developing scalable body area sensor networks (BASNs) for remote health applications [1]. Instigated by the limited power source and small form factors, BASNs has an outstanding design and operational challenges, mainly focusing on energy optimization. The wearable health monitoring systems (WHMS) [2] aim a larger scale telemedicine system for ambulatory health monitoring. Advances in wireless sensing and wearable sensors have made BASNs a promising technology for healthcare monitoring without constraining the activities of the patient [3].

During the last few years, most of the research in the area of remote monitoring has focused on the concerns regarding wireless sensor designs, signal compression techniques, sensor miniaturization, and low-power hardware design [4–6]. The state-of-the-art for hardware, technologies, and standards for BASN have been discussed in multiple papers [7], and different encoding algorithms have been demonstrated. However, most of them ignore the trade-off between encoding and transmission energy. The cross-layer design of energy minimization that addresses the time-frequency allocation under misuse constraints for delay-sensitive transmission in BASNs is an ignored subject in the research field. In addition to this, other papers [8] studied the trade-offs between the compression ratio and distortion for lossy EEG compression without considering the energy consumption. The properties of compressed ECG data for energy-saving utilizing a selective encryption mechanism have also been studied [9]. Other research works focused on reducing power consumption at MAC layer by avoiding idle listening and collision [10] or by presenting latency-energy optimization [11].

From the energy consumption point of view, the effect of source encoding is two-fold. First, increasing the compression ratio reduces the amount of data transmission, which results in saving a large amount of transmission energy. An inversely proportional scenario is when efficient data compression often needs higher computational complexity, leading to more significant energy consumption and increasing in encoding distortion [12]. These two adverse effects indicate that in practical system design, there is always a trade-off between the transmitted rate, energy consumption, and encoding distortion. To find the best trade-off solution, it needs to develop an analytic framework to model the energy-rate-distortion (E-R-D) nature of the ECG monitoring encoding system. The E-R-D analysis proposes a theoretical and practical guideline in system design and performance optimization for wireless EEG monitoring system under delay and distortion constraints.

4.2   ENERGY MANAGEMENT ON MEDICAL WEARABLE DEVICES

Wearable medical devices are available even in micro/nanoscale, and the miniaturization of the devices is often limited due to the size and shape of the power supply. An aspect that significantly influences the ability to use a device is the battery life, and it is essential to maintain a low power consumption of the sensor. Thus, a reduction of the energy consumption of the devices is desired, which implies a continuous decrease in the number of integrated circuits and improved efficiency in energy scavenging technologies, until the design and implementation of solutions without batteries, with a zero-energy consumption [13]. These technologies include those that are wirelessly powered by dedicated sources, such as inductive power—for example, a wearable stethoscope that is both wireless and battery-less [14], or those that harvest energy from motion and the environment, such as capturing ambient R.F. energy and solar power. An innovative technology that has as its first advantage the low energy consumption is represented by BLE (Bluetooth low energy), which generally differs from the others in that the RF (radio frequency) mode is activated only to transmit data or to receive data, then deactivates to save energy consumption. Study results [14] show that current consumption is on average reduced by BLE about two times more than by ZigBee and 1.5 times more than ANT+ (adaptive network topology). It is a challenge to establish exact numbers that indicate the current average consumption for each low-power wireless standard, as it depends on many vendor-specific parameters, some of which are software-related, others are hardware-related. Wearable devices [15] can only use batteries as power supply rather than stationary power, and thus it is a tedious process to charge such devices. Frequently recharging the batteries or replacing them can inevitably reduce the practicability and affect the preference and satisfaction of users. Additionally, large amounts of energy consumption can provoke high heat. If the cooling system can't keep up with overheating, the whole situation could damage the experience of the user or even cause low-temperature scald. Therefore, the energy consumption of wearable devices is a problem that must be paid attention to.

Wearable devices [16] autonomy depends on the battery type. For example, conventional batteries lithium-ion (Li-ion) coin cells are adequate for sensors and low-power wearable devices. Depending on the battery design, these can be flexible or rigid. Some constructs utilize high-energy-density batteries that should be able to resist folding, bending, and stretching [17], while still being cooperative with integration and miniaturization. The property of elasticity in wearable is in direct relation to the mechanical properties of the electrodes. A rechargeable Li-ion type of battery comprises pouch cells containing 100 photos lithographically patterned Al and Cu disks. These disks are connected in parallel with molded pads of cathode and anode that consist of $Li_4Ti_5O_{12}$ and $LiCoO_2$. This battery has been manufactured from low-modulus silicone elastomers as substrates, segmented active components, and interconnected structures. The system includes a spacer where a gel electrolyte can be injected in order to act as a medium for ionic transportation. The battery has a stretchability of 300% with a capacity density of approx. 1.1 mA h/cm^2. Currently, lithium polymer technology is the standard for commercial products. However, Li-air and Zn-air batteries may be able to offer longer energy storage time. Zn-air batteries need the atmospheric oxygen to diffuse into a porous carbon electrode and comprise an anode made of zinc metal and a cathode made of the air electrode. Additionally, Li-air batteries are likely contenders because of their energies can go up to 12,000 and 4000 W h/kg.

Wireless power transfer can be achieved thanks to the stretchable microfluidic devices. Liquids in microfluidic channels can deform without significant hysteresis and flow without discontinuity upon bending. The maximum power efficiency reached around 10% in correspondence to a receiving power of 0.47 W from a transmitting power of 4.6 W. The device that resulted is capable of functioning after 1000 cycles of strain (25%). In a different embodiment, the transfer of wireless energy was managed through modified conductive yarns of silver-coated copper wire and polyester filaments. The Ag-plated copper filament with a diameter of 40 μm was wrapped around polyester yarn of 150 T/m (twists per meter). The copper filament surrounding the polyester yarn was then turned 700 T/m. For the final yarn, three strands of the twisted yarn were piled together with 550 T/m in order to create a yarn density of 547 deniers and a diameter of 300 μm. Because of such a twisting structure, part of the copper wire was left exposed on the surface of the conductive yarn. Therefore, the conductive yarn will be in contact with the skin after being integrated into clothing. Consequently, current leakage to the skin could occur. However, the conductive yarn keeps insulation from the skin because the exposure area of the copper wire is insignificant. The resulting measurements on the resistance of the conductive yarn presented a resistance of 89 mΩ/cm, even after the conductive yarn was touched by the hand. The number of coil turns was designed to have the resonant frequency of 6.78–13.56 MHz, which is included in the industrial, scientific, and medical (ISM) band.

Triboelectric nanogenerators (TENGs) are capable of converting mechanical energy from the human body motions into electric power with a coupled effect of triboelectric effect and electrostatic induction. Their properties of high flexibility, light weight, and pollution-free make the TENGs optimal applicants for powering wearable devices. The main aspect that can improve the current and output voltage is increasing surface charge induced by triboelectrification of TENGs. On the one hand, nanostructures, such as nanoparticles, nanorods, nanowires, and nanopatterns, were introduced to TENGs in order to boost the friction and contact area for a higher output performance. However, the manufacturing and decoration of these nanostructures are usually complicated and time consuming. On the other hand, increasing charge density through material optimization and surface functionalization, like fluorocarbon plasma treatment and ionized-air injection, could improve the output performance from the materials aspect. Evidently, the approaches of plasma treatment and ionized-air injection were consuming energy and always implied extreme environmental conditions, like a vacuum.

A type of energy source for wearable devices is represented by environmental sources which energy is available in realistic scenarios typically comes from light, vibration (kinetic, wind, and other vibrations), thermal heat, and electromagnetic radiation. After that, the energy must be converted into electrical energy. This step can be done by using the energy transducers.

4.3 LOW POWER CONSUMPTION MICROCONTROLLERS AND HARVESTING ENERGY

In the last decades, development in the field of electronics has been focused on miniaturization, flexibility, functionality. Most of the time, a nano/microdevice has a power consumption in the range of micro-milliwatt. In this case, it is imperative, for high-quality power sources, to match in size of these devices, thus forming self-powered nano/microsystems [18].

Some applications require long-time usage, such as continuously monitoring for specific parameters. Even though batteries could work somewhat in most of the applications, human intervention for replacing them must be considered. A better solution would be to use the energy coming from the environment, such as thermal, light, and mechanical energy. This type of energy is called harvesting energy (HE) [19]. The properties that have to be considered when choosing such an approach are power density, maximum voltage and current, the size, shape and weight, water resistance, and operating temperature [20].

Due to the importance of HE from the environment to power a device of tiny dimensions (nano/microdevices), different forms of mechanical energy (which is all around us) have been discovered as being capable of generating electricity. This category includes the following: air blowing/vibration, ultrasonic waves, pressure, small friction, and also body movement. Several types of nanogenerators have been fabricated using piezoelectric materials, such as ZnO, InN, ZnS, PZT, BaTiO3, and PVDF [21].

4.3.1 Piezoelectric Materials

Piezoelectric materials can detect vibrations and convert them into electrical energy. Thus, they generate an electrical charge when a mechanical load is applied. For example, polycrystalline ceramic is a common piezoelectric material, having a high efficiency of mechanical to electrical energy conversion [22].

As for other sources, the harvested energy can be stored in rechargeable batteries, rather than capacitors. According to [23], a thick film of piezoelectric material is placed on a thin steel beam, and when the beam resonates, the material is deformed, and electrical energy is generated. In [24], a circuit that uses step-down converters and HE is stored more efficiently; even four times more power is stored compared to other circuits. However, a simplified converter was later used to increase even more efficiency.

Using piezoelectric materials has the advantages that no external voltage source is needed, considering the material itself can directly generate the desired voltage. On the other hand, these materials might have leakages of charge over time [25].

Regarding the personal and wearable nano/microsystems that are self-powered, they require flexible and lightweight piezoelectric materials. One of these materials can be the tough lead zirconate titanate (PZT) textile that is made of aligned parallel nanowires. This material is transferred on the surface of a thick polyethylene terephthalate (PET) film to create flexible and wearable nanogenerators. This type of nanogenerator has a maximum output voltage and current of 6 V and 45 nA, respectively. It has been proven that such nanogenerators can power a ZnO nanowire UV sensor to detect UV light quantitatively, and also it can light a commercial LCD.

As mentioned before, we can obtain flexible nanogenerator based on a PZT textile (FTNG). Usually, an effective PZT textile in a nanogenerator has the length, width, and thickness of 1.5 cm, 0.8 mm, and 5 μm, respectively. The flexibility and soft properties of the PZT textile is a significant advantage in the fabrication of a wearable nanogenerator by combining the textile with a chemical fabric. Taking this into account, we can say that this type of wearable nanogenerator can work by

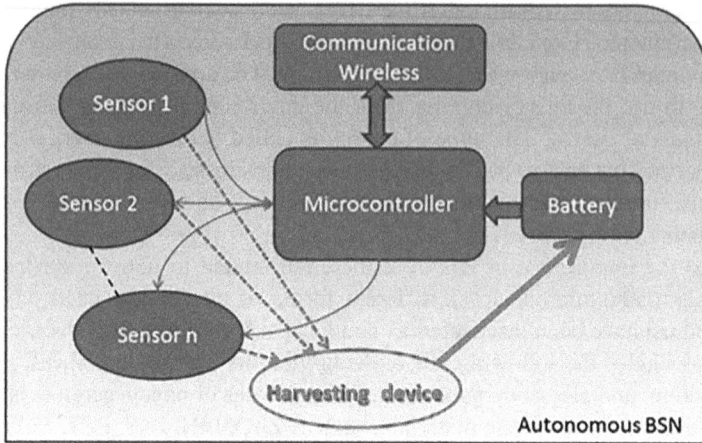

FIGURE 4.1 Autonomous BSN-based harvesting energy.

stretching and releasing the chemical fabric. The stress that appears in the fabric can be transferred to nanogenerator due to the close bond between the substrate-free generator and the fabric. In this way, at the stretching state of the textile, a piezoelectric potential appears along the PZT nanowires. When it is released, this difference in potential disappears. By doing this process periodically, the electrons will flow back and forth continuously and the wearable nanogenerator will be capable of generating alternating current (ac) electricity. In Figure 4.1 is presented the schematic of a system that is self-powered and contains a UV sensor and a FTNG (a), the UV photoresponse from a ZnO nanowire UV sensor powered by a FTNG (b), and the graph of the voltage drop on a UV sensor versus UV light intensity (c) [26].

4.3.2 PHOTOVOLTAIC CELLS

A photovoltaic cell is a device that converts light energy into electrical energy. Sunlight is the primary source of light energy that photovoltaic cells work with, mainly because there are many convenient areas such as roadways or marine locations, where the sunlight has high availability [27]. Usually, an array of photovoltaic cells is used to produce enough electrical power for a specific application. There also needed batteries or other power supply solutions to store the converted energy.

On a large scale, a microcontroller is an embedded device composed of a central processing unit (CPU), memory, and resources that allow the interaction with the external environment. Nowadays, they are used with multiple power sources, but for longer-lasting applications, energy consumption must be taken into consideration. Low-power consumption microcontrollers are devices that use as low energy as possible to process the input data. Depending on the application, two things must be taken into consideration: performance and energy saving. However, a compromise between these two is always an appropriate choice. Using low-power microcontrollers along with HE is the most common solution for Internet of Things applications. This is possible by using rechargeable batteries, which are powered by the energy coming from the systems described previously in this subchapter. A disadvantage could be that the status of the devices is highly dependent on the environment, and there could be interruptions on specific periods.

4.4 EDGE COMPUTING USED TO MITIGATE THE HIGH POWER CONSUMPTION PROCESSES

Power consumption is the measuring of vitality utilized per unit time. Power consumption is critical in computerized frameworks. The battery life of portable systems, for example, mobile phones and PCs constrained by power utilization. Influence is likewise critical for frameworks that are connected, on the grounds that power costs cash and in light of the fact that the framework will overheat on the off chance that it draws an excessive amount of intensity [28].

An efficient way to process the mass data, which is generated by IoT devices, is cloud computing. It can provide elastic computational resources on demand. There may be two problems when it comes to cloud computing for IoT: high transmission latency and resource demand mismatch. The mass data transmission, generated by IoT devices, requires quite a lot of time, energy, and bandwidth. Also, for long-distance transmission can be in a more significant delay. To solve these problems, the edge (or fog) computing is proposed. Edge computing distributes at the edge of the network, a part of VMs coming from cloud computing centers. With the help of the radio access network (RAN) together with the fog computing (FC) paradigm, some of the functions of data and control planes can be processed at the local baseband unit (BBU). The end-to-end delay can be reduced because the fog node is close to IoT devices. The high energy consumption is produced by the fog nodes because they have to transmit and process a large amount of data, in ECIoT. To save power in fog nodes, the power must be controlled. There are four types of schemes in which power is allocated for different optimization objectives: allocation of water-filling power, constant and proportional power allocation, and channel inversion power allocation.

Edge computing is a networking philosophy focused on bringing computing as close as possible to the source of data to reduce latency and bandwidth use. In less complicated terms, edge computing means running fewer procedures in the Cloud and moving those procedures to nearby places, for example, on a client's PC, an IoT gadget, or an edge server. Carrying calculation to the system's edge limits the measure of long-separation correspondence that needs to occur between a customer and server. It aims to deliver context-aware storage and distributed computing at the edge of the networks.

There are difficulties associated with edge computing, quite revolving around availability, which can be irregular or described by low transfer speed as well as high inactivity at the system edge. That represents an issue if enormous quantities of savvy edge gadgets are running programming (AI applications, for instance) that necessities to speak with central cloud servers [29]. Edge computing has likewise been created to take care of the bottleneck issue of system assets in IoT. By offloading the information calculation and capacity to end users, the reaction time and traffic stream will be fundamentally decreased [30].

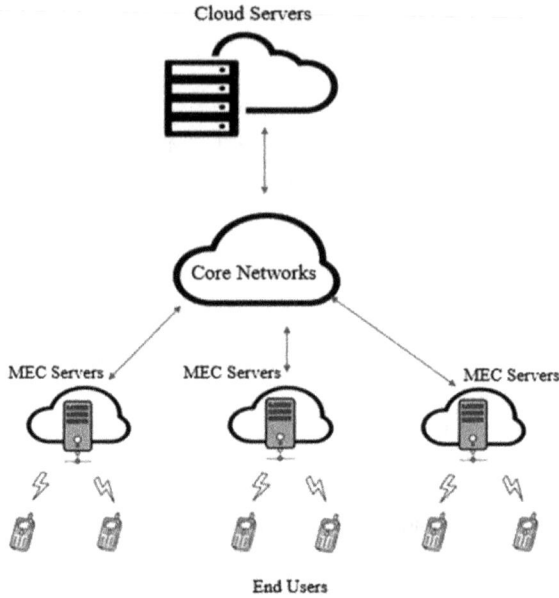

FIGURE 4.2   The basic edge computing architecture.

In Figure 4.2 is presented the architecture of edge computing. As we can see in this image, the edge computing servers are closer to the end user than cloud servers. The edge computing servers provide better Quality of Service (QoS) and also lower latency to the end users, than the cloud servers, even though their computation power is smaller than the one of the cloud servers. Figure 4.3 presents the structure of edge

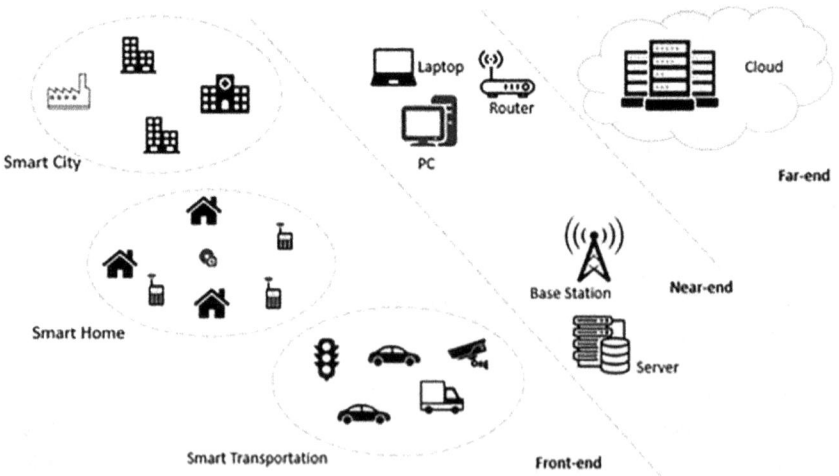

FIGURE 4.3   Typical structure of edge computation networks.

computation. As it is illustrated, this kind of process can be split into three main components: the front-end, near-end, and far-end.

The front-end aspect of edge computing is responsible for providing better responsiveness and more interaction for the end users. Also, it is the stage where the end devices, such as sensors or actuators, are deployed. In near-end environment, data storage and computation are migrated. The servers which are in the far-end environment can provide more computing power and data storage.

The most popular models of edge computing are hierarchical model, and software-defined model. The first of the models mentioned above, the hierarchical one, is a very appropriate method to describe the network structure of edge computing. The software-defined model is constructive in making the management of the edge computing for IoT more simple.

To solve the problem of the large quantity of data that come from many sources which can easily hit millions of tuples per second, the edge computing technology was proposed in order to use computing resources near IoT sensors for storing locally the data and processing preliminary the data. By using edge computing, it is possible to decrease network congestion and also to accelerate the process of analysis and resulting decision-making.

The Edge layer is positioned between the Cloud and the end devices. Moreover, there are several Edge implementation models, depending on the devices that can be used as intermediate edge nodes, the networks, the communication protocols, and also on the services that the Edge layer can offer. These models can be structured in three main categories: mobile edge computing (MEC), FC, and cloudlet computing (CC).

FC is characterized by the fact that there is a computing layer which is leveraging devices such as M2M gateways and wireless routers, that are called FC nodes (FCNs). These are utilized for the computation and storage of data that are gathered from end devices locally before sending them forward to the Cloud. MEC process is used for the deployment of the nodes, which are intermediate and can store and process data in the base stations of cellular networks. This will provide Cloud computing capabilities inside the radio area network (RAN) [31].

When it comes to power consumption, it is higher if the LTE or radio networks are used rather than the case where we use Wi-Fi. More than this, the access of mobile Edge nodes provides a higher power consumption than of Cloudlets. FC implementation leads to low consumption of energy like 802.15.4 and BLE, because it allows access to the nodes through access mediums.

4.5 BLUETOOTH NETWORK AND BODY AREA NETWORK (BAN) TOPOLOGIES FOR WEARABLE MEDICAL DEVICES

In a Bluetooth network, the Pico Network (piconet) is the basic unit where a master device allows/requests a connection from various slave devices (Figure 4.4).

The combination of two piconets represents a Scatter Network (scatternet). In scatternet, a device can act as a slave in the first Pico network and as a master in the second one (Figure 4.5).

In general, BAN is used low costs sensors with low power to create the network.

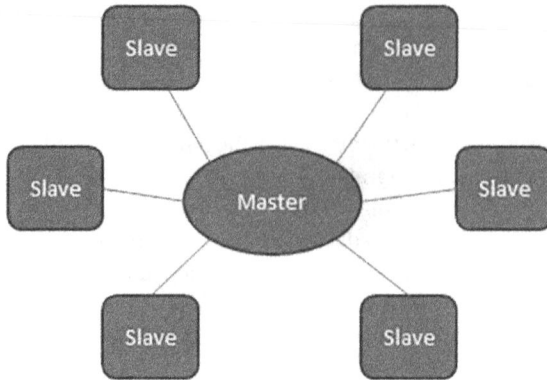

FIGURE 4.4 Bluetooth-piconet.

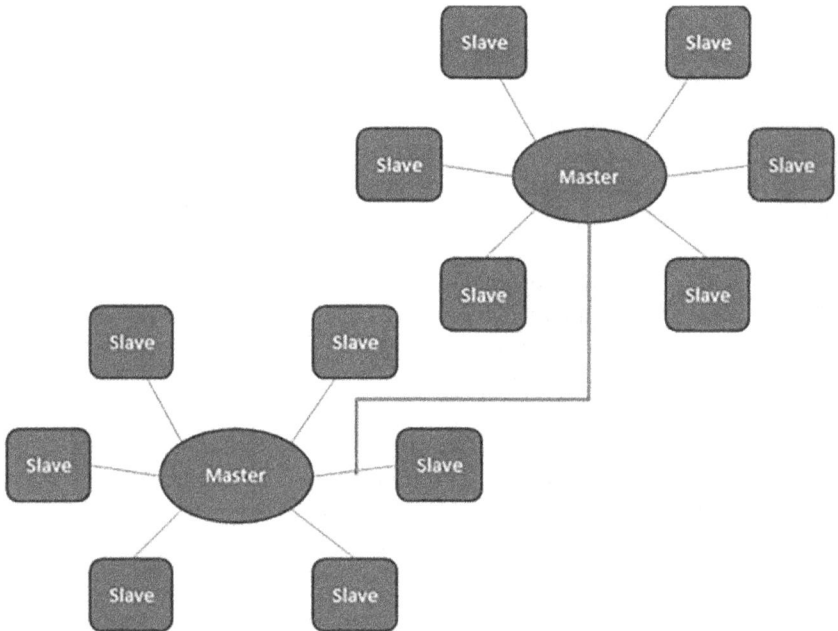

FIGURE 4.5 Bluetooth-scatter network.

A BAN mainly is based on sensors working on star or mesh topologies:

- In a star network topology, the sensors are connected to a BAN coordinator (microcontroller) that receives data from sensors and sends these data
  to another aggregator (mobile phone). The star topology is based on the
  Master-Slave model (Figure 4.6).

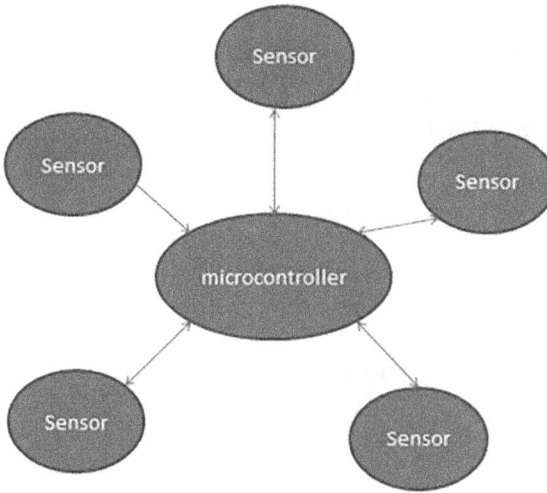

FIGURE 4.6 BAN—start topology.

- A mesh topology is based on the peer-to-peer model. Here, the presence of the BAN coordinator is necessary, but devices can connect directly to each other (Figure 4.7).

In general, BAN has used star topologies because it is easier to coordinate the device and to manage the power.

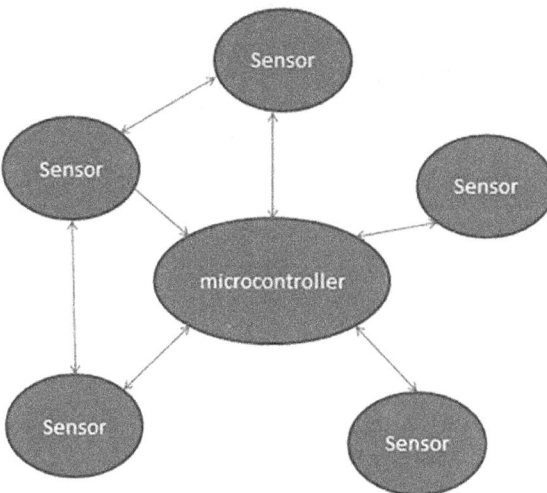

FIGURE 4.7 BAN—mesh topology.

4.6 PERSPECTIVE IN ENERGY MANAGEMENT IN THE CONTEXT OF CONDITIONS IMPOSED BY SECURITY, RELIABILITY, ACCURACY, AND FUNCTIONALITIES OF WEARABLE DEVICES

In health remote monitoring based on wearable devices, data security includes secure patient authentification, authorization, and data protection.

In general, the necessity of wearable device security comes from the aspects concerning personal data protection and avoiding an intrusion or unauthorized access to sensitive data if these are sent to another aggregator (phone, tablet, PC) or Cloud.

In general, wearable reliability and quality can be evaluated by test to high-low temperature (−10°C to +50° C), temperature changes, temperature shocks, high/low humidity, humidity cycling, mechanical resistance, ESD/EMC, flammability, resistance to chemicals.

Some scientific studies [32–34], concerning the six models of wearable devices used for heart rate, indicate the accuracy of 30% on average when subjects were walking compared to when seated at rest.

The accuracy of wearable (e.g., hearth rate device based on photoplethysmography [PPG] method) is variable between specific devices and activity types, but the sensors' accuracy is not affected by the skin tone of the wearer [34, 35], according to NPJ Digital Medicine study [32].

Management of energy dissipation and battery lifetime remains a challenge for wearable devices used. In addition, low-quality data collection, a non-reliable monitoring process, and missing important health events can be the effect on the wrong power management or weak algorithms used for selecting the information to be sent to the aggregators or Cloud. It is already known that a considerable volume of data sent repetitively can generate power consumption very quickly. However, an investigation conducted by Duke University researchers among 53 individuals, highlighted consistent over-reporting of heart rate during low-intensity activity, accuracy, and output differences between consumer- and research-grade devices, and the results were that the researchers and clinicians should be aware of when choosing devices for clinical research and clinical decision support.

The wearable medical devices request IoT scalable applications that can allow energy consumption optimization, data management, security, and privacy with low energy consumption requirements.

The challenges for medical wearable devices are:

- low power consumption;
- miniaturization;
- respond to the European regulation for medical devices;
- to be adaptive to patients' needs; and
- to ensure security, reliability, sensitive data privacy through edge software applications.

REFERENCES

1. Aileni, R.M., Suciu, G., Balaceanu, C.M., Beceanu, C., Lavinia, P.A., Nadrag, C.V., Pasca, S., Sakuyama, C.A.V. and Vulpe, A., 2019. Body area network (BAN) for healthcare by wireless mesh network (WMN). In Body Area Network Challenges and Solutions (pp. 1–17). Springer, Cham.
2. Cao, H., Leung, V., Chow, C. and Chan, H., 2009. Enabling technologies for wireless body area networks: A survey and outlook. IEEE Communications Magazine, 47(12), pp. 84–93.
3. Patel, M. and Wang, J., 2010. Applications, challenges, and prospective in emerging body area networking technologies. IEEE Wireless Communications, 17(1), pp. 80–88.
4. Kim, H., Kim, Y. and Yoo, H.J., 2008, August. A low cost quadratic level ECG compression algorithm and its hardware optimization for body sensor network system. In 2008 30th Annual International Conference of the IEEE Engineering in Medicine and Biology Society (pp. 5490–5493). IEEE.
5. Abdulghani, A.M., Casson, A.J. and Rodriguez-Villegas, E., 2012. Compressive sensing scalp EEG signals: Implementations and practical performance. Medical & Biological Engineering & Computing, 50(11), pp. 1137–1145.
6. Daou, H. and Labeau, F., 2012, August. Pre-processing of multi-channel EEG for improved compression performance using SPIHT. In 2012 Annual International Conference of the IEEE Engineering in Medicine and Biology Society (pp. 2232–2235). IEEE.
7. Chen, M., Gonzalez, S., Vasilakos, A., Cao, H. and Leung, V.C., 2011. Body area networks: A survey. Mobile Networks and Applications, 16(2), pp. 171–193.
8. Cárdenas-Barrera, J.L., Lorenzo-Ginori, J.V. and Rodríguez-Valdivia, E., 2004. A wavelet-packets based algorithm for EEG signal compression. Medical Informatics and the Internet in Medicine, 29(1), pp. 15–27.
9. Ma, T., Shrestha, P.L., Hempel, M., Peng, D., Sharif, H. and Chen, H.H., 2012. Assurance of energy efficiency and data security for ECG transmission in BASNs. IEEE Transactions on Biomedical Engineering, 59(4), pp. 1041–1048.
10. Omeni, O., Wong, A.C.W., Burdett, A.J. and Toumazou, C., 2008. Energy efficient medium access protocol for wireless medical body area sensor networks. IEEE Transactions on Biomedical Circuits and Systems, 2(4), pp. 251–259.
11. Alam, M.M., Berder, O., Menard, D. and Sentieys, O., 2012, May. Latency-energy optimized MAC protocol for body sensor networks. In 2012 Ninth International Conference on Wearable and Implantable Body Sensor Networks (pp. 67–72). IEEE.
12. He, Z., Liang, Y., Chen, L., Ahmad, I. and Wu, D., 2005. Power-rate-distortion analysis for wireless video communication under energy constraints. IEEE Transactions on Circuits and Systems for Video Technology, 15(5), pp. 645–658.
13. Huang, H., Chen, P.Y., Hung, C.H., Gharpurey, R. and Akinwande, D., 2016. A zero power harmonic transponder sensor for ubiquitous wireless μL liquid-volume monitoring. Scientific Reports, 6(1), pp. 1–10.
14. Cho, K.J. and Asada, H.H., 2002, April. Wireless, battery-less stethoscope for wearable health monitoring. In Proceedings of the IEEE 28th Annual Northeast Bioengineering Conference (IEEE Cat. No. 02CH37342) (pp. 187–188). IEEE.
15. Jiang, H., Chen, X., Zhang, S., Zhang, X., Kong, W. and Zhang, T., 2015, July. Software for wearable devices: Challenges and opportunities. In 2015 IEEE 39th Annual Computer Software and Applications Conference (Vol. 3, pp. 592–597). IEEE.
16. Yetisen, A.K., Martinez-Hurtado, J.L., Ünal, B., Khademhosseini, A. and Butt, H., 2018. Wearables in medicine. Advanced Materials, 30(33), p. 1706910.
17. Kwak, S., Kang, J., Nam, I. and Yi, J., 2020. Free-form and deformable energy storage as a forerunner to next-generation smart electronics. Micromachines, 11(4), p. 347.

18. Wu, W., Bai, S., Yuan, M., Qin, Y., Wang, Z.L. and Jing, T., 2012. Lead zirconate titanate nanowire textile nanogenerator for wearable energy-harvesting and self-powered devices. ACS Nano, 6(7), pp. 6231–6235.

19. Amirtharajah, R. and Chandrakasan, A.P., 1997, June. Self-powered low power signal processing. In Proceedings of Symp. on VLSI Circuits Digest of Technical Papers (pp. 25–26).

20. Fry, D.N., Holcomb, D.E., Munro, J.K., Oakes, L.C. and Matson, M.J., 1997. Compact portable electric power sources (No. ORNL/TM-13360). Oak Ridge National Lab., Oak Ridge, TN (United States).

21. Fan, H.H., Jin, C.C., Wang, Y., Hwang, H.L. and Zhang, Y.F., 2017. Structural of BCTZ nanowires and high performance BCTZ-based nanogenerator for biomechanical energy harvesting. Ceramics International, 43(8), pp. 5875–5880.

22. Joseph, A.D., 2005. Energy harvesting projects. IEEE Pervasive Computing, 4(1), pp. 69–71.

23. Glynne-Jones, P., El-Hami, M., Beeby, S.P., James, E.P., Brown, A.D., Hill, M. and White, N.M., 2000. A vibration-powered generator for wireless microsystems.

24. Ottman, G.K., Hofmann, H.F. and Lesieutre, G.A., 2003. Optimized piezoelectric energy harvesting circuit using step-down converter in discontinuous conduction mode. IEEE Transactions on Power Electronics, 18(2), pp. 696–703.

25. Wang, L. and Yuan, F.G., 2007, April. Energy harvesting by magnetostrictive material (MsM) for powering wireless sensors in SHM. In Sensors and Smart Structures Technologies for Civil, Mechanical, and Aerospace Systems 2007 (Vol. 6529, p. 652941). International Society for Optics and Photonics.

26. Zhang, M., Gao, T., Wang, J., Liao, J., Qiu, Y., Yang, Q., Xue, H., Shi, Z., Zhao, Y., Xiong, Z. and Chen, L., 2015. A hybrid fibers based wearable fabric piezoelectric nanogenerator for energy harvesting application. Nano Energy, 13, pp. 298–305.

27. Raghunathan, V., Kansal, A., Hsu, J., Friedman, J. and Srivastava, M., 2005, April. Design considerations for solar energy harvesting wireless embedded systems. In IPSN 2005. Fourth International Symposium on Information Processing in Sensor Networks, 2005. (pp. 457–462). IEEE.

28. Vereecken, W., Van Heddeghem, W., Deruyck, M., Puype, B., Lannoo, B., Joseph, W., Colle, D., Martens, L. and Demeester, P., 2011. Power consumption in telecommunication networks: overview and reduction strategies. IEEE Communications Magazine, 49(6), pp. 62–69.

29. Mao, Y., You, C., Zhang, J., Huang, K. and Letaief, K.B., 2017. A survey on mobile edge computing: The communication perspective. IEEE Communications Surveys & Tutorials, 19(4), pp. 2322–2358.

30. Yu, W., Liang, F., He, X., Hatcher, W.G., Lu, C., Lin, J. and Yang, X., 2017. A survey on the edge computing for the Internet of Things. IEEE Access, 6, pp. 6900–6919.

31. Dolui, K. and Datta, S.K., 2017, June. Comparison of edge computing implementations: Fog computing, cloudlet and mobile edge computing. In 2017 Global Internet of Things Summit (GIoTS) (pp. 1–6). IEEE.

32. Bent, B., Goldstein, B.A., Kibbe, W.A. and Dunn, J.P., 2020. Investigating sources of inaccuracy in wearable optical heart rate sensors. NPJ Digital Medicine, 3(1), pp. 1–9.

33. Stahl, S.E., An, H.S., Dinkel, D.M., Noble, J.M. and Lee, J.M., 2016. How accurate are the wrist-based heart rate monitors during walking and running activities? Are they accurate enough?. BMJ Open Sport & Exercise Medicine, 2(1), p. e000106.

34. Wearable heart rate sensor accuracy didn't vary across skin tones in small Duke study, online available: https://www.medtechdive.com/news/wearable-heart-rate-sensors-did-not-vary-across-skin-tones-duke-study/572073

35. Chow, H.W. and Yang, C.C., 2020. Accuracy of optical heart rate sensing technology in wearable fitness trackers for young and older adults: validation and comparison study. JMIR mHealth and uHealth, 8(4), p. e14707.

5 Collaborative Data Analytics for a Smart World

Rani Deepika Balavendran Joseph
University of North Texas,
Denton, TX, USA

CONTENTS

5.1 SIGNIFICANCE OF SMART BUILDINGS

Smart buildings are a combination of advanced technologies with well-designed architecture. The primary goal of these buildings is to create comfort, safety, and sustainability by considering all social and economic needs. For instance, a smart system in The Edge Building in Holland knows the car model and even the sugar quantity needed by an employee. In this building, light and temperature are automatically controlled based on employee's preference. In addition, Duke Energy Center in North Carolina, Capital Tower in Singapore, Glumac in China, Hindmarsh Shire Centre in Australia, and Doug Peter Ron (DPR) construction firm in the United States are popular for efficient energy and water management systems in smart buildings. The Crystal Building in London is an example of an energy efficient architecture that has zero annual heating bill [1]. These are a few examples of existing and popular buildings that showcase the significance of smart buildings in providing social and economic needs of a human.

5.1.1   PRIMARY COMPONENTS IN SMART BUILDINGS

A smart building can have a system that controls temperature based on number of people present in a room, an LCD screen that updates on events that are happening in a building, an app that sends reminder on change of printer cartridge, voice navigator that directs to a required room, a sensor that turn off lights, and so on. These all applications need to be connected through internet with a proper control management system. For this purpose, one needs to use Internet of Things (IoTs), where one can control the entire functioning of the building by connecting each device through internet. In recent times, the usage of internet has been increasing from multiple set of people for multiple purposes. Humans use internet for multiple purposes such as receiving and sending emails, using social networks, watching web content, and even working from home [2]. Next generation of computers are outside of the traditional computers, whereas all things around can be connected in a network in form or the other. Revolution in internet has exceptionally connected people around the world. Next pace will be connecting millions or billions of devices to create an automated or a smart environment. This connection of things in a network is termed as IoTs [3].

Kevin Ashton was the first person to come up with the word "Internet of Things" (IoTs). In other words, he was a technology pioneer who invented IoTs in 1999. According to him, like internet, IoT has a capability to change the world. However, a human-centric data captured by computers and internet can limit their performance. This is because human beings have limited time, accuracy, and attention, and they are not good in gathering data on things. In other hand, economy, society, and survival are based on things in the real world rather than information or ideas. Therefore, it is needed to have computers that can make decisions from the gathered data without any limits. Gathered data from Sensors and Radio Frequency Identification (RFID) technology can enable computers to sense the world to reduce the cost, waste, and loss. In simple words, an individual can consider IoT as a system that connects anything and everything through wireless network [4].

IoT is defined based on the network vision and objects connected. Based on the network-oriented vision this is classified into three visions: Things-, internet-, and Semantic-oriented vision. Things-oriented vision is based on the sensors or objects connected, Internet vision is considered based on the interaction of physical devices, and semantic represents the huge data that is collected from the connected sensors and devices. Considering RFIDs things, the IoT is focused on connecting smart object using a suitable architecture, and in internet view, IoT is considered as a wireless network. Finally, from the semantic view, it is collection of data from different devices. The main challenges include storage, management, and integration of data collected from multiple devices. For such huge amount of information captured by IoTs, solutions are provided by semantic technologies [2].

IoT can connect millions or billions of devices to create an automated or a smart environment. Vision of next-generation computing is possible with RFID and sensor technologies [3]. Smart technologies can be incorporated in any building structure like healthcare center, corporate tower, houses, and so on. These buildings are characterized based on sharing information efficiently among multiple devices. Smart buildings are incorporated with sensors, actuators, meters, and digital networks to operate intelligently. The main sources for the collection of data in smart buildings are meters and sensors. In smart buildings, while IoTs collects the information from the sensors, a storage environment helps to store and process data that is obtained from sensors [5]. Smart buildings are built with an automated BMS by integrating multiple devices that can ensure efficiency and security by monitoring different operations and controlling energy and water consumption. BMS and data analyzed can comfort the life of an individual by automatically controlling multiple functions by integrating all devices in a building. Henceforth, two main components to be discussed while designing smart buildings are IoTs and data storage environments. Focusing on these two components can improve quality, increase productivity, and provide safety [6–8].

5.1.2 Integration of Smart Buildings

The key factor in smart buildings is integration, which allows connecting multiple building systems and then link automation to enterprise. This integration helps facilities executives and organizations to make use of smart building benefits both in constructing new buildings and renovating the existing one with smart technologies. An advanced building automation system (BAS) is commonly used for integration. BAS helps in integrating each building systems in a network that allows facility executives to take advantage of energy in building systems. Therefore, integrating individual building systems through BAS is helpful to organizations and facility executives. After integrating building systems, the next step is linking BAS to enterprise system; this can enhance the performance of a building and improve the decision-making. Performance of smart buildings depends on integrating BAS and managing facility's systems like heat, ventilation, air conditioning (HVAC), and lighting. There are no predefined competencies of smart buildings, but these buildings use the concept of integration to tap systems energy that may lead to for future enhancement [9].

Usually, integration in smart building links HVAC, meters, and lighting systems; however, advanced building may even connect elevators, fire alarms, and access control systems. Duke Energy Center is a 48-floor smart building that was chosen as the smartest building in 2010. In Duke's building, there are 16 different building systems and three BAS that are linked with Internet Protocol (IP) network. BAS in Duke's building is customized to support different protocols like Open Platform Communication (OPC), LonWorks, Programmable Logic Controller

(PLC), Modbus, and Building Automation and Control network (BACnet), which can help for efficient system operation and data collection from multiple building systems. Integrating center in Duke's building includes building systems like emergency intercom systems, light gathering blinds, seven generators, security cameras, uninterrupted power supplies (UPS), water filtration systems, sub-submeters, elevator, and video monitoring systems. Sub-submeters in integration system can observe energy consumption from HVAC, load, and lighting based on each floor [9].

Smart objects are considered as small computers with combination of sensors, actuators, and communication devices that can be installed in cars, light switches, thermometers, and in machines that cannot be imagined as computers. Extensive use of smart objects in the state-of-art applications needs a reliable and compatible communicating framework. IP has proven capability that supports broad applications, tools, and technologies. Smart objects capability is limited to devices like RFIDs, but with the help of sensors and bidirectional communication, these smart objects can provide data such as temperature, energy, and pressure measurement. Broad range of smart objects applications are in fields such as transportation, smart energy management, healthcare monitoring, home, and building automation. This IP stack takes small space, which can be easily installed in a small battery-operated device. Hence, characteristics of IP can make the IoTs dream come true by connecting billions of devices. IoT can be possible with the recent advancement in low-cost portable devices, which can be helpful in in wide range of application such as health, safety, and energy [10].

SixFox and narrow band IoT (NB-IoT) are examples of low-power wide area networks (LPWAN) Wireless technologies. Compared to short-range technologies like Zigbee, LPWAN are popular technologies for information exchange among multiple devices in smart buildings. SixFox and NB-IoT can be used as suitable long-range wireless technologies for communication or interaction in smart buildings. For operating devices in smart buildings, using low-cost microcontrollers and low-power network shows a considerable impact on IoT [11]. A communication protocol called BACnet can manage the interaction among the devices in smart buildings. Efficient communication or interaction can be possible by knowing the overall structure and relationship among subsystems. For this purpose, it is difficult for the developers to design a framework that can manage and control the existing physical resources in various subsystems is not possible. Therefore, developing a framework that can manage and control resources is one of the challenges faced in designing smart buildings. Building information model for providing building geometric information and indicating sensors and systems in a building can be used to represent building [9, 12, 13].

Managing and controlling the privileges accessed by occupant is another challenge faced in smart building. Buildings with multiple organizations and thousands of people's access to those buildings make traditional role-based access control unreliable. This traditional method needs to have information on roles of each individual and their relative authority. There are issues based on role-based access, for instance, this will provide access to same resources to all occupants in a building with different workstations. Hence, it is required to automate the process by granting and revoking fine-grain permissions for users in the building. A framework is built based on block chain smart contract that can overcome the drawback of traditional method. This framework provide authority to users of the building to access privileges and flexible in managing access between occupants and visitors. Two application programming interfaces (API) developed are query API and BACnet API, which can serve the purpose in multiple buildings. A small modification to the BACnet API can make this framework to connect to BMS and serve the needs in real-time scenario [14].

5.2 DATA ANALYSIS FOR EFFICIENT DESIGN

As discussed in the previous sections, recent technologies allow advancement in a building functionality. Information and Communication Technologies (ICT) are used extensively to improve the performance of a building. An individual comfort and less energy consumption are the primary goals of a smart building. These goals can be achieved with the combination of advanced intelligence technologies and architecture models. Cyber-Physical Systems in smart buildings are BMS that are used to monitor and control building operations. These management systems are embedded with different sensors and meters. Sensors and meters are the main sources for generating huge amount of data. Processing and analyzing this data can have multiple benefits such as observing energy usage or fault discovery and diagnosis (FDD) [15]. This section discusses on significance of analyzing data collected from sensors and meters, and advantages of sharing data to perform collaborative data analysis.

5.2.1 BENEFITS OF DATA ANALYSIS

Benefits of analyzing data for BMS are generating ground truth data, validating data, predicting energy consumption, generating a schedule for maintenance, estimating comfort level of occupant, building simulation, detecting, and identifying fault. These benefits can be fulfilled by using collaborative data analysis. Ground truth data collection is a challenging task as it relates to the operating data of a BMS. Due to complexity of BMS, collecting ground truth data and labeling that data is a very difficult task. For instance, if the operation pattern of one building is identified different from another buildings, then it is difficult to state that one building confidently as different from other buildings. Another purpose of validating data is to ensure that data collected from multiple sources is error free. This process includes careful FDD as fault can be occurred due to incorrect data. Fault identification and building a model is a very time taking process, hence collaborating, and sharing data among different building can make this process easy and fast [16].

Intelligent smart buildings require microelectromechanical systems (MEMS), communication systems, and data analytics for comfortable and safety living of an individual. Implementing the value-driven model is used for predicting the factors, computing, and providing optimal solutions. This model is built focusing on multi-objective optimization for building intelligent buildings. Another benefit of interactive building structure is monitoring the crowd in commercial buildings to make decision on positioning of businesses and advertisements. The ideas and applications are evaluated by using the simulators. With the help of object-oriented simulator, the experimental cost is reduced. Recent time popularity of the CAD tools for researching in different fields is evident. For example, an electrical engineer can reduce fabrication cost by thorough design and testing of application using CAD tools or simulators. Similarly, the floor plan of the smart building is transformed to CAD tools and data is saved using XML. By using this XML structure, a user can save or export files for future reference. The record is saved in the lag from loading a file to successfully simulating it. Therefore, this can do simulation in two forms: manually and predefined form [17, 18].

5.2.2 Growth of Smart Cities

Efficient design of smart buildings can lead to a development of smart cities, which eventually can provide better living socially and economically. Systematic Literature Review projected that 29% of researchers' literature were focused on the smart cities, whereas next importance is given to healthcare, commercial, and environment applications. IoT is a framework that integration of smart objects, wireless network, and processing technologies, which helps to provide smart services to customers or humans. Different problems that may be faced while dealing with the smart cities are IoT platforms scalability, heterogenous data streams management, monitoring vehicles, Quality of Service (QoS) aware composing, and finding location [19, 20]. Formal verification is a challenging task in the IoT devices, due to the interconnection between the different devices. Verification of the services compatibility is a main challenge faced by IoT; this can be solved by considering a proper framework that verifies the compatibility of multiple services that are integrated with IoT. An efficient architecture can be built by processing and analyzing data collected from these devices that are connected in a network.

Multinational companies like IBM, Cisco, and General Electric had shown their contributions on developing smarter systems that can benefit companies and communities [21]. Global economic crisis in 2008 made IBM to think about the challenges and problems that were controlling the world. It infused computing power into things like car, phones, water, food, and power lines. In addition, advanced analytics and algorithms helped to understand the captured information and data from these things. IBM's campaign on Smarter Planet explained how smarter systems and technologies can become more powerful and accessible in coming years. The launch of Smarter Planet has become a framework for IBM's growth strategy and encouraged world to think of smarter systems that can plan transportation. Control electric power, manage healthcare, provide public safety, water, and food. Revenue generated within a year of Smart planet launch has showcased the significance of building smart systems that can be a benefit to industries and communities [22]. Similarly, Cisco's intelligent urbanization helps to integrate management of a city using a network and General Electric's Ecomagination is helpful to solve today's environment problems and help society.

Smart homes or smart buildings are basic blocks of developing smart cities. There is a need of standard codes for buildings to provide features of smart homes like solar energy, waste management, plantation, rainwater harvesting, video surveillance, LED lighting, sustainability, ventilation, communication, and integrated safety. Interventions like architectural and engineering are proposed for standard building codes in both locally and nationally. Building coed maintains an optional and mandatory clause for the individual homes and large apartments. Such codes can solve some issues related to infrastructure in homes and apartments [23].

There is an innumerable growth of population in cities because of the opportunities, in future, we can even see more than 70% of population living in smart cities, which are considered as hub for businesses. Most famous smart city in korero is New Songdo City, which has taken care of aspects such as transportation, open space, utilities, and central parks. Examples of newly build smart cities and modernizing exiting ones are King Abdullah Economic City (Saudi Arabia), Gujarat International Finance Tec-City (India), San Francisco TechConnect (the United States), and Santander (Spain). These cities are built using IoTs that can connect all devices that surround us through wireless communication network. IoTs with wireless network and capturing data from different devices such as temperature, pressure, and energy measurement can be used for designing multiple application, which can serve industry and society.

5.2.3 COLLABORATIVE ANALYSIS

For meeting the need of BMS and efficiently using the data captured by the sensors and meters, it is required to know the potential and a model that is suitable for of the collaborative data analytics. Instead of taking a data for a single building, gathering the data, or integrating the data of multiple building can provide efficient results. This type of collaborative analysis can meet the needs of BMS, which mainly focus on the energy savings and occupant comfort. The huge amount of data which is gathered by the meters and sensors is used for various purposes like FDD and observing power consumption. Sharing data among different fields and integrating it for making decision may show synergic effect. Therefore, collaborative data analysis may also benefit or show a positive effect on BMS control. Cloud computing and IoT technologies provide a supportive mechanism for sharing of data and taking decisions. This type of analysis is required to obtain accurate predictions for making high-quality decisions. Factors that need to be considered for the collaborative data analysis are sharing of data, specialization, credit, access, expertise, concurrency, usability, flexible semantics, and motivation. These factors required for collaboration of data are similar to human collaboration. For instance, humans need to get motivated to participate; similarly building owners need to know the importance or potential of collaborating data for taking efficient decisions. However, we can exclude the one relating to behavior and psychology of humans [24].

Processing multiple buildings data instead of a single building can help to solve a relevant business problem more accurately. Similarly, processing data from multiple sources also helps to design a better building model. A collaborative data collected can help to create a framework that suits the environment and comforts lives. Collaborative framework is built based on the data that is collected from multiple sources. Efficient framework for smart buildings can ease the life of humans and save their time from concentrating on trivial matters [15]. Increased devices in the IoT may need to concentrate on interface and convergence subsystem. This convergence system helps to integrate sensors and actuators into a common network. Special hardware and software are required that can support these two subsystems [25].

Collaborative data analytics can be classified as offline and online analytics. Offline analytics allows collecting data of individual building periodically and stored in a local BMS of a building. This data later can be transfer to a storage unit where we can use uploaded data for processing and predicting. In other hand, online analytics allows collecting data continuously from required buildings and predict or draw insights from the collected data at any time. For this purpose, we need to design a special BMS which can be connected to networks like internet or cloud [16]. Collaborative analysis can be provided in public and private scenarios. In public scenario, where the data from building owners of similar buildings is shared and processed to take decisions that can be useful for organizations and owners. Collaborative data analytics in private scenario can help to a building owner to improve energy consumption. In addition to online and offline analytics, building owners can perform their own analytics using own software and hardware. In addition, the cloud computing services provide a cost-effective solution to building owners [26]. More security and privacy issues need to be considered for building a collaborative data analytics model for controlling BMS. This collaborative analytics is suitable for controlling smart grids, as some of the smart grid issues can be solved by the sharing and integrating data.

5.2.3.1 Methods for Collaborative Data Analysis

This section discusses on two methods: aggregation and disaggregation, for collaboration the operation of the smart grid. Bi-level aggregation method is used to collaborate the smart grid operations in smart buildings. Operating constraints are considered in the first level of this aggregation method and formulating the optimization problems are considered in the second level. The operating issues in a distribution system like phase imbalance, capacity of a transformer, and thermal limit. This can also be implemented using three directions. Three directions considered are loads for storing, optimal power flow in a building, and estimating the thermal dynamic model accuracy, complexity, and formulating optimization of aggregation problem in secondary level [27].

Disaggregation method for end users was developed for commercial smart buildings to provide efficient energy management system. Standard disaggregation methods were suitable for highly generated data, whereas end-user disaggregation method is based on the low frequency data generated from the BASs. This automation data consists of operational information on fans and pumps. This automation system data helps to disaggregate low-frequency data into distribution, chillers, lighting, and pumps, which are considered as end uses. For instance, model build based on the automation system data is used for control state of fans and electricity used by fans using air handling unit with variable speed option. Genetic algorithms were used to evaluate the model parameters to avoid the misfit between the measured and observed values. The hourly commercial building data is used for building the model that can control the electricity level [13].

5.3 DATA ANALYSIS FOR EFFICIENT DESIGN

Huge amount of data is generated because of installing communication and information systems in the environment. Storing, processing, and analyzing of captured data is a challenging task. The framework or an architecture of a building developed needs to focus on the sectors like computing, networking, storing, and visualizing. Building automation and management systems provide large amount of data that need to be analyzed for providing insights for making decisions. BAS and BMS are installed in both public and commercial buildings to monitor and control multiple applications. Therefore, data mining is a popular approach for handling large amount of data collected by these systems.

Data mining techniques are used in various fields such as retail, finance, and communication. These techniques are classified as supervised and unsupervised techniques. Supervised learning technique learns from the historic data and develops a relation between input and output, whereas unsupervised learning technique directed by an explicit target. The primary motive of this unsupervised learning is to recognize unknown structures of data. This will help in the growth of individuals sectors to provide better shared environment. Virtual cloud model framework can be used for proper storing, computing, and visualizing of data. It is visualizing and interpreting the analyzed data by considering constraints like security and management of data [3]. There are a few common steps that need to be followed for analyzing and interpreting data.

5.3.1 DATA PRIVACY AND MANIPULATION

Smart meters provide information on energy consumption by storing, collecting, and processing of data, which can be helpful for application such as detecting anomaly and avoiding peak response. Primary risk in smart meter data is leaking of privacy information of occupancy and users' motion patterns. In addition to techniques used for preserving the privacy of data, there are data manipulation techniques. For example, down sampling and addition of noise are two techniques for data manipulation that can help to preserve the privacy of data. These techniques can be easily implemented with meters and do not need to install batteries. Using these techniques can provide a better protection to the users' privacy, and this can be used in fault detection. Though nonintrusive approach may provide better service, but this can enable data curator to extract sensitive information of a user. Encryption techniques can help to protect the data from untrusted parties, but this cannot stop the data curators from performing data analysis on the extracted data [28, 29]. Aggregator encryption is used for the collected data to protect the privacy of the smart meter. This encryption scheme provides finer encryption times and smaller cipher texts.

Techniques like battery-based load hiding (BHL) can help to protect data by installing batteries in smart meter to hide load curves. Smart meter data can provide information on the individual appliances load in a house. Technique that can be used to calculate the load of the appliances or activity of the appliances from outside is using BHL technique. That is, a battery is installed in each appliance at home and controlling those appliances using smart techniques. Such techniques can help to track the activities of appliances but cannot provide privacy protection. To overcome this problem in BHL techniques, researchers have proposed randomized and multi-tasking BHL algorithms for providing protection. There are other factors like battery charging and discharging, and economic benefits need to be considered while using BHL technique [30, 31].

Using batteries, we can not only provide the privacy but also can reduce the electricity bill. The main challenge is a trade-off between the privacy and reducing electricity bill. This dynamic optimization problem depends on future consumption of electricity. Hence, a dynamic optimization framework needs to be designed that can control the privacy and the electricity bill of a consumer. An online control algorithm is proposed by researchers that can provide privacy and also reduce electricity bill of the battery controlled devices [31]. Online control algorithm is developed using the Lyapunov optimization technique. This online algorithm can provide an optimized solution without information on load and prices of electricity. Interpreting the numerical results obtained, proposed algorithm has better privacy protection than the existing algorithm.

Embedding smart devices is a cost-effective task; hence, the control of the smart buildings is obtained by updating in their smart phones or devices. The primary issue in such flexibility is cyber security. Knowledge management is needed to manage and handle the issues obtained with cybersecurity issues of sensor network and other devices that are installed in the smart buildings. The main idea of capturing the forensic data is to reduce the cost of investigation after the incident or a crime. There is a ten-step process proposed by researcher for forensic readiness [32]. Forensic readiness is a digital evidence that is used by the organization when it is required. The primary goal of forensic readiness is to reduce the investigation cost by gathering and using the digital evidence. This is an enhancement for existing security activities where we need determine the crimes and disputes that need electronic evidence. Organizations need to communicate with the IT services to implement framework that capture the digital evidence. Introducing forensic readiness in business and capturing of digital evidence can help organizations to cooperate with the law enforcement agencies. Both forensic readiness and cyber security testing are needed to secure the information which is shared among the smart devices.

Adding noise can manipulate data, and this technique can protect the data collected in smart meter that can be used to estimate energy cost [33]. Researchers have examined lightweight approach is used for privacy and utility by adding noise. Nonintrusive devices are used for evaluating privacy, whereas benefits of smart meter were considered for validating utility. This lightweight approach is efficient and reliable compared to the other approaches that can fulfill customer and power provider needs. This approach is mainly focused on adding noise and using real data of consumers. Data manipulation techniques provide coarse grained protection compared to encryption techniques.

5.3.2 DATA STORAGE FRAMEWORK

IoT in smart building provide comfort, safety, and energy efficiency, and improve life quality of an individual. The main objective of smart building is to improve cities and infrastructure facilities. Intelligent system in the smart buildings is considered in three levels: input data, system, and services level. Infrastructure level based on input data shows data collected from different sources by connecting different devices. Different types of data may include temperature, humidity, and energy consumption. System-level infrastructure represents collecting, storing, processing, and analyzing of data. Services level represents the system that offers services to suppliers, managers, and residents. Smart building ecosystem components include sensors, networks, big data pipeline, and analytics. All these components together may allow an Individual to efficiently utilize smart building services. Applications of IoT in smart buildings include providing facilities, detecting occupancy, tracking resource's location, improving the comfort, and managing energy [5].

Framework to collect, store, manage, and visualize data. For integrating information and making efficient utilization of services, a framework that can collect, store, process, and visualize data that has been collected from multiple devices. This WattDepot consists of sensors, servers, and clients to perform analysis on data and store results in repository. WattDepot is used in kuikui nut energy competition. Activities of energy literacy are carried to gain kuikui points, and competition is based on the total points gained. WattDepot framework helps to collect energy data, analyze, and visualize that is required for kuikui cup. Another application includes carbon intensity control by using spotlight visualization and use analytical tool for common time stamp data sets in a repository to solve timestamp problem. Need to implement a suitable privacy system and create a database that can effectively use the services provided by WattDepot [34].

In addition, cloud computing frameworks are also used for processing, storing, and analyzing data that can provide efficient solutions. A huge number of resources are required for the processing and storing of gathered data. Therefore, cloud computing is an emerging platform for commercial purposes and efficient utilization of resources. This cloud model is popular for providing scalable and cost-efficient solutions for the smart buildings. This cloud computing platform is focused in improving the functionalities of fault diagnosis and detection. Smart Building Diagnosis as a Service (SBDaaS) is a cloud service proposed by the researchers [35]. SBDaaS provides efficient services like storing, processing, and analyzing the collected data from different building systems. However, being service based, SBDaaS allows adding, removing, and modifying services as required. In addition to the scalable and cost-efficient framework, developing a suitable model based on the data type is a significant task. Machine and deep learning models for accurate data predictions are discussed in the next subsection.

5.3.3 MACHINE AND DEEP LEARNING MODELS

In smart building, various algorithms and methods are using for processing and analyzing data. Methods and algorithms such as statistical and conventional learning, neural networks (NNs), probabilistic graphical modeling, data mining, and clustering algorithms are used. Selection of these methods and algorithms depends on the application, type, quality, and size of the data. Hadoop and Spark are the famous data pipeline architectures for data processing. Hadoop framework is suitable for batch processing, whereas Spark is a suitable framework for the streaming data [36]. Selecting a suitable machine learning model for predicting the collected data plays a significant role. Therefore, this section provides a detailed explanation on machine learning models for smart buildings.

5.3.3.1   Deep Learning Techniques

Machine learning models for smart buildings are used in automation systems are built with model-based solutions using control algorithms. These solutions are not suitable for recent designs because of increase in customer demand and complex building structure. Therefore, learning from the experience or data is the appropriate solution that can provide energy saving solution and efficient control system. Machine learning algorithms can help to infer the information from occupancy, devices, and profiles [37]. IoT has provided space for new technologies like big data and cloud computing. The primary objective of comforting and improving life of an individual is achieved by efficiently using the sensors' data in smart buildings. For processing and monitoring huge amount of data, proper sensor management systems and cloud servers are required. Green smart building or an energy-efficient smart building is possible by developing a network that can collect and manage sensors' data [38].

For instance, unemployment is the main problem faced by country's economy. Being unemployed is producing less economy, which effects the social and political disruption. There are many models built to predict the unemployment levels. Another approach can be using the smart meter data. A machine learning model can classify whether an individual is employed or unemployed based on behavioral patterns and energy consumption. Sixteen variables are considered for building classification model. In addition, with a smart meter, unemployment rates can be predicted using GPS data, call records, and Google searches. Logistic regression techniques are popular for determining unemployment. Data analysis is performed using techniques like generalized linear models (GLM) and NNs. Especially machine learning models like random forest (RF), gradient boosting machines (GBM), distance weighted discrimination (DWD) with kernel < decision tree, and multiplayer perception are used. The metrics like area under curve, sensitivity, and specificity are measure for all the machine learning models built. MLP and DWD models made accurate predictions on unemployment compared to all other machine learning models. These predictions may affect based on the data set considered for analysis [39].

Linear time-invariant techniques are based on parametric or nonparametric methods. A room occupancy is considered a deconvolution problem and is mathematically formulated including the parameters that can help to estimate the occupancy. Support-vector machine (SVM) and NN-based models are built to estimate or detect occupancy. These models are built assuming building occupancy signal for that period is available. Data is categorized as occupied and empty space based on vector machine techniques. Smart meter data is used to build the vector machine model in three categories: polynomial, linear, and radial basis function. Smart meter data is used to observe the pattern of occupancy and by using k-nearest neighbor (KNN) and vector machine for prediction. In addition to smart meter data, sensor data can also be added to provide accurate and efficient results [40–42].

In addition to this model, using blind identification method can help to estimate the occupancy by ignoring the assumption on availability of signal. Various factors are considered while predicting the occupancy of a building. Factors like type of the building, position of the building, type of the business performed in commercial building, and time. Descriptive statistical summary and statistical test were performed on each month and type of office data. Due to extensive advancement in sensor and wireless technology, a framework for analyzing unstructured data is in demand. Hence, unsupervised learning algorithms also need to be considered for building models or interpreting the data captured from different sources. Exploring the occupancy of the building with nonintuitive mechanism may need clustering algorithms. In addition, nonparametric clustering algorithms may need for exploring the occupancy at large areas. Three-step data mining models include transformation decision tree to clustering techniques for accurate efficient detection of occupancy [43–46].

5.3.3.2 Deep Learning Techniques

Deep learning technique is used to identify the occupancy that can control the cost of energy generation. Metrics like F2 score, accuracy or association, and correlation coefficient are considered for evaluating this deep learning method. This deep CNN model for detecting or classifying the occupancy is named as DeepECO. For feature selection, the model is compared with the principal component analysis (PCA) and SHapeley Additive exPlanation (SHAP). Data captured from the smart meters are considered for building this model. CNN model built with SHAP feature selection technique has outperformed the base CNN model and PCA-CNN model. All three metrics are compared to evaluate these three models developed with different feature select techniques. An ensemble model was built with the combination of PCA, SHAP, and correlation-based feature selection (CFS). That is, a model is built by considering PCA for feature extraction, CFS for ranking, and SHAP for feature selection [47, 48].

The Bayesian models are fast computational model compared to the numerical model that takes hours to simulate. These models are used to generate the indirect variables calculated or generated with the direct variables, which are captured from the meters installed in buildings. Software programs that are computationally demanding cannot be used for predicting but can be used for generating the data sets that can be transferred to Bayesian networks. The future values that will descript the energy and thermal values are interfered from the Bayesian network by considering input values. This step of generating the future values is repeated thousands of times to make network values feasible to model-based prediction control (MPC). This MPC will help to improve performance of BMS with optimal behavior [49].

Two efficient approaches of machine learning are deep and reinforcement learning. Deep learning is an efficient approach that has capabilities such as approximation, prediction, and classification. However, for decision-making and optimal controlling a reinforcement learning is a suitable approach. The combination of these two approaches is used for a few applications like video games [50]. Deep reinforcement learning (DRL) is gaining popularity for different applications such as dosage of a patient, vehicle classification, visual navigation, traffic signaling, and localizing objects [51–55]. Smart IoT services can be supported by proposed DRL framework that used both labeled and unlabeled data [56]. This semisupervised model was built using environmental data, which has both labeled and unlabeled data. The proposed model was to infer positioning policy, when compared to the model build with only labeled data.

The probabilistic graphical model helps to find the correlation of variable based on the condition. Various variables extracted from the sensors is examined by using conditional algorithms. This technique involved in two parts, one is determining dependency and another part is observing the probability distribution of another variables. Accurate prediction of occupancy in commercial building is possible with the help of Markov models. A simulator is used to analyze data of water-based heating system using DRL. Priority is given to selection of variable for developing a controller need to be developed that can provide a adapt to the environment change. These deep learning reinforcement algorithms have the capability of adapting to high dimensional and complex objective. These deep learning algorithms are combined with NNs to explain nonlinear and complex relation of an environment which is controlled [57, 58].

HVAC controlling strategy is obtained by applying machine learning models. Current shallow machine learning models and nonintrusive approaches failed to detect occupancy that can efficiently enhance building-to-grid integration. Researchers have come up with occupancy detection using deep learning model by using the advanced metering infrastructure (AMI) data. Deep model developed is a combination of the convolutional NN and bidirectional long short-term memory (BiLSTM). The deep learning architecture consists of four CNN layers and three BiLSTM layers network. This CNN-BiLSTM model has predicted the occupancy with 90% accuracy when compared to other classifier model such as KNN, SVM, a Gaussian process model, RF, and adaptive boosting. This model has shown better results compared to the model CNN, LSTM, and CNN-LSTM [59]. In addition, to these models, there are time series models which are used for predictions based on the type of the data.

5.3.4   TIME SERIES MODELS

Time series models are useful in smart buildings services or in IoT applications like detecting occupancy, identifying nonintrusive activities, and predicting energy usage. The main advantage of the forecast is we can use the past time observations or data to predict the future time observations or activities. Forecasting is a process of predicting the future activities at time $t+1$ using past activities at time t. Advanced sensor and network technologies has provided huge time series data related to temperature, energy consumption, and humidity measurement of devices. This time series data reflects different events in a building and helpful for designing efficient smart buildings services. Therefore, a quantitative analysis on this time series data can reveal correlation among multiple events. This analysis can help in managing and controlling multiple applications like energy consumption and detecting nonintrusive activities.

A model can examine the relationship among time series that are observed from sensors. The data captured by sensor and wireless networks are highly nonstationary and traditional correlation analysis is not suitable. Hodrick Prescott filter with granger causality is used to address the issue of nonstationarity and to capture components. Furthermore, ARIMAX model is built and F statistics is interpreted that can provide the casual relationship among time series data. Causality measured is different from the data with symmetrical measures. This causality can exploit potential of more dynamic and asymmetric relation among time series data. And, it is important to consider the pairwise causality or cross casual dependency need to be performed to build a reliable model [60].

Occupancy detection is the key component for providing efficient smart building services, which are limited by the sensor technology and algorithms. Occupancy is detected using a time series models using the Wi-Fi network data. Auto-regressive integrated moving average (ARIMA) and LSTM models are built considering three different intervals of occupancy, building, and access point levels. First model is built individual for different intervals and then combined model for three intervals is built. Finally, model metric like RSTM is compared with the ARIMA model for accuracy. Though RMSE is lower for LSTM compared to ARIMA, this model will take more time to run [61].

Researchers have used data reduction, transformation, and applying machine learning models to solve the unusual patterns in the data obtained energy meters. Detection of anomalies in the energy consumption can decrease energy waste and avoid abnormal events in the time series data. Anomaly patterns after mining were examined to depict the impact of fault identification in the early stage. These anomalies obtained may be due to abnormal functioning of control systems installed in the building. Therefore, study of change in energy consumption based on the time series data generated by the meters is important to identify the anomalies. Symbolic aggregate approximation (SAX) is used to measure the abnormal patterns in consumption of energy time series data. This SAX method helps the stakeholders to identify the behavior of the building, to explain the suitable strategy for energy management, and to send alerts based on abnormal energy consumption. The regression tree model was applied for segmentation of data set using a splitting criterion. Self-tuning is the special characteristic of a model to reduce time series effectively by finding suitable length and number of windows. Pattern recognition techniques were applied after encoding the energy consumption time series data [62].

5.3.5 Metrics for Model Performance

Root-mean-squared error (RMSE) is an example of such metrics. Suitable model depends on the type of data, quality of data, and features combination. Thorough research is needed to include the features before predicting performance. RMSE is used to measure the performance of the model built. This is the deviation of the predicted value from the observed or original values. The individual difference in the value is called residual and the difference performed on overall data sample is called as RMSE [63]. RMSE is mostly used for the models that predict air quality, geoscience, and air quality. Another metric is MAE. There is no agreement on which metric to use. MAE is always greater than the RMSE value and these both are used for evaluation the model performance. RMSE depends on sample size, magnitude of error, and MAE value. More reliable conclusions are made by using RMSE than MAE when the sample data is normally distributed. RMSE metric reliability depends on the outliers. Therefore, it is advisable to treat outliers when evaluating the model using RMSE value. Considering absolute value in mathematical calculation is not required in all cases. This avoiding of absolute value in calculation is one main difference in RMSE over MAE. Sometimes, it is required to calculate sensitivity and adjust model parameters by minimizing the cost function in data assimilation field. For such applications, RMSE is preferred than MAE.

$$\text{RMSE} = \sqrt{\frac{1}{n}\sum_{i=1}^{n}\left|e_i^2\right|} \tag{5.1}$$

$$\text{MAE} = \frac{1}{n}\sum_{i=1}^{n}\left|e_i\right| \tag{5.2}$$

Equations (5.1) and (5.2) represent the mathematical formula for calculating RMSE and MAE metrics. This formula is not showing the difference in the predicted values from observed values. The n in this formula represents the sample size, e is the model errors, where $i = 1, 2, 3, \ldots, n$. Unbiased errors with normal distributed are a basic assumption in performing RMSE metric calculation [63].

Determining the sensitivity of model using only one metrics for evaluating same model will not depict any change in error distribution. Error distribution is an important factor to be considered while measuring performance among different models with a single metric. We need to consider the remaining metrics like variance, mean, skewness, and kurtosis while measuring model error variation, as distribution of errors is normally distributed while considering RMSE. Therefore, it is always better to consider multiple metrics to evaluate performance of model as considering the single metrics will not provide information on certain aspects of model errors.

Model selection is based also identification techniques. There are techniques called gray box and black box. Gray box techniques no need full information to build a model, whereas black box technique need full information on the data obtain meaningful results. Examples of gray box techniques are MRC relevant identification (MRI), deterministic and probabilistic semi-physical modeling (DSPM and PSPM). There are two stages following to select model complexity. First one is minimizing the input variables that can improve the quality of model built. And next stage is evaluating the model improvement based on the input sets by using criteria like KS test, fit factor, T-criterion, whiteness, and coefficient of determination. Selection of input states is stopped once there is no qualitative improvement in the model selected. Following these various criterions for model selection can provide technical and economic savings with a smaller number of input states, which eventually means less number of sensors [64].

5.4 DATA ANALYSIS FOR DIFFERENT APPLICATIONS

Recent trend includes designing an interactive structure or architecture of a building based on the climatic or environmental changes. Such interactive structures may be considered esthetics of buildings or advancement in functionality of buildings. Technology for building interactive structures used automated louvers and self-dimming glasses that respond for varying outdoor lighting and heating patterns [6]. This section discusses on multiple applications in smart building such as observing energy consumption, occupancy detection, health monitoring, automated lighting, and HVAC systems. As mentioned in the previous section that there is no well-defined model that suits each application in smart buildings. Therefore, this section, rather than providing different models for each application, provides an idea on requirement of each application and building systems that need to be considered to provide accurate predictions.

5.4.1 ENERGY CONSUMPTION

Smart metering is the initial point for the conversion of traditional building structure to a smart building structure. This smart meter deployed in the smart buildings need to automatically monitor, control, interact, and optimize energies. Another important step is designing a smart lighting system that has sensors, controllers, drivers, communication, and logic systems. This smart metering and lighting systems can turn the conventional buildings to smart buildings that can provide a smart world with 100% renewable energies. Carbon footprint can be reduced with the help of renewable energy sources. The automatic systems can provide an optimized solution with the proper data usage and efficiently use the renewable sources [65].

Smart meters have become popular for measuring energy consumption. Power systems with smart meters can measure power consumptions in an hourly period or even less, whereas traditional meters can provide information on cumulative power consumption over a billing period. Therefore, smart meters data provides more information on electricity usage pattern to both electricity end users and electric utilities management. End user can analyze this energy usage patterns to save energy cost.

However, the one who manages electric utilities can use smart meter data to improve efficiency and robustness of a power grid. In addition to smart meters, AMI provides data that can control voltage in distributed power grids. Smart meters are gaining popularity with increase in electricity prices and decrease in electronics cost. Like any other network equipment, smart meters are also susceptible to errors. Cause for missing data in smart meters is due to human errors made during installation, issues in data storage, and failures in communication. Hence, it is important to develop methods that can correctly analyze missing data to increase the capability of smart meters. Collaborative inference method is proposed with 85% accuracy by considering the exogenous factors [66].

ARIMA model is used for inferencing missing values of target building data by considering the data of building with similar characteristics. Other factors like temperature and number of people in that building are considered for building model. Seasonal auto-regression integrated moving average model with exogenous variables (SARIMAX) model is considered for the inferencing missing data in smart meters, and cross-correlation function is used to relate a building with missing data to the building with known data [66]. ARIMA is the most popular model for analyzing time series data. ARIMA model provides better results for analyzing different times series data. This family of models are called Box-Jenkins models as they are first invented in 1976 by Box and Jenkins [67]. A model built with transfer functions is considered the ARIMAX model. Another model in this ARIMA family is called ARIMAX model, also called as transfer function model. ARIMA model deals with only one time series variable, whereas ARIMAX model deals with two time series variables. Researchers have used ARIMA and ARIMAX for forecasting gross domestic product (GDP) per capita. From metric-like mean absolute percentage error

(MAPE) and RMSE, it is observed that ARIMA showed slight accurate predictions than the ARIMAX model [68, 69].

Drawbacks in intuitive approach have given rise to nonintrusive approach, where there is no need of installation cost and no false-positive predictions. One approach is using GPS data of mobile phones and another approach is using power meter reading. This smart meter reading is called as load curve data. Applying supervised machine learning algorithms such as SVM, KNN, and HMM is used for building model. Using this supervised learning is a tedious process as this requires information from occupancy and cannot be generalized. The sparse human action recovery with knowledge of appliances (SHARK) model is trained based on the appliance switching mode and consumption constraints. The major problem is in maintaining the privacy of the human in detecting the occupancy [70].

Smart cities are based on the IoT implemented in the infrastructure. In addition, including IoT in services and residents provide better services. Collection of massive amounts of data eventually turned into dealing with big data, which need huge number of resources and computational resources. Though the existing data analysis techniques works, a huge amount of data processing time is longer for smart cities compared to the individual buildings. Therefore, for proper processing of big data for smart cities, quick actions are required. Extensive research is conducted to manage the fog and clod technologies.

The processes of decision-making for efficient energy are also considered as a service. Hence, researchers have developed a service-based framework that can help monitoring and controlling energy consumption [71]. A framework which is based on the service may consider all the activities related energy as services. These services are used for integration for any application by registering, advertising, and making available. Hence, this type of framework can integrate technologies that can perform big data analytics for managing energy in smart cities.

There are some potential barriers need to be considered in the smart grid technologies, which are great hindrance for efficient reduction in energy cost. The challenges faced by smart grids are device network, detection, security, and data dependency. Important things that need to be considered to remove potential barriers are to develop a data storage and visualization tools that can serve the purpose of smart grid, to design security mechanism that can control and integrate smart grid data, to improve optimization and organization mechanisms in wireless technologies, to develop QoS aware network, to develop a heterogenous environment for optimal solution, and to design a novel simulator for robust modeling of smart grid parameters [72]. Increase in the power consumed electronic devices has increased the electricity consumption in industrial, commercial, and residential buildings. Presence detection or occupancy detection is considered are an effective way to control energy consumption by controlling HVAC and lighting systems [70].

5.4.1.1 Economic Dispatch

The main objective of an optimization problem called economic dispatch is to reduce the cost of the energy generated by considering generator constraints. Technology used for smart grid makes power providers and operators to release resources economically. The primary objective is to optimize the energy consumption by considering the privacy constraints. This finally leads to save finances and reduces the cost of building the costly infrastructures that can meet the highly variable load demands. A particle swarm optimization (PSO) is a method that helps to solve this economic dispatch problem. In this method, the generators characteristics like cost functions, ramp rate limits and operating zones are considered. And the efficiency of the method is compared with the generic algorithms. Hence, researchers have showcased that PSO provide better solution for economic dispatch problem than generic algorithms. This method is better in a few aspects like providing high quality solution, efficient computation, and stable characteristics. These aspects are compared with the generic algorithms to estimate the performance of PSO method [73, 74].

In an extension to this research, an effective PSO algorithm with social weight is proposed. All constraints like cost functions, ramp rate limits, and operating zones are considered while performing operations using this method. This method is faster in finding quality solution for the economic dispatch problem in power systems [74]. An event triggered method is used for economic dispatch to reduce information exchange. In this event, triggered-based distributed approach, initially cost function is formulated using the θ-logarithmic barrier and balanced supply demand is allocating using distributed algorithm. This hybrid event triggered method is used to solve the problem in a distributed architecture. This method has achieved a tradeoff between the quality of solution and the number of resources for communication; this can be taken care of operators by following proper network conditions [75].

Recent economic dispatch is deviated from the traditional centralized one and more focused on distributed architecture with smart entities. Hence, researchers have proposed method for solving such complex economic dispatch problem is self-organizing and decentralizing framework. This framework effectively solves economic dispatch problem even in the case of variable loads by using dynamic agents. The primary objective in this method is to solve the problem by using the network of agents that can exchange and process data in a distributed environment. Though, this is developed on few assumptions, this method can be more efficient on solving the other problems during economic dispatch analysis [76].

It is the main constraint in economic dispatch providing the privacy. Hence, researchers have come up with a distributed and effective heterogenous privacy-preserving consensus (DisEHPPC)-based approach. This approach includes demand response server, manager, and controller. The privacy in this method is obtained in two phases. This method is carried in three stages: customer privacy is focused on first stage, convergent point is obtained for measuring optimality of solution, and the last stage is providing the theoretical analysis of the convergent points and privacy degree that can provide a quality solution for the economic dispatch problem [77].

Building management systems (BMSs) for energy monitoring are called building energy management system (BEMS). This same management system can be also used to control HVAC systems. Several sensor nodes are deployed for monitoring utilization of energy and observing environmental conditions that can eventually save energy. Intelligent algorithms are implemented by the control system to control the energy consumption. However, building management can also be facilitated using cloud computing. This type of computing model can reduce the requirement of organizations to provide infrastructure required for smart buildings. In other words, this cloud computing model have expensive software, hardware, and network framework. In addition, this model also reduces the cost of manpower to operate developed framework. For processing, storing, and analyzing data that is gathered for smart meters, a suitable infrastructure is needed. This cloud computing provides a solution for such problem. Increase amount of data requires more storage space, computation, and sophisticated software for implementing intelligent algorithms. Cloud computing is suitable in providing economical and scalable solutions for smart buildings' needs. In addition, this allows collecting huge amount of data from multiple sources and process data with high speed. Therefore, cloud-based BEMS are more effective in integrated data analysis for monitoring and controlling energy consumption [16, 78, 79].

5.4.2 OCCUPANCY DETECTION

Humans' activity can produce multiple patterns in energy consumption and in the renewable energy generation. Hence, it is important to obtain the information on the occupancy of a building equipped with renewable energy generation systems. Predicting the occupancy of a building can be provide multiple benefits such as controlling the energy consumption, efficient utilization of grid, and researching on human's mobility. Especially, population mobility and social distancing have been slogans for past one year due to COVID-19 pandemic. Initially infrared motion or carbon dioxide sensors are used for detecting the occupancy in smart buildings [80]. These sensors and dedicated devices need hardware installation and more prone to false-positive results. That is, sensors are costly and find hard to differentiate between human and pet movement [70]. Therefore, using sensors or cameras has not been an efficient approach when there is renewable energy generation that can cause bidirectional energy flow. Hence, collecting the data from the smart meters for detecting the occupancy in smart buildings is proposed by researchers [39, 48, 70, 81, 82].

As a supplement to sensor and smart meter data, smart phone data can also be used to detect the occupancy. This low-cost technique for determining the occupancy using Bluetooth of a smart phone can reliable and accurate results. In addition, there are some drawbacks with this technique. This smart phone technique can generate false-positives due to Bluetooth signal range, latency rate of Bluetooth response, power consumed with Bluetooth, and wrong placing of mobile phones [83]. For detecting the footsteps and occupant localization, a sparse passive sensing system is used. Detecting occupancy and crowd information can help to respond in emergency and manage crowd. For monitoring crowd dynamics, a sensor network is constructed using sensor nodes with interfacing circuits that consume less power. Through the sensor network or array, surface vibrations are collected to detect the footsteps. Monitoring crowd dynamics and human flow achieved by using passive seismic sensors in a building lead to Internet of Ears (IoEs). IoEs can be used to detect footsteps, walking directions and position of occupant. Position of occupant or occupant localization can be identified by using the least square method [84].

The researchers have investigated the need for occupancy detection in the sensor networks and for building management. Considering a single measurement point can lead to wrong predictions of occupancy, whereas considering the whole sensors data and performing extensive analysis on captured data can efficiently detect occupancy in a building. There are many applications for the occupancy detection in smart buildings. An example of such application is a video filmed by French brothers Jules and Naudet on firefights regarding twin towers incidence showcases the importance of occupancy detection mechanism in elevators. Another example can be false alarms which are the result of single sensitive detectors [85].

Oxnard police department explained that in 1995, only 2% of alarms are related to crime and remaining 98% are false alarms, which cost average of $62. This is due sensors detected inanimate substances that cause false alarms. Therefore, developing home security systems with sensor networks that can accurately detect occupancy can in turn reduced the cost due to false alarms. In addition, another application is for controlling the HVAC system in buildings based on occupancy. A properly developed framework can control the air conditioning in homes that can provide air only to occupied rooms. Most of the houses have hot or cold rooms, irrespective of the temperature. This can falsely control the thermostat based on the occupancy. Hence, in line duct fans are used in such room that can lower or raise the ambient temperature [85].

A sensor networks designed can control air flow to room by controlling air conditioners and fans. Lastly, another application is providing a considerable economic, mental, and health benefits to the elderly people who are alone at homes. People who are living may have some sudden health issues, this can be addressed by sensor monitoring systems. Such benefits include turning off stove when there is no human activity in the kitchen, sending messages or emails to family members or care takers if there is unusual activity is detected. These unusual patterns are detected by developing a statistical criterion that has been implemented based on the probability distribution which is associated with daily events [85].

5.4.3 HVAC CONTROLLING

Researchers even came up with HVAC controlling strategy based on detection of occupancy in a building. This strategy proposed by the researcher has controlled 20% energy based on the occupancy in a building. Markov chain model is used to train the data and detected occupancy to control heat and ventilation. Though this researcher has not mentioned the control HVAC in cold region, by considering even more data that is collected by system, further research can be done that can provide better controlling strategy. Energy plus is a standard tool, which is used by US industries to estimate the energy that is controlled by the occupancy strategies. Air handler unit and air volume vent in each room are considered as attributes HVAC system. When developing a suitable model for colder regions, one can use air handler unit and solar gain variables. CNN-BiLSTM model is used for predicting the occupancy. This model depicted 90% accuracy when compared to SVM, KNN, and RF models [59].

There are personalized models that can provide accurate results based on the data points collected and the HVAC system will not control till these predictions are obtained. A thermal environment that is self-tuned is obtained by knowing an individual thermal comfort and HVAC systems that is controlled by personalized models. This self-tuned HVAC system can improve the comfort level of an occupant by controlling the energy consumption and deploying where it is in demand. We need to select the proper evaluation methods and suitable learning methods that can provide meaningful results in the BMS. Different control approaches were obtained by properly making using data that is collected from the systems installed in the buildings. Complex computational problems can be solved by diving into small problem and providing the suitable solution for simple problem to solve the complex one. This can provide the optimum control solution for the existing control systems in buildings [86].

Increase in air infiltration can lead to decrease in materials of hygrothermal insulation, degrades acoustic insulation, increase in heating and cooling. The rate at which the air is filtered is defined by the airtightness. Measuring this air tightness is a primary issue, and no model is developed that can accurately predict this air tightness. Airtightness in a building and its uncertainty can be measured by using the regression models. For predicting the air rate based on the mean average models like ordinary least square (OLS), iterative weighted least square (IWLS) and weighted line of organic correlation (WLOC) method were used. These models like IWLS and WLOC provide an accurate uncertainty when compared to OLS model [87].

5.4.4 HEALTHCARE

A framework, single-input or single-output model predictive control is installed in smart grids for electricity managing and pricing in commercial and conventional buildings. A model is built by considering the single input with minimum variable transformations and depicting performance index of the single optimum output. This kind of framework can be helpful when the pricing and demand of electricity is in peaks. Through connection of smart object, the society is connected unprecedently. Infrastructure developed for smart home has become possible with the sensor and actuator technologies. This smart home technology has created interest in fields like engineering, clinical, social, and computer sciences to meet requirements of humans. Design of smart home technologies can be evolved with life of humans. While designing automated homes or buildings, developers need to focus on two groups of people. First groups can be people who are interested on technical or general interest of technology or innovations, whereas second kind may be people who are health oriented [88].

For instance, depression is a major health problem faced in current world, which has been not easily recognized and very complicate treatment monitoring system. Depression is not well treated at early stages because an individual may not feel very comfortable to visit doctor and take proper treatment because of public disgrace. Hence, researchers have developed a depression monitoring system that can monitor patients' movements 24/7. For collecting the information, devices like sensors, touch screens, mobiles, and supporting software are used. These devices give collective data based on the parameters like sleep, speech, weight, and day-to-day activities of a patient. This data collected from multiple behaviors of a patient is aggregated and used for better treatment and diagnosis. These depression monitoring systems are installed at home for treatment at the early stages of depression. With proper care and attention, this professionally and personally debilitating disorder can be treated with positive results. Additionally, depression shows a dangerous impact on mood, which increases alcohol consumption, decreases individual hygiene, affects social relationships, and reduces interest in taking existing medications.

A special tool called symptom monitoring tool is used to track an individual report that includes past symptoms that helps to improve the treatment and diagnosis. Empath is an example of the architecture that is installed at home for monitoring depression by considering sleep, weight, speech, and other activities with retroactive individual report. Though the researchers have considered only speech, sleep, weight, and occupancy for providing better information for treatment and diagnosis of depression, there can be other factors too that can enhance depression treatment.

Empath is specially designed focusing on monitoring elderly people who are living alone. And this researcher has focused only considered data collected at home, but data that is generated outside is ignored. For example, by considering exercise and social relationships outside home can be useful for better treatment. Also, this research ignored mood effects like alcohol consumption, social interaction, missing regular medication, and exercise. Especially, detecting activities that are taken place outside home is a challenging task. One can link smart watches data also to track these activities but creating a suitable framework that can integrate all these devices data is a challenging task. Empath is also used for monitoring soldiers' post-traumatic stress disorder (PTSD) after war. This disorder has symptoms like problem in concentrating, difficulty in sleeping or sometimes over sleeping, and lack of interest in performing day-to-day activities and living [89].

5.4.5 TRANSPORTATION

Accurate travel time prediction is necessary for better transportation management. Travel time is a key attribute for evaluating road network performance, which helps to come up with better management strategies and tools. With this travel time prediction, better planning and scheduling of transportation is possible for public transport system. However, for an individual, predicting accurate travel time can help to make better travel plans by considering features like route selection, mode of transport, and departure time. GPS data of vehicles can be used for prediction of travel time. Most important thing to solve the GPS data storage problem is feature extraction, which can reduce the storage size and time cost for data analysis. Decision tree ensemble methods, which have efficiency and effectiveness, are popular methods to solve prediction and classification problems. RF and gradient boosting regression trees (GBRT) are examples of such methods. Main problem in feature selection is combinatorial effect that needs to be considered while building a model using these methods. It is important to consider the model fitting by considering different strategies and using suitable evaluation metric [90].

5.5 CONCLUSION

Smart buildings are developed based on combination of smart technologies and well-defined architectures. The primary objective of these buildings is to provide comfortability, sustainability, safety, and security. Popular buildings like Duke Energy

Center, The Edge Building, DPR Constructions, Glumac, and Hindmarsh showcase the significance of smart buildings. Efficiently controlling temperature, observing energy consumption, monitoring health, detecting number of occupants, controlling HVAC, and lighting systems are few applications in smart buildings. Devices performing these applications need to be connected in a network. This is possible with the concept of IoTs. IoT allows connecting each device through internet. BMS is used to connect different building systems that can control and monitor each application. The devices like sensors and actuators are the main sources of the data. BEMS is a management system that can be used to control HVAC systems.

Several sensor nodes are deployed for monitoring utilization of energy and observing environmental conditions that can eventually save energy. Intelligent algorithms are implemented by the control system to control the energy consumption. However, building management can also be facilitated using cloud computing. Developing a framework that can process, store, and analyze gathered data from different devices helps to design s smart building efficiently. Cloud computing models are scalable and cost-effective platforms for processing, storing, and analyzing collected data. WattDepot and SBDaaS are suitable cloud platform for providing cloud services. In addition to creating a suitable framework, it is important to predict the accurate results from the given data. In smart buildings, different data mining, clustering, NNs, and statistical and conventional model are used to process and analyze data. Selecting a suitable machine learning model for predicting the collected data plays a significant role. Humans' activity can produce multiple patterns in energy consumption and in the renewable energy generation. Hence, it is important to obtain the information on the occupancy of a building equipped with renewable energy generation systems.

Processing multiple buildings data instead of a single building can help to solve a relevant business problem more accurately. Similarly, processing data from multiple sources also helps to design a better building model. A collaborative data collected can help to create a framework that suits the environment and comforts lives. Collaborative framework is built based on the data which is collected from multiple sources. Efficient framework for smart buildings can ease the life of humans and save their time from concentrating on trivial matters. Predicting the occupancy of a building can be provide multiple benefits such as controlling the energy consumption, efficient utilization of grid, and researching on human's mobility. Machine learning and deep learning models such as PCA, CNN, SVM, KNN, GLM, RF, and Bayesian models are used. Deep learning and reinforcement learning are becoming popular due to customer demand and complexity of a building structure. These models provide solutions based on the previous experience. These models are applied to the data collected based on smart meters, number of occupants, temperature meters, and HVAC systems. Therefore, extensive literature provided in this chapter on models built for different applications showcase the significance of data analysis in smart buildings.

REFERENCES

1. Planet Technology USA, "7 Incredible Examples of Smart Buildings (And What Makes Them Smart)," *Planet Technology USA*, 2019. https://planetechusa.com/7-incredible-examples-of-smart-buildings-and-what-makes-them-smart/ (accessed Feb. 09, 2021).

2. R. Mehta, J. Sahni, and K. Khanna, "Internet of Things: Vision, applications and challenges," *Procedia Comput. Sci.*, vol. 132, pp. 1263–1269, 2018, doi: 10.1016/j.procs. 2018.05.042.

3. J. Gubbi, R. Buyya, S. Marusic, and M. Palaniswami, "Internet of Things (IoT): A vision, architectural elements, and future directions," *Futur. Gener. Comput. Syst.*, vol. 29, no. 7, pp. 1645–1660, 2013, doi: 10.1016/j.future.2013.01.010.

4. K. Ashton, "That 'Internet of Things' Thing," *RFID Journal*, 2009. https://www. rfidjournal.com/that-internet-of-things-thing.

5. A. Daissaoui, A. Boulmakoul, L. Karim, and A. Lbath, "IoT and big data analytics for smart buildings: A survey," *Procedia Comput. Sci.*, vol. 170, pp. 161–168, 2020, doi: 10.1016/j.procs.2020.03.021.

6. M. El-Shafie and L. Fakeih, "The smart environment of commercial buildings," 2018 15th Learn. Technol. Conf. L T 2018, pp. 161–166, 2018, doi: 10.1109/LT.2018.8368501.

7. A. E. M. Taha and A. Elabd, "IoT for certified sustainability in smart buildings," *IEEE Netw.*, pp. 1–7, 2020, doi: 10.1109/MNET.011.2000521.

8. B. Ramprasad, J. Mcarthur, M. Fokaefs, C. Barna, M. Damm, and M. Litoiu, "Leveraging existing sensor networks as IoT devices for smart buildings," *IEEE World Forum Internet Things, WF-IoT 2018 - Proc.*, vol. 2018-Janua, pp. 452–457, 2018, doi: 10.1109/WF-IoT.2018.8355121.

9. Siemens, "Improving Performance with Integrated Smart Buildings," 2012. [Online]. Available: http://www.usa.siemens.com/intelligent-infrastructure/assets/pdf/smart-building-white-paper.pdf.

10. A. Dunkels and J. Vasseur, "The Internet of Things : IP for Smart Objects," *IPSO Alliance White Pap.*, 2008.

11. S. Karg and B. Lucia, "Towards low-energy, low-cost and high-performance IoT-based operation of interconnected systems," IEEE 4th World Forum Internet Things, pp. 706–711, 2018.

12. T. M. Lawrence *et al.*, "Ten questions concerning integrating smart buildings into the smart grid," *Build. Environ.*, vol. 108, pp. 273–283, 2016, doi: 10.1016/j. buildenv.2016.08.022.

13. H. Burak Gunay, Z. Shi, I. Wilton, and J. Bursill, "Disaggregation of commercial building end-uses with automation system data," *Energy Build.*, vol. 223, p. 110222, 2020, doi: 10.1016/j.enbuild.2020.110222.

14. L. Bindra, C. Lin, E. Stroulia, and O. Ardakanian, "Decentralized access control for smart buildings using metadata and smart contracts," Proc. - 2019 IEEE/ACM 5th Int. Work. Softw. Eng. Smart Cyber-Physical Syst. SEsCPS 2019, pp. 32–38, 2019, doi: 10.1109/SEsCPS.2019.00013.

15. S. Lazarova-Molnar and N. Mohamed, "Collaborative data analytics for smart buildings: Opportunities and models," *Cluster Comput.*, vol. 22, no. November 2017, pp. 1065–1077, 2019, doi: 10.1007/s10586-017-1362-x.

16. N. Mohamed, S. Lazarova-molnar, and J. Al-jaroodi, CE-BEMS : A Cloud-Enabled Building Energy Management System, pp. 1–6, 2016.

17. C. Lin *et al.*, "Multi-objective value-driven smart building management solutions," Proc. - 2015 Int. Conf. Cloud Comput. Big Data, CCBD 2015, pp. 260–263, 2016, doi: 10.1109/CCBD.2015.52.

18. M. K. Bidhandi, M. K. Bidhandi, and S. A. H. R. Ebrahimi, "Introduce an object-oriented simulator for analyzing discrete events in smart buildings," 2014 Int. Congr. Technol. Commun. Knowledge, ICTCK 2014, no. Ictck, pp. 26–27, 2015, doi: 10.1109/ ICTCK.2014.7033497.

19. A. Souri, P. Asghari, and R. Rezaei, "Software as a service based CRM providers in the cloud computing: Challenges and technical issues," *J. Serv. Sci. Res.*, vol. 9, no. 2, pp. 219–237, Dec. 2017, doi: 10.1007/s12927-017-0011-5.

20. A. Souri, A. M. Rahmani, and N. Jafari Navimipour, "Formal verification approaches in the web service composition: A comprehensive analysis of the current challenges for future research," *Int. J. Commun. Syst.*, vol. 31, no. 17, pp. 1–28, 2018, doi: 10.1002/dac.3808.

21. S. Sundmaeker, H. Guillemin, P. Friess, and Woelfflé, S. "Vision and Challenges for Realizing the Internet of Things", *Cluster of European research projects on the internet of things, European Commision*, vol. 3, no. 3, 2010.

22. IBM100, "Smarter Planet", Retrieved from https://www.ibm.com/ibm/history/ibm100/us/en/icons/smarterplanet/ *IBM*, p. 1, 2011.

23. S. Ghosh, "Smart homes: Architectural and engineering design imperatives for smart city building codes," Int. Conf. Technol. Smart City Energy Secur. Power Smart Solut. Smart Cities, ICSESP 2018 - Proc., vol. 2018-Janua, pp. 1–4, 2018, doi: 10.1109/ICSESP.2018.8376676.

24. S. Lazarova-Molnar and N. Mohamed, "Towards collaborative data analytics for smart buildings," *Lect. Notes Electr. Eng.*, vol. 424, no. March, pp. 459–466, 2017, doi: 10.1007/978-981-10-4154-9_53.

25. A. Mejías, R. S. Herrera, M. A. Márquez, A. Calderón, I. González, and J. Andújar, "Easy handling of sensors and actuators over TCP/IP networks by open source hardware/software," *Sensors (Switzerland)*, vol. 17, no. 1, 2017, doi: 10.3390/s17010094.

26. "IEEE Xplore Full-Text PDF:" https://libproxy.library.unt.edu:2301/stamp/stamp.jsp?tp=&arnumber=707933 (accessed Jan. 18, 2020).

27. Y. Liu *et al.*, "Coordinating the operations of smart buildings in smart grids," *Appl. Energy*, vol. 228, no. July, pp. 2510–2525, 2018, doi: 10.1016/j.apenergy.2018.07.089.

28. F. Benhamouda, M. Joye, and B. Libert, "A new framework for privacy-preserving aggregation of time-series data," *ACM Trans. Inf. Syst. Secur.*, vol. 18, no. 3, pp. 1–21, 2016, doi: 10.1145/2873069.

29. H. Yang, L. Cheng, and M. C. Chuah, "Evaluation of utility-privacy trade-offs of data manipulation techniques for smart metering," 2016 IEEE Conf. Commun. Netw. Secur. CNS 2016, pp. 396–400, 2017, doi: 10.1109/CNS.2016.7860526.

30. J. Zhao, T. Jung, Y. Wang, and X. Li, "Achieving differential privacy of data disclosure in the smart grid," Proc. - IEEE INFOCOM, pp. 504–512, 2014, doi: 10.1109/INFOCOM.2014.6847974.

31. H. V. P. Lei Tang, Xu Chen, Junshan Zhang, "Optimal Privacy-Preserving Energy Management for Smart Meters," 2014.

32. R. Rowlingson, "Ten step process for forensic readiness," *Int. J. Digit. Evid.*, vol. 2, no. 3A, pp. 1–28, 2004, [Online]. Available: https://www.utica.edu/academic/institutes/ecii/publications/articles/A0B13342-B4E0-1F6A-156F501C49CF5F51.pdf.

33. P. Barbosa, A. Brito, and H. Almeida, "Defending against load monitoring in smart metering data through noise addition," Proc. ACM Symp. Appl. Comput., vol. 13-17-Apri, pp. 2218–2224, 2015, doi: 10.1145/2695664.2695800.

34. R. S. Brewer and P. M. Johnson, WattDepot: An Open Source Software Ecosystem for Enterprise-Scale Energy Data Collection, Storage, Analysis, and Visualization, pp. 91–95, 2010, doi: 10.1109/smartgrid.2010.5622023.

35. N. Mohamed, S. Lazarova-Molnar, and J. Al-Jaroodi, "SBDaaS: Smart building diagnostics as a service on the cloud," Proc. 2nd Int. Conf. Intell. Green Build. Smart Grid, IGBSG 2016, 2016, doi: 10.1109/IGBSG.2016.7539417.

36. V. Diaconita, A. R. Bologa, and R. Bologa, "Hadoop oriented smart cities architecture," *Sensors (Switzerland)*, vol. 18, no. 4, 2018, doi: 10.3390/s18041181.

37. D. Djenouri, R. Laidi, Y. Djenouri, and I. Balasingham, "Machine learning for smart building applications: Review and taxonomy," *ACM Comput. Surv.*, vol. 52, no. 2, 2019, doi: 10.1145/3311950.

38. A. P. Plageras, K. E. Psannis, C. Stergiou, H. Wang, and B. B. Gupta, "Efficient IoT-based sensor BIG Data collection–processing and analysis in smart buildings," *Futur. Gener. Comput. Syst.*, vol. 82, pp. 349–357, 2018, doi: 10.1016/j.future.2017.09.082.
39. C. A. C. Montaklez and W. Hurst, "A machine learning approach for detecting unemployment using the smart metering infrastructure," *IEEE Access*, vol. 8, pp. 22525–22536, 2020, doi: 10.1109/ACCESS.2020.2969468.
40. A. Akbar, M. Nati, F. Carrez, and K. Moessner, "Contextual occupancy detection for smart office by pattern recognition of electricity consumption data," *IEEE Int. Conf. Commun.*, vol. 2015-Septe, pp. 561–566, 2015, doi: 10.1109/ICC.2015.7248381.
41. C. Duarte, K. Van Den Wymelenberg, and C. Rieger, "Revealing occupancy patterns in an office building through the use of occupancy sensor data," *Energy Build.*, vol. 67, pp. 587–595, 2013, doi: 10.1016/j.enbuild.2013.08.062.
42. A. Ebadat, G. Bottegal, D. Varagnolo, B. Wahlberg, and K. H. Johansson, "Estimation of building occupancy levels through environmental signals deconvolution," Proceedings of the 5th ACM Workshop on Embedded Systems For Energy-Efficient Buildings - BuildSys'13, 2013, pp. 1–8, doi: 10.1145/2528282.2528290.
43. J. Zhao, B. Lasternas, K. P. Lam, R. Yun, and V. Loftness, "Occupant behavior and schedule modeling for building energy simulation through office appliance power consumption data mining," *Energy Build.*, vol. 82, pp. 341–355, 2014, doi: 10.1016/j.enbuild.2014.07.033.
44. N. T. Nguyen, R. Zheng, and Z. Han, "UMLI: An unsupervised mobile locations extraction approach with incomplete data," *IEEE Wirel. Commun. Netw. Conf. WCNC*, pp. 2119–2124, 2013, doi: 10.1109/WCNC.2013.6554890.
45. D. Chen, S. Barker, A. Subbaswamy, D. Irwin, and P. Shenoy, Non-Intrusive Occupancy Monitoring using Smart Meters, pp. 1–8, 2013, doi: 10.1145/2528282.2528294.
46. S. D'Oca and T. Hong, "Occupancy schedules learning process through a data mining framework," *Energy Build.*, vol. 88, no. 510, pp. 395–408, 2015, doi: 10.1016/j.enbuild.2014.11.065.
47. S. Sasikala, S. A. A. Balamurugan, and S. Geetha, "An efficient feature selection paradigm using PCA-CFS-Shapley values ensemble applied to small medical data sets," 2013 4th Int. Conf. Comput. Commun. Netw. Technol. ICCCNT 2013, pp. 4–8, 2013, doi: 10.1109/ICCCNT.2013.6726773.
48. N. Pal, P. Ghosh, and G. Karsai, "DeepECO: Applying deep learning for occupancy detection from energy consumption data," Proc. - 18th IEEE Int. Conf. Mach. Learn. Appl. ICMLA 2019, pp. 1938–1943, 2019, doi: 10.1109/ICMLA.2019.00311.
49. A. Carbonari, M. Vaccarini, and A. Giretti, "Bayesian networks for supporting model based predictive control of smart buildings," *Dynamic Programming and Bayesian Inference, Concepts and Applications*, InTech Publications, 2014, doi: 10.5772/58470.
50. Y. LeCun, Y. Bengio, and G. Hinton, "Deep learning," *Nature*, vol. 521, no. 7553, pp. 436–444, May 2015, doi: 10.1038/nature14539.
51. V. Mnih *et al.*, "Playing Atari with Deep Reinforcement Learning," pp. 1–9, 2013, [Online]. Available: http://arxiv.org/abs/1312.5602.
52. Q. Zhao, C. Xu, and S. Jin, "Traffic signal timing via parallel reinforcement learning," *Smart Innov. Syst. Technol.*, vol. 149, no. 3, pp. 113–123, 2019, doi: 10.1007/978-981-13-8683-1_12.
53. Y. Zhu *et al.*, "Target-driven visual navigation in indoor scenes using deep reinforcement learning," *Proc. - IEEE Int. Conf. Robot. Autom.*, no. 1, pp. 3357–3364, 2017, doi: 10.1109/ICRA.2017.7989381.
54. S. Nemati, M. M. Ghassemi, and G. D. Clifford, "Optimal medication dosing from suboptimal clinical examples: A deep reinforcement learning approach," *Proc. Annu. Int. Conf. IEEE Eng. Med. Biol. Soc. EMBS*, vol. 2016-October, pp. 2978–2981, 2016, doi: 10.1109/EMBC.2016.7591355.

55. T. Zhang, N. Li, J. Huang, J.-X. Zhong, and G. Li, "An active action proposal method based on reinforcement learning," 2018 25th IEEE International Conference on Image Processing (ICIP), Oct. 2018, vol. 53, no. 9, pp. 4053–4057, doi: 10.1109/ICIP.2018.8451741.

56. M. Mohammadi, A. Al-Fuqaha, M. Guizani, and J. S. Oh, "Semi-supervised deep reinforcement learning in support of IoT and smart city services," *arXiv*, vol. 5, no. 2, pp. 624–635, 2018.

57. S. Brandi, M. S. Piscitelli, M. Martellacci, and A. Capozzoli, "Deep reinforcement learning to optimise indoor temperature control and heating energy consumption in buildings," *Energy Build.*, vol. 224, p. 110225, 2020, doi: 10.1016/j.enbuild.2020.110225.

58. M. Leyli-Abadi *et al.*, "Predictive classification of water consumption time series using non-homogeneous Markov models," Proc. - 2017 Int. Conf. Data Sci. Adv. Anal. DSAA 2017, vol. 2018-January, pp. 323–331, 2017, doi: 10.1109/DSAA.2017.32.

59. V. L. Erickson and A. E. Cerpa, "Occupancy based demand response HVAC control strategy," BuildSys'10 - Proc. 2nd ACM Work. Embed. Sens. Syst. Energy-Efficiency Build., pp. 7–12, 2010, doi: 10.1145/1878431.1878434.

60. Y. Zhou, Z. Kang, L. Zhang, and C. Spanos, "Causal analysis for non-stationary time series in sensor-rich smart buildings," *IEEE Int. Conf. Autom. Sci. Eng.*, pp. 593–598, 2013, doi: 10.1109/CoASE.2013.6654000.

61. B. Qolomany, A. Al-Fuqaha, D. Benhaddou, and A. Gupta, "Role of deep LSTM neural networks and Wi-Fi networks in support of occupancy prediction in smart buildings," Proc. - 2017 IEEE 19th Intl Conf. High Perform. Comput. Commun. HPCC 2017, 2017 IEEE 15th Intl Conf. Smart City, SmartCity 2017 2017 IEEE 3rd Intl Conf. Data Sci. Syst. DSS 2017, vol. 2018-January, pp. 50–57, 2018, doi: 10.1109/HPCC-SmartCity-DSS.2017.7.

62. F. A. Qureshi and C. N. Jones, "Hierarchical control of building HVAC system for ancillary services provision," *Energy Build.*, vol. 169, pp. 216–227, 2018, doi: 10.1016/j.enbuild.2018.03.004.

63. T. Chai and R. R. Draxler, "Root mean square error (RMSE) or mean absolute error (MAE)? – Arguments against avoiding RMSE in the literature," *Geosci. Model Dev.*, vol. 7, no. 3, pp. 1247–1250, 2014, doi: 10.5194/gmd-7-1247-2014.

64. S. Prívara, Z. Váňa, E. Žáčeková, and J. Cigler, "Building modeling: Selection of the most appropriate model for predictive control," *Energy Build.*, vol. 55, pp. 341–350, 2012, doi: 10.1016/j.enbuild.2012.08.040.

65. F. M. Bhutta, "Application of smart energy technologies in building sector - Future prospects," ICECE 2017 - 2017 Int. Conf. Energy Conserv. Effic. Proc., vol. 2018-January, pp. 7–10, 2017, doi: 10.1109/ECE.2017.8248820.

66. N. Duan, J. Cadena, P. Sotorrio, and J. Y. Joo, "Collaborative inference of missing smart electric meter data for a building," IEEE Int. Work. Mach. Learn. Signal Process. MLSP, vol. 2019-October, 2019, doi: 10.1109/MLSP.2019.8918698.

67. G. E. P. Box, G. M. Jenkins, G. C. Reinsel, *Time Series Analysis: Forecasting and Control*, 5th Edition. John Wiley & Sons, 2015, 215AD.

68. D. Peter and P. Silvia, "ARIMA vs. ARIMAX – Which approach is better to analyze and forecast macroeconomic time series?," *Int. Conf. Math. Methods Econ.*, Vol 2, pp. 136–140, 2012.

69. A. Pankratz, *Forecasting with Dynamic Regression Models*. Indianapolis, IN: John Wiley & Sons, 2012.

70. G. Tang, K. Wu, J. Lei, and W. Xiao, "The meter tells you are at home! Non-intrusive occupancy detection via load curve data," 2015 IEEE Int. Conf. Smart Grid Commun. SmartGridComm 2015, pp. 897–902, 2016, doi: 10.1109/SmartGridComm.2015.7436415.

71. N. Mohamed, J. Al-Jaroodi, and I. Jawhar, "Service-oriented big data analytics for improving buildings energy management in smart cities," 2018 14th Int. Wirel. Commun. Mob. Comput. Conf. IWCMC 2018, pp. 1243–1248, 2018, doi: 10.1109/IWCMC.2018.8450469.

72. M. Yesilbudak and I. Colak, "Main barriers and solution proposals for communication networks and information security in smart grids," 6th IEEE Int. Conf. Smart Grid, icSmartGrids 2018, pp. 58–63, 2019, doi: 10.1109/ISGWCP.2018.8634478.

73. Z. L. Gaing, "Particle swarm optimization to solving the economic dispatch considering the generator constraints," *IEEE Trans. Power Syst.*, vol. 18, no. 3, pp. 1187–1195, 2003, doi: 10.1109/TPWRS.2003.814889.

74. J. Guo, C. Jin, W. Liu, and W. Zhou, "An effective particle swarm optimization algorithm with social weight in solving economic dispatch problem considering network losses," Proc. - 2012 3rd Glob. Congr. Intell. Syst. GCIS 2012, no. 4, pp. 80–83, 2012, doi: 10.1109/GCIS.2012.83.

75. C. Li, X. Yu, W. Yu, T. Huang, and Z. W. Liu, "Distributed event-triggered scheme for economic dispatch in smart grids," *IEEE Trans. Ind. Informatics*, vol. 12, no. 5, pp. 1775–1785, 2016, doi: 10.1109/TII.2015.2479558.

76. V. Loia and A. Vaccaro, "Decentralized economic dispatch in smart grids by self-organizing dynamic agents," *IEEE Trans. Syst. Man, Cybern. Syst.*, vol. 44, no. 4, pp. 397–408, 2014, doi: 10.1109/TSMC.2013.2258909.

77. A. Wang, W. Liu, T. Dong, X. Liao, and T. Huang, DisEHPPC : Enabling Heterogeneous Smart Grids, pp. 1–12, 2020.

78. S. Lazarova-Molnar and N. Mohamed, "Challenges in the data collection for diagnostics of smart buildings," *Lect. Notes Electr. Eng.*, vol. 376, pp. 941–951, 2016, doi: 10.1007/978-981-10-0557-2_90.

79. E. M. Hanna, N. Mohamed, and J. Al-Jaroodi, "The cloud: Requirements for a better service," 2012 12th IEEE/ACM International Symposium on Cluster, Cloud and Grid Computing (ccgrid 2012), May 2012, pp. 787–792, doi: 10.1109/CCGrid.2012.93.

80. C. Luppe and A. Shabani, "Towards reliable intelligent occupancy detection for smart building applications," *Can. Conf. Electr. Comput. Eng.*, pp. 0–3, 2017, doi: 10.1109/CCECE.2017.7946831.

81. M. Jin, R. Jia, and C. J. Spanos, "Virtual occupancy sensing: Using smart meters to indicate your presence," *IEEE Trans. Mob. Comput.*, vol. 16, no. 11, pp. 3264–3277, 2017, doi: 10.1109/TMC.2017.2684806.

82. C. Feng, A. Mehmani, and J. Zhang, "Deep learning-based real-time building occupancy detection using AMI data," *IEEE Trans. Smart Grid*, vol. 11, no. 5, pp. 4490–4501, 2020, doi: 10.1109/TSG.2020.2982351.

83. W. S. G. Newsham, "Smart phone based occupancy detection in office buildings," Proc. IEEE 20th Int. Conf. Comput. Support. Coop. Work Des., pp. 4–8, 2016.

84. X. Tang, M. C. Huang, and S. Mandal, "An 'Internet of Ears' for crowd-aware smart buildings based on sparse sensor networks," *Proc. IEEE Sensors*, vol. 2017-December, pp. 1–3, 2017, doi: 10.1109/ICSENS.2017.8234263.

85. X. Guo, D. K. Tiller, G. P. Henze, and C. E. Waters, "The performance of occupancy-based lighting control systems: A review," *Light. Res. Technol.*, vol. 42, no. 4, pp. 415–431, 2010, doi: 10.1177/1477153510376225.

86. S. Lee and P. Karava, "Towards smart buildings with self-tuned indoor thermal environments – A critical review," *Energy Build.*, vol. 224, p. 110172, 2020, doi: 10.1016/j.enbuild.2020.110172.

87. M. Prignon, C. Delmotte, A. Dawans, S. Altomonte, and G. van Moeseke, "On the impact of regression technique to airtightness measurements uncertainties," *Energy Build.*, vol. 215, 2020, doi: 10.1016/j.enbuild.2020.109919.

88. S. Mennicken, A. Hwang, R. Yang, J. Hoey, A. Mihailidis, and E. M. Huang, "Smart for life: Designing smart home technologies that evolve with users," Conf. Hum. Factors Comput. Syst. - Proc., vol. 18, no. April 2015, pp. 2377–2380, 2015, doi: 10.1145/2702613.2702631.

89. R. F. Dickerson, E. I. Gorlin, and J. A. Stankovic, "Empath: A continuous remote emotional health monitoring," Proc. 2nd Conf. Wirel. Heal. - WH '11, p. 1, 2011, [Online]. Available: http://dl.acm.org/citation.cfm?id=2077552%5Cnhttp://dl.acm.org/citation.cfm?doid=2077546.2077552%0Ahttp://dl.acm.org/citation.cfm?doid=2077546.2077552.

90. X. Li, R. Bai, P. O. Siebers, and C. Wagner, "Travel time prediction in transport and logistics: Towards more efficient vehicle GPS data management using tree ensemble methods," VINE J. Inf. Knowl. Manage. Syst., vol. 49, no. 3, pp. 277–306, 2019, doi: 10.1108/VJIKMS-11-2018-0102.

6 Machine Learning-Based Optimal Consensus Networked Control with Application to Van der Pol Oscillator Systems

Luy Nguyen Tan
Industrial University of Ho Chi Minh City,
Ho Chi Minh, Vietnam

CONTENTS

6.1 INTRODUCTION

In recent years, thanks to IoT technology and machine learning, consensus control for multiple agents has been widely researched for a smarter world, in which the stabilization of multi-agent system (MAS) has gained the attention of the scientific community. Currently, the distributed control methods mainly focus on affine nonlinear

DOI: 10.1201/9781003268796-8

system dynamics [1–6]. However, it is more favorable to design the controllers for strict feedback nonlinear systems, because they are popularly used in industrial applications, such as a leader and follower mechanism or a distributed formation. For more details, please see [5] and references therein.

The conventional control methods only consider the single-agent strict-feedback systems (SASS) [7, 8]. Adaptive control algorithms, based on back-stepping techniques, can handle the uncertain parameters but ignore the external disturbances and cost functions. For multi-agent strict-feedback systems (MASS), by using Lyapunov theory, adaptive control algorithms are performed in [9]. The algorithms can maintain the desired formations while guaranteeing the track performance between agents and leaders. Moreover, observers are employed to estimate states for agents. In the light of the fast development of H_∞ control theory [10], H_∞ adaptive control algorithms are proposed [11, 12]. Although the robust control methods can deal with external disturbances and model errors, they do not minimize/maximize performance index functions defined by users.

The objective of optimal control for a nonlinear system is to seek optimal control strategies deriving from the Pontryagin maximum principle. However, the principle may be only a necessary condition. It is required that a sufficient condition of Hamilton-Jacobi-Bellman (HJB). Unfortunately, there do not exist analytical HJB solutions [13, 14] due to differential nonlinear. Recently, reinforcement learning (RL) techniques [14], a basic core of the machine learning theory [15], has emerged as one of the most well-known methods being employed to approximate the HJB solutions for MAS problems [16]. For example, RL can learn the solutions of differential, stochastic, and Markov games, two-player or multiplayer games, or Nash Q-learning. A branch of RL, adaptive dynamic programming (ADP) [5], is widely researched for optimal control designs [5, 17–24]. Inspired by naturally behavioral psychology, almost ADP-based algorithms use policy iteration techniques, which use two or three neural networks (NN) for control structures [17–24], called actor-disturber-critic (ADC). The critic approximates a value function while the others tune the optimal control strategy and disturbance rejection policy via the critic. The NN-weight training process is executed in two sequential steps: evaluation of actor's policy and improvement of disturber's policy. Unfortunately, with the ADC structure, the training algorithms have the disadvantages of sequential update, and hence, they require stabilized initial weights [18].

To overcome the disadvantages of an ADC structure, Dierks and Jagannathan [18] presented a control algorithm with a structure of a single NN. The unknown internal dynamics is identified online with a state observer. The algorithm could minimize the predefined performance in the absence of available states and initially stabilizing controllers. In [22–24], control structures with only one NN are utilized with application to multi-mobile robots. Inspired from [18], this chapter presents an ADP control technique using a structure with a single NN with application to MASS of Van der Pol oscillators.

In many industrial applications, there exists the phenomenon of saturation, backlash, hysteresis, or dead-zone. The phenomenon causes the input constraints and unstable closed-loop dynamics. As actuators are physically limited by the bounds of voltages, currents, flows, and torques [25–27], the saturation constraint of systems causes more and more difficulties in finding the optimal controllers. In optimal control with ADP, the constraints are integrated into the HJB equation, which is, therefore, detrimental in approximating the HJB equation online while requiring the asymptotical stability of closed-loop systems. Recently, control algorithms for multiple agents with saturated inputs have been reported [27, 28]. However, only adaptive control has been devoted. In contrast, saturated optimal control schemes were presented in [29–31] using ADP, but unfortunately, the researchers considered SASS in affine form without disturbances. Meanwhile, for nonlinear systems with saturated inputs, strict-feedback form, Chen et al. [31] provided an adaptive control algorithm; of course, it did not optimize any the predefined performance functions.

Recently, the Van der Pol oscillator system has been intensely focused on since it has been increasingly encountered in real-life applications. The system was originally derived from a dynamical system of a triode-valve electrical circuit [32] then extended to the dynamical systems of chaos, relaxation oscillations, elementary bifurcations [33–37]. The dynamical systems have been used to design various practical applications in robotics, power systems, combustion processes, biomedical engineering, environmental monitoring.

In [38–40], adaptive control schemes via NN were developed for uncertain Van der Pol oscillator systems. Meanwhile, a feedback linearizing control scheme via NN was designed for output tracking [40]. In [41], a standard higher-order sliding-mode differentiator was designed for an observer applying to experimental Van der Pol oscillator systems. Despite the modern control techniques approach, the above studies only address the systems in SASS and provide adaptive controllers without optimizing any predefined performances.

Van der Pol oscillator systems in [42] was controlled by adaptive optimal. The systems were presented as strict-feedback nonlinear systems, and a data-driven control approach was employed to design control laws, of which HJB equation was solved online. For Van der Pol oscillator systems with input constraints, Vamvoudakis et al. [29], based on ADP, designed the optimal control algorithm, by which the states approach to the origin or are uniformly ultimately bounded (UUB). However, optimal tracking control algorithms may only be applied to SASS.

It is worth emphasizing that MASS of Van der Pol oscillator dynamics is rarely considered even in the absence of input constraint and disturbance. To develop applications with overcoming the disadvantages of the existing methods mentioned above, a novel method that provides a basis framework to design distributed optimal control algorithm for MASS is studied in this chapter. The agent dynamics includes multi-input and multi-output (MIMO) Van der Pol oscillator systems with the uncertain strict-feedback nonlinear, drift internal parameters, external disturbances, and input constraints. The main contribution in this chapter includes:

1. Extending ADP theory [14, 19, 30, 43, 44] from the simple single-agent systems to multi-agent MIMO systems with uncertain strict-feedback nonlinear, drift internal parameters, external disturbances, and constrained inputs. A novel algorithm is developed to avoid system identification unlike [44, 45] used for MAS with strict-feedback dynamics.

2. By utilizing MAS configuration presented by graph theory and the ideal of structures with a single NN, this chapter extends the existing works in [22, 23, 46] to design ADP-based distributed H_∞ optimal consensus control algorithm with application to MASS of electronic Van der Pol oscillator systems. For relaxing the drift unknown dynamics, the integral reinforcement leaning (IRL) technique [43] is also used.
3. The stability and convergence are rigorously analyzed by Lyapunov theory to show that the approximation and consensus errors UUB. In addition, predefined performance functions also approach to the suboptimality.

The chapter is organized as follows. Communication graph and Van der Pol oscillator dynamics are described in Section 6.2, control algorithms are built in Section 6.3, numerical simulation is studied in Section 6.4, and the conclusion is given in Section 6.5.

6.2   GRAPH THEORY AND VAN DER POL OSCILLATOR DYNAMICS

6.2.1   DEFINITION

Definition 1 [47]: If $\bar{\sigma}_i = \sigma_i / (\sigma_i^T \sigma_i + 1)$ satisfies the following inequality

$$\beta_{1i} I \leq \int_{t-T}^{t} \bar{\sigma}_i(\tau) \bar{\sigma}_i^T(\tau) d\tau \leq \beta_{2i} I \tag{6.1}$$

where $T > 0$, β_{2i}, β_{1i} are positive constants and $\beta_{2i} > \beta_{1i}$. Then it is persistently exciting (PE).

Definition 2 [47]: Given a compact set $\Omega \in \mathbb{R}^n$ and $\dot{x} = f(x,u)$, an equilibrium point $x_0, \forall x_0 \in \Omega$, is UUB if there exists $\|x - x_0\| \leq B$, $\forall t > t_0 + T(B, x_0)$ for $B \geq 0$.

Definition 3 [28]: Consider dynamics $\dot{x} = f(x,u)$, $Q(x) > 0$, $Q(x) = 0$ if only if $x = 0$, $R = R^T > 0$, a control strategy u is admissible on Ω, if u is continuous on Ω, keeps the system stable on Ω and $V(x_0) = \int_0^\infty \left(Q(x) + u^T R u \right) dt$ finite, $\forall x_0 \in \Omega$.

6.2.2   COMMUNICATION GRAPH THEORY

A graph, often employed to form a topology among agent in MAS, is notated $\Delta(\mathcal{V}, \Xi, \mathcal{A})$, $\mathcal{V} = \{s_1, \ldots, s_N\}$, and $\Xi \subseteq \mathcal{V} \times \mathcal{V}$ are the vertex and edge sets. $\mathcal{A} = [\alpha_{ij}]$ is the connection matrix, $\alpha_{ij} = 0$ if $\alpha_{ij} \notin \Xi$, $\alpha_{ij} = 1$ if $\alpha_{ij} \in \Xi$. If agent j receives data from i, it is in the set $N_i = \{j : s_j \in \mathcal{V}, (s_i, s_j) \in \Xi\}$. Denote $\beta_i = \sum_{j \in N_i} \alpha_{ij}$, $\mathcal{B} = \text{diag}(\beta_i)$, then $\mathcal{M} = \mathcal{B} - \mathcal{A}$ is a Laplacian. The connection from the leader s_0 to all followers specified by $\mathcal{C} = \text{diag}[c_1, c_2, \ldots, c_N]$, where $c_i = \{0,1\}$ according to the connection between node i and node 0. There has a strongly connected graph if $\forall (s_i, s_j) \in \mathcal{V}$, $s_i \neq s_j$, s_i directly connects to s_j exists. The matrices $\mathcal{M}$ and $\mathcal{A}$ of the directed graph are irreducible [46].

6.2.3 Agent Nonlinear Dynamics

Consider the Van der Pol oscillator agent as strict-feedback nonlinear system with external disturbances and constrained input given by

$$
\begin{cases}
\dot{x}_{i,1} = x_{i,2} + k_{i,1}(x_{i,1})d_{i,1} \\
\quad\vdots \\
\dot{x}_{i,n} = -x_{i,n} - \dfrac{1}{2}x_{i,n}\left(1 - x_{i,n-1}^2\right) - x_{i,n-1}^2 x_{i,n} + g_{i,n}\left(x_{i,1}, x_{i,2}, \ldots, x_{i,n}\right)u_{i,n} \\
\qquad + k_{i,n}(x_{i,1}, x_{i,1}, \ldots, x_{i,n})d_{i,n} \\
y_i = x_{i,1}
\end{cases}
\tag{6.2}
$$

where $x_{i,l} \in \mathbb{R}, l = 1, \ldots, n$ are state vectors, which are available for full state feedback, $y_i \in \mathbb{R}$ is an output vector, $g_{i,n}(.) \in \mathbb{R}$, $u_i \in \mathbb{R}$ is a control input vector, which satisfies $u_i \leq \lambda_i$ where λ_i is a positive value, $k_{i,l}(.) \in \mathbb{R}, l = 1, \ldots, n$ are state-dependent functions, $d_{i,l} \in \mathbb{R}, l = 1, \ldots, n$ external disturbances. The dynamics (6.2) is presented by an overall form as

$$
\begin{cases}
\dot{\mathcal{X}}_i = \mathcal{F}_i(\mathcal{X}_i) + \mathcal{G}_i(\mathcal{X}_i)\mathcal{U}_i + \mathcal{K}_i(\mathcal{X}_i)\mathcal{D}_i \\
\mathcal{Y}_i = \mathcal{T}_i \mathcal{X}_i
\end{cases}
\tag{6.3}
$$

where $\mathcal{X}_i = [x_{i,1}, x_{i,2}, \ldots, x_{i,n}]^T \in \mathbb{R}^n$, $\mathcal{U}_i = [0, 0, \ldots, u_{i,n}]^T \in \mathbb{R}^n$, $\mathcal{D}_i = [d_{i,1}, d_{i,2}, \ldots, d_{i,n}]^T$ $\in \mathbb{R}^n$, $\mathcal{T}_i = [1 \quad 0 \quad \cdots \quad 0] \in \mathbb{R}^{1 \times n}$,

$$
\mathcal{F}_i(\mathcal{X}_i) =
\begin{bmatrix}
x_{i,2} \\
x_{i,3} \\
\vdots \\
-x_{i,n} - \dfrac{1}{2}x_{i,n}(1 - x_{i,n-1}^2) - x_{i,n-1}^2 x_{i,n}
\end{bmatrix}
\in \mathbb{R}^n, \quad
\mathcal{G}_i(\mathcal{X}_i) =
\begin{bmatrix}
0 & 0 & \cdots & 0 \\
0 & 0 & 0 & 0 \\
\vdots & \vdots & \ddots & \vdots \\
0 & 0 & \cdots & y_i
\end{bmatrix}
\in \mathbb{R}^{n \times n},
$$

$$
\mathcal{K}_i(\mathcal{X}_i) =
\begin{bmatrix}
k_{i,1} & 0 & \cdots & 0 \\
0 & k_{i,2} & 0 & 0 \\
\vdots & \vdots & \ddots & \vdots \\
0 & 0 & \cdots & k_{i,n}
\end{bmatrix}
\in \mathbb{R}^{n \times n},
$$

Assumption 1: $\mathcal{F}_i(x_i)$ is Lipschitz continuous, $\mathcal{G}_i(x_i)$, and $\mathcal{K}_i(x_i)$ are bounded such that $\mathcal{G}_i(x_i) \leq g_i^{max}$, $\mathcal{K}_i(x_i) \leq k_i^{max}$, $\mathcal{D}_i \in L_2[0, \infty)$, where g_i^{max}, k_i^{max} are some unknown positive constants.

Remark 1: The subsystem (6.3) with Assumption 1 is practical in many industrial applications [30, 42, 43], where internal dynamics is Lipschitz and the measured output is bounded. Note that each subsystem has uncontrolled unstable limit cycles and unstable equilibrium points at the origin [29].

6.2.4 DISTRIBUTED CONSENSUS DYNAMICS

For each agent, the consensus error $\mathcal{Z}_i \in \mathbb{R}^n$ is established as

$$\mathcal{Z}_i = \sum_{j \in N_i} \alpha_{ij} (\mathcal{X}_i - \mathcal{X}_j) + c_i (\mathcal{X}_i - \mathcal{X}_0) \tag{6.4}$$

where $\mathcal{X}_0 \in \mathbb{R}^n$ is the state vector of a leader. Define $\mathcal{Z} = [\mathcal{Z}_1^T, \mathcal{Z}_2^T, ..., \mathcal{Z}_N^T]^T$, $\mathcal{X} = [\mathcal{X}_1^T, \mathcal{X}_2^T, ..., \mathcal{X}_N^T]^T \in \mathbb{R}^{Nn}$ as overall vector, $\bar{\mathcal{X}}_0 = 1_N \otimes I_n \mathcal{X}_0 \in \mathbb{R}^{Nn}$, the global consensus error vector in the communication graph $\Delta(\mathcal{V}, \Xi, \mathcal{A})$ can be written as

$$\mathcal{Z} = \left((\mathcal{M} + \mathcal{C}) \otimes I_n \right) (\mathcal{Z} - \bar{\mathcal{X}}_0) = \left((\mathcal{M} + \mathcal{C}) \otimes I_n \right) \mathcal{E} \tag{6.5}$$

where $\mathcal{E}$ is global synchronization error between the agents and the leader,

$$\mathcal{E} = \mathcal{X} - \bar{\mathcal{X}}_0 \in \mathbb{R}^{Nn} \tag{6.6}$$

The following lemma shows the relationship of performance index between the global and local consensus errors.

Lemma 1 [46]: The following inequality is satisfied if $\mathcal{C} \neq 0$

$$\|\mathcal{E}\| \leq \frac{\|\mathcal{Z}\|}{\underline{\sigma}(\mathcal{M} + \mathcal{C})} \tag{6.7}$$

where $\underline{\sigma}(.)$ is the minimum singular value. Then, the bounded local consensus error implies the bounded global consensus error.

Later, instead of finding the global controllers, it is required to design the distributed local controllers. The consensus error dynamics from (6.4) is rewritten as

$$\dot{\mathcal{Z}}_i = \sum_{j \in N_i} \alpha_{ij} (\dot{\mathcal{X}}_i - \dot{\mathcal{X}}_j) + c_i (\dot{\mathcal{X}}_i - \dot{\mathcal{X}}_0) \tag{6.8}$$

Using the communication graph and (6.3), consensus dynamics (6.8) is transformed into

$$\dot{\mathcal{Z}}_i = \mathcal{F}_{\mathcal{Z}_i}(t) + (\beta_i + c_i)(\mathcal{G}_i \mathcal{U}_i + \mathcal{K}_i \mathcal{D}_i) - \sum_{j \in N_i} \alpha_{ij} (\mathcal{G}_j \mathcal{U}_j + \mathcal{K}_j \mathcal{D}_j) \tag{6.9}$$

where $\mathcal{F}_{\mathcal{Z}_i}(t) = (\beta_i + c_i) \mathcal{F}_i (\mathcal{X}_i) - \sum_{j \in N_i} \alpha_{ij} \mathcal{F}_j (\mathcal{X}_j)$.

The objective is to build an algorithm to guarantee that the consensus error (6.8) of each agent is UUB and the value function is L_2-bounded regardless of external disturbances and saturated inputs.

6.3 DISTRIBUTED SATURATED OPTIMAL CONTROL ALGORITHM

In the section, the differential graphical game theory and ADP are extended to approximate distributed saturated optimal control strategies and disturbance rejection policies.

6.3.1 PERFORMANCE INDEX FUNCTION OF BOUNDED L_2-GAIN

For all admissible policies $\mathcal{U}_i$, $\bar{\mathcal{U}}_i = \{\mathcal{U}_j, j \in N_i\}$, define an output vector $\xi_i = \left[\mathcal{Z}_i^T(t), \mathcal{U}_i(t), \bar{\mathcal{U}}_i(t) \right]^T$, the performance index of (6.9) has bounded L_2-gain, if the following inequality is satisfied for $\gamma > \gamma^* > 0$:

$$\int_0^\infty \|\xi_i^2\| dt = \int_0^\infty \left(Q_{ii}(\mathcal{Z}_i) + U_{ii}(\mathcal{U}_i) + \sum_{j \in N_i} U_{jj}(\mathcal{U}_j) \right) dt \leq \gamma^2 \int_0^\infty \left(\mathcal{D}_i^T \mathcal{D}_i + \sum_{j \in N_i} \mathcal{D}_j^T \mathcal{D}_j \right) dt \quad (6.10)$$

with γ^* is the smallest γ for which dynamics (6.9) is stable, $Q_{ii}(\mathcal{Z}_i)$ is positive definite functions. In the case of unconstraint, the nonnegative functions $U_{ii}(\mathcal{U}_i)$ can be chosen as $U_{ii}(\mathcal{U}_i) = \mathcal{U}_i^T R_{ii} \mathcal{U}_i$, $R_{ii} = R_{ii}^T > 0$, otherwise, it can be chosen as [28]

$$U_{ii}(\mathcal{U}_i) = 2\lambda_i \int_0^{\mathcal{U}_i} \tanh^{-T}\left(s_i / \lambda_i \right) R_{ii} ds_i \quad (6.11)$$

A local infinite horizontal function, satisfying (6.10), is given by

$$J_i \left(\mathcal{Z}_i(0), \mathcal{D}_i, \bar{\mathcal{D}}_i, \mathcal{U}_i, \bar{\mathcal{U}}_i \right)$$

$$= \int_0^\infty \left(Q_{ii}(\mathcal{Z}_i) + U_{ii}(\mathcal{U}_i) + \sum_{j \in N_i} U_{jj}(\mathcal{U}_j) - \gamma^2 \mathcal{D}_i^T \mathcal{D}_i - \gamma^2 \sum_{j \in N_i} \mathcal{D}_j^T \mathcal{D}_j \right) dt \quad (6.12)$$

where $\bar{\mathcal{D}}_i = \{\mathcal{D}_j, j \in N_i\} \neq 0$.

6.3.2 OPTIMAL CONTROL AND DISTURBANCE REJECTION POLICIES

It is shown in [12] that the game theories [48] are combined with ADP to approximate the optimal value $V_i^*(\mathcal{Z}_i(0))$ subject to (6.9). Approximation of the optimal value satisfies the min-max law as follows

$$V_i^*(\mathcal{Z}_i(0)) = \min_{\mathcal{U}_i} \max_{\mathcal{D}_i} J_i(\mathcal{Z}_i(0), \mathcal{D}_i, \bar{\mathcal{D}}_i, \mathcal{U}_i, \bar{\mathcal{U}}_i) \quad (6.13)$$

Equation (6.13) provides the saddle point $(\mathcal{U}_i^*, \mathcal{D}_i^*)$, from which the optimal control strategy $\mathcal{U}_i^*$ (first player) and the disturbance rejection policy $\mathcal{D}_i^*$ (second player) are derived. If there exists a saddle point, the following condition will be satisfied

$$V_i^*\left(\mathcal{Z}_i(0)\right) = \min_{\mathcal{U}_i}\max_{\mathcal{D}_i} \mathcal{J}_i\left(\mathcal{Z}_i(0), \mathcal{D}_i^*, \bar{\mathcal{D}}_i, \mathcal{U}_i^*, \bar{\mathcal{U}}_i\right)$$

$$= \max_{\mathcal{U}_i}\min_{\mathcal{D}_i} \mathcal{J}_i\left(\mathcal{Z}_i(0), \mathcal{D}_i^*, \bar{\mathcal{D}}_i, \mathcal{U}_i^*, \bar{\mathcal{U}}_i\right) \tag{6.14}$$

V_i^* is the Nash equilibrium if there exists the following condition for the control strategies $\mathcal{U}_i$, $\bar{\mathcal{U}}_i$ and disturbance rejection policies $\mathcal{D}_i, \bar{\mathcal{D}}_i$

$$\mathcal{J}_i\left(\mathcal{D}_i, \bar{\mathcal{D}}_i^*, \mathcal{U}_i^*, \bar{\mathcal{U}}_i^*\right) \leq \mathcal{J}_i\left(\mathcal{D}_i^*, \bar{\mathcal{D}}_i^*, \mathcal{U}_i^*, \bar{\mathcal{U}}_i^*\right) \leq \mathcal{J}_i\left(\mathcal{D}_i^*, \bar{\mathcal{D}}_i^*, \mathcal{U}_i, \bar{\mathcal{U}}_i^*\right) \tag{6.15}$$

Consider the state-dependent strategies, the consensus value function is described as

$$V_i(\mathcal{Z}_i(t)) = \int_t^\infty \left(Q_{ii}(\mathcal{Z}_i) + U_{ii}(\mathcal{U}_i) + \sum_{j \in N_i} U_{jj}(\mathcal{U}_j) - \gamma^2 \left(\mathcal{D}_i^T \mathcal{D}_i + \sum_{j \in N_i} \mathcal{D}_j^T \mathcal{D}_j \right) \right) d\tau \tag{6.16}$$

Using Leibniz formula, the Hamiltonian is derived from (6.15) as

$$H_i\left(Z_i, \nabla V_i(Z_i), D_i, \bar{D}_i, U_i, \bar{U}_i\right) \equiv \left(\nabla V_i(Z_i)\right)^T \left(F_{Z_i} + (\beta_i + c_i)(G_i U_i + K_i D_i)\right)$$

$$- \sum_{j \in N_i} \alpha_{ij}\left(G_j U_j + K_j D_j\right) - \gamma^2 D_i^T D_i - \gamma^2 \sum_{j \in N_i} D^T{}_j D_j + Q_{ii}(Z_i) + U_{ii}(U_i)$$

$$+ \sum_{j \in N_i} U_{ii}(U_i) = 0 \tag{6.17}$$

From (6.17), the disturbance rejection policy and the optimal control strategy are derived from the stationary condition as

$$\mathcal{D}_i^* = \frac{1}{2\gamma^2}(\beta_i + c_i)\mathcal{K}_i^T(\mathcal{X}_i)\nabla V_i(\mathcal{Z}_i)$$

$$\mathcal{U}_i^* = -\lambda_i \tanh\left(\frac{1}{2\lambda_i}(\beta_i + c_i)R_{ii}^{-1}\mathcal{G}_i^T(\mathcal{X}_i)\nabla V_i(\mathcal{Z}_i)\right) \tag{6.18}$$

Remark 2: Unsaturated control strategy [12] is chosen as $\mathcal{U}_i^* = (\beta_i + c_i)R_{ii}^{-1}\mathcal{G}_i^T\nabla V_i/2$. Since $\|\mathcal{U}_i^*\| \leq \lambda_i$ is required, (6.11) is utilized to define a new policy dealing with saturation. Substituting u_i^* from (6.18) into (6.11) we obtain that

$$U_{ii}(\mathcal{U}_i^*) = 2\lambda_i \int_0^{u_i^*} \tanh^{-T}(\zeta_i/\lambda_i)R_{ii}d\zeta_i$$

$$= 2\lambda_i \tanh^{-T}\left(\mathcal{U}_i^*/\lambda_i\right)R_{ii}\mathcal{U}_i^* + \lambda_i^2 \bar{R}_{ii}\ln\left(\bar{1} - \left(\mathcal{U}_i^*/\lambda_i\right)^2\right)$$

$$= \lambda_i(\beta_i + c_i)(\nabla V_i)^T \mathcal{G}_i \tanh\left(\frac{1}{2\lambda_I}(\beta_i + c_i)R_{ii}^{-1}\mathcal{G}_i^T\nabla V_i\right)$$

$$+ \lambda_i^2 \bar{R}_{ii}\ln\left(\bar{1} - \tanh^2\left(\frac{1}{2\lambda_i}(\beta_i + c_i)R_{ii}^{-1}\mathcal{G}_i^T\nabla V_i\right)\right) \tag{6.19}$$

where $\bar{1} = \bar{1}_m$, and $\bar{R}_{ii} \in \mathbb{R}^{1 \times n}$ contains diagonal elements of R_{ii}. Using (6.17), we obtain the Hamilton-Jacobi-Isaac (HJI) equation:

$$H_i^*(\mathcal{Z}_i, \nabla V_i(U_i), D_i^*, U_i^*) = Q_{ii}(\mathcal{Z}_i) + U_{ii}(U_i^*)$$

$$+ \sum_{j \in N_i} U_{jj}(\mathcal{U}_j^*) + (\nabla V_i(\mathcal{Z}_i))^T (\mathcal{F}_{\mathcal{Z}_i} + (\beta_i + c_i)(\mathcal{G}_i \mathcal{U}_i^* + \mathcal{K}_i \mathcal{D}_i^*))$$

$$- \sum_{j \in N_i} \alpha_{ij}(\mathcal{G}_j \mathcal{U}_j^* + \mathcal{K}_j \mathcal{D}_j^*) - \gamma^2 \mathcal{D}_i^{*T} \mathcal{D}_i - \gamma^2 \sum_{j \in N_i} \mathcal{D}_j^{*T} \mathcal{D}_j = 0 \tag{6.20}$$

According to [13], the coupled HJI equation (6.20) does not has analytical solution. However, there always exists a smooth minimal nonnegative solution [12] such that the policies (6.18), depended on the optimal value V_i^*, satisfy the Nash equilibrium solutions.

Followed by the Weierstrass theorem [28], the smooth optimal value function $V_i^*(\mathcal{Z}_i)$ is presented by

$$V_i^*(\mathcal{Z}_i) = \mathcal{W}_i^T \Phi_i(\mathcal{Z}_i) + \mathcal{E}_i(\mathcal{Z}_i) \tag{6.21}$$

for NN activation functions $\Phi_i(\mathcal{Z}_i): \mathbb{R}^n \to \mathbb{R}^\pi$, approximation errors $\mathcal{E}_i(\mathcal{Z}_i)$, true weights $\mathcal{W}_i \in \mathbb{R}^\pi$.

Property 1 [49]: If vector $\Phi_i(\mathcal{Z}_i)$ is chosen as complete independent basis set, $\pi \to \infty$, $\mathcal{E}_i(\mathcal{Z}_i) \to 0$, $\nabla \mathcal{E}_i(\mathcal{Z}_i) \to 0$. On the other hand, for a fixed number π, $\|\mathcal{E}_i(\mathcal{Z}_i)\| < \varepsilon_i^{\max}$, $\|\nabla \mathcal{E}_i(\mathcal{Z}_i)\| \le \varepsilon_{\mathcal{Z}_i}^{\max}$ for $\varepsilon_i^{\max}$ and $\varepsilon_{\mathcal{Z}_i}^{\max}$ are positive constants.

Using (6.21) for (6.18) and then substituting the result to (6.20), we obtain the HJI equation:

$$H_i^*\left(Z_i, W_i, \nabla \Phi_i(Z_i), D_i^* U_i^*\right)$$

$$= Q_{ii}(Z_i) + W_i^T \nabla \Phi_i(Z_i)\left(F_{Z_i} + (\beta_i + c_i)\left(G_i U_i^* + K_i D_i^*\right) - \sum_{j \in N_i} \alpha_{ij}\left(G_j U_j^* + K_j D_j^*\right)\right)$$

$$+ U_{ii}\left(U_i^*\right) + \sum_{j \in N} U_{jj}\left(U_j^*\right) - \gamma^2\left(D_i^{*T} D_i + \sum_{j \in N} D_j^{*T} D_j^*\right) - \mathcal{E}_{H_i} = 0 \tag{6.22}$$

where

$$\mathcal{E}_{H_i} = H_i^* - H_i = -\nabla \mathcal{E}_i^T\left(\mathcal{F}_{\mathcal{Z}_i} + (\beta_i + c_i)\left(\mathcal{G}_i \mathcal{U}_i^* + \mathcal{K}_i \mathcal{D}_i^*\right) - \sum_{j \in N_i} \alpha_{ij}\left(\mathcal{G}_j \mathcal{U}_j^* + \mathcal{K}_j \mathcal{D}_j^*\right)\right) \tag{6.23}$$

Remark 3: Note Property 1, $\mathcal{E}_{H_i}$ is bounded: $\exists b_{i,\varepsilon} > 0 : \sup_{\mathcal{Z}_i \in \Omega} \|\mathcal{E}_{H_i}\| \le b_{i,\varepsilon}$. If $\pi \to \infty$, $\mathcal{E}_{H_i}$ will uniformly converge to zero [28].

Since the ideal NN weights $\mathcal{W}_i$ (6.21) are not available, the value function is approximated by

$$\hat{V}_i = \hat{\mathcal{W}}_i^T \Phi_i \tag{6.24}$$

Accordingly, the disturbance rejection and optimal control policies are estimated as

$$
\begin{cases}
\hat{\mathcal{D}}_i = \dfrac{1}{2\gamma^2}(\beta_i + c_i)\mathcal{K}_i^T \nabla \Phi_i^T \hat{\mathcal{W}}_i \\[3mm]
\hat{\mathcal{U}}_i = -\lambda_i \tanh\left(\dfrac{1}{2\lambda_i}(\beta_i + c_i) R_{ii}^{-1} \mathcal{G}_i^T \nabla \Phi_i^T \hat{\mathcal{W}}_i \right)
\end{cases}
\tag{6.25}
$$

The approximate Hamilton is yielded by substituting (6.21) and (6.25) into (6.22)

$$
\hat{\mathcal{H}}_i(\mathcal{Z}_i, \hat{\mathcal{W}}_i) = Q_{ii}(\mathcal{Z}_i) + W_i^T \nabla \Phi_i(\mathcal{Z}_i)\left(\mathcal{F}_{\mathcal{Z}_i} + (\beta_i + c_i)\left(\mathcal{G}_i \hat{\mathcal{U}}_i + \mathcal{K}_i \hat{\mathcal{D}}_i \right) \right)
$$

$$
- \sum_{j \in N_i} \alpha_{ij}\left(\mathcal{G}_j \hat{\mathcal{U}}_j + \mathcal{K}_j \hat{\mathcal{D}}_j \right) + U_{ii}(\hat{\mathcal{U}}_i) + \sum_{j \in N_i} U_{jj}(\hat{\mathcal{U}}_j) - \gamma^2 \left(\hat{\mathcal{D}}_i^T \hat{\mathcal{D}}_i + \sum_{j \in N_i} \hat{\mathcal{D}}_j^T \hat{\mathcal{D}}_j \right)
\tag{6.26}
$$

where

$$
U_{ii}(\hat{\mathcal{U}}_i) = \rho(\beta_i + c_i)\hat{\mathcal{W}}_i^T \nabla \Phi_i \mathcal{G}_i \tanh\left(\frac{1}{2\lambda_i}(\beta_i + c_i) R_{ii}^{-1} \mathcal{G}_i^T \nabla \Phi_i^T \hat{\mathcal{W}}_i \right)
$$

$$
+ \lambda_i^2 \bar{R}_{ii} \ln\left(\bar{1} - \tanh^2\left(\frac{1}{2\lambda_i}(\beta_i + c_i) R_{ii}^{-1} \mathcal{G}_i^T \nabla \Phi_i^T \hat{\mathcal{W}}_i \right) \right)
$$

From (6.26) and (6.22), it can be seen that if $\hat{\mathcal{W}}_i$ is tuned to minimize a residual error function of $\hat{\mathcal{H}}_i\left(\mathcal{Z}_i, \hat{\mathcal{W}}_i \right)$ to obtain $\hat{\mathcal{H}}_i \to \mathcal{H}_i^*$, then $\hat{\mathcal{W}}_i \to \mathcal{W}_i$. In addition, to avoid identifying unknown parameters, the IRL technique is used [43]. The residual error function $E_i = \dfrac{1}{2}\mathcal{E}_{\mathcal{H}_i}^T \mathcal{E}_{\mathcal{H}_i}$ is chosen to minimize, where

$$
\mathcal{E}_{\mathcal{H}_i} = \int_{t-T}^{t} \hat{\mathcal{H}}_i\left(\mathcal{Z}_i, \hat{\mathcal{W}}_i \right) d\tau
\tag{6.27}
$$

where $T > 0$ is an optional parameter. The tuning law for weights is proposed via the normalized gradient descent

$$
\dot{\hat{\mathcal{W}}}_i = -\chi_i \Theta_i \Psi_i
\tag{6.28}
$$

for update rate $\chi_i > 0$, $\Theta_i = \sigma_i / \left(\sigma_i^T \sigma_i + 1 \right)^2$, where

$$
\sigma_i = \Delta \Phi_i\left(\mathcal{Z}_i(t) \right) = \int_{t-T}^{t} \nabla \Phi_i \left(\mathcal{F}_{\mathcal{Z}_i} + (\beta_i + c_i)\left(\mathcal{G}_i \hat{\mathcal{U}}_i + \mathcal{K}_i \hat{\mathcal{D}}_i \right) - \sum_{j \in N_i} \alpha_{ij}\left(\mathcal{G}_j \hat{\mathcal{U}}_j + \mathcal{K}_j \hat{\mathcal{D}}_j \right) \right) d\tau
$$

$$
= \int_{t-T}^{t} \nabla \Phi_i \dot{\mathcal{Z}}_i d\tau = \Phi_i(\mathcal{Z}_i(t)) - \Phi_i\left(\mathcal{Z}_i(t - T) \right)
\tag{6.29}
$$

$$\Psi_i = \sigma_i^T \hat{\mathcal{W}}_i + \int_{t-T}^{t} \left(Q_{ii} + U_{ii}(\hat{\mathcal{U}}_i) + \sum_{j \in N_i} U_{jj}(\hat{\mathcal{U}}_j) - \frac{1}{4\gamma^2}(\beta_i + c_i)^2 \hat{\mathcal{W}}_i^T \nabla \Phi_i \mathcal{K}_i \mathcal{K}_i^T \nabla \Phi_i^T \hat{\mathcal{W}}_i \right.$$

$$\left. - \frac{1}{4\gamma^2} \sum_{j \in N_i}(\beta_j + c_j)^2 \hat{\mathcal{W}}_j^T \nabla \Phi_j \mathcal{K}_j \mathcal{K}_j^T \nabla \Phi_j^T \hat{\mathcal{W}}_j \right) d\tau \qquad (6.30)$$

If $\|\mathcal{Z}_i(t)\| = \|\mathcal{Z}_j(t)\| = 0$, $\hat{\mathcal{W}}_i$ is not further updated. To make sure $\hat{\mathcal{W}}_i$ converges, the PE condition [47] in Definition 1 is used.

Algorithm 1: Distributed saturated optimal consensus control

$\hat{\mathcal{W}}_i^{(0)} \leftarrow 0$; $\hat{V}_i^{(0)} \leftarrow 0$; $\hat{\mathcal{U}}_i^{(0)} \leftarrow 0$; $\hat{\mathcal{D}}_i^{(0)} \leftarrow 0$; $T > 0$; $Q_{ii} > 0$, $R_{ii} > 0$, $\gamma > 0$, Φ_i, χ_i, PE probe noise ϖ_i; $\kappa \in \mathbb{R}^+$, $\lambda_i > 0$; stop step k_{max}; $k \leftarrow 0$;
   Repeat

$$\hat{V}_i^{(k)} = \hat{\mathcal{W}}_i^{(k)T} \Phi_i;$$

$$\hat{\mathcal{U}}_i^{(k)} = -\lambda_i(\beta_i + c_i)\tanh\left(\frac{1}{2}\lambda_i^{-1}R_{ii}^{-1}\mathcal{G}_i^T(\mathcal{X}_i)\nabla\Phi_i^T\hat{\mathcal{W}}_i^{(l)}\right) + \varpi_i(t);$$

$$\hat{\mathcal{D}}_i^{(k)} = \frac{1}{2\gamma^2}(\beta_i + c_i)\mathcal{K}_i^T\nabla\Phi_i^T\hat{\mathcal{W}}_i^{(l)} + \varpi_i(t);$$

$$\Theta_i = \Delta\Phi_i/(\Delta\Phi_i^T\Delta\Phi_i + 1)^2; \ k \leftarrow k+1; \ \dot{\hat{\mathcal{W}}}_i^{(k)} = -\chi_i\Theta_i\Psi_i;$$

If $\left\|\hat{\mathcal{W}}_i^{(k)} - \dot{\hat{\mathcal{W}}}_i^{(k-1)}\right\| \leq \kappa$ then $\varpi_i(t) \leftarrow 0$
   Until $(k >= k_{max})$
   From the definitions of the NN weight update law, the disturbance rejection policy the optimal control policy, a pseudo code of distributed control algorithm is written (see Algorithm 1).
   Remark 4: In Algorithm 1, knowledge of drift dynamics $\mathcal{F}_{\mathcal{Z}_i}(t)$ does not identify and the initial weights can be assigned to zeros without system stability loss. In other words, the stabilized initial controller is not required.

6.3.3 STABILITY AND CONVERGENCE

While the proposed algorithm is executed, the stability and convergence are guaranteed through the stability and convergence analysis in this section.
   Theorem 1: Let Van der Pol oscillator agents in MASS be defined in (6.3) with external disturbances and constraint inputs. Let the consensus value function be (6.16), the coupled HJI be (6.26), the parameter update law be designed in (6.28), the disturbance rejection policy, and the control policy be designed in (6.25). Then, the consensus and approximation errors are UUB. Moreover, when the algorithm

executes on a limited number of iterations, the L_2-gain function converges to the near-optimal value.

Proof: Consider the overall Lyapunov candidate:

$$L = V^* + \frac{1}{2}\sum_{i=1}^{N} \text{trace}\left(\tilde{W}_i^T \tilde{W}_i\right) \tag{6.31}$$

$V^* = \sum_{i=1}^{N} V_i^*$, V_i^* is a nonnegative smooth optimal HJI solution to (6.22). Taking the derivative of (6.31) yields

$$\dot{L} = \dot{V}^* + \sum_{i=1}^{N}\tilde{W}_i^T \dot{\tilde{W}}_i \tag{6.32}$$

Now, the overall stability and convergence are proven. The overall tracking dynamics, based on (6.9) and graph $\bar{G}(V, \Xi, A)$, is written as

$$\dot{Z} = \mathcal{F}_{Z} - I_n \otimes (\mathcal{M} + \mathcal{C})(\mathcal{G}\mathcal{U} + \mathcal{K}\mathcal{D}) \tag{6.33}$$

$\mathcal{Z} = \left[\mathcal{Z}_1^T, ..., \mathcal{Z}_N^T \right]^T$, $\mathcal{F}_Z = \left[\mathcal{F}_{Z_1}^T, ..., \mathcal{F}_{Z_N}^T \right]^T$, $\mathcal{U} = \left[\mathcal{U}_1^T, ..., \mathcal{U}_N^T \right]^T$, $\mathcal{D} = \left[\mathcal{D}_1^T, ..., \mathcal{D}_N^T \right]^T$
$\mathcal{G} = \text{diag}[\mathcal{G}_1, ..., \mathcal{G}_N]$, $\mathcal{K} = \text{diag}[\mathcal{K}_1, ..., \mathcal{K}_N]$. According to the disturbance rejection and optimal control policies, the overall policies is written as

$$\begin{cases} \mathcal{D}^* = \dfrac{1}{2\gamma^2} \mathcal{K}^T(\mathcal{X})(I_n \otimes (\mathcal{M}+\mathcal{C}))^T \nabla V^* \\[2mm] \mathcal{U}^* = -\lambda \tanh\left(\dfrac{1}{2\lambda} R^{-1}\mathcal{G}^T(\mathcal{X})(I_n \otimes (\mathcal{M}+\mathcal{C}))^T \nabla V^* \right) \end{cases} \tag{6.34}$$

where $\mathcal{X} = [\mathcal{X}_1, ..., \mathcal{X}_N]^T$, $\mathcal{U}^* = \left[\mathcal{U}_1^{*T}, ..., \mathcal{U}_N^{*T} \right]^T$, $\mathcal{D}^* = \left[\mathcal{D}_1^{*T}, ..., \mathcal{D}_N^{*T} \right]^T$,
$R = \text{diag}[R_{11}, ..., R_{NN}]$, $\lambda = \text{diag}[\lambda_1, ..., \lambda_N]$. Using (6.32) and the gradient $\nabla \hat{V} = \nabla \Phi^T \hat{W}$, $\nabla \Phi = \left[\nabla \Phi_1^T, ..., \nabla \Phi_N^T \right]^T$, $\hat{W} = \left[\hat{W}_1^T, ..., \hat{W}_N^T \right]^T$, the approximate optimal strategy $\hat{\mathcal{U}} = \left[\hat{\mathcal{U}}_1^T, ..., \hat{\mathcal{U}}_N^T \right]^T$ and the policy $\hat{\mathcal{D}} = \left[\hat{\mathcal{D}}_1^T, ..., \hat{\mathcal{D}}_N^T \right]^T$ are computed as

$$\begin{cases} \hat{\mathcal{U}} = -\lambda \tanh\left(\hat{N}_1\right) \\[2mm] \hat{\mathcal{D}} = \dfrac{1}{2\gamma^2}\mathcal{K}^T(\mathcal{X})(I_n \otimes (\mathcal{M}+\mathcal{C}))^T \nabla \Phi^T \hat{W} \end{cases} \tag{6.35}$$

where $\hat{N}_1 = \lambda^{-1}R^{-1}\mathcal{G}^T(I_n \otimes (\mathcal{M}+\mathcal{C}))^T \nabla \Phi^T \hat{W}$. Then, $\dot{V}^*$ can be presented as

$$\dot{V}^* = \left(\nabla V^*\right)^T \left(\mathcal{F}_Z + I_n \otimes (\mathcal{M}+\mathcal{C})(\mathcal{G}\hat{\mathcal{U}} + \mathcal{K}\hat{\mathcal{D}}) \right)$$

$$= W^T \nabla \Phi \left(\mathcal{F}_Z - I_n \otimes (\mathcal{M}+\mathcal{C}) \left(\lambda \mathcal{G}\tanh(\hat{N}_1) - \frac{1}{2\gamma^2}\mathcal{K}\mathcal{K}^T \Gamma^T \nabla \Phi^T \hat{W} \right) \right) + \varepsilon_V \tag{6.36}$$

$$\mathcal{W}=\left[\mathcal{W}_1^T,...,\mathcal{W}_N^T\right]^T, \quad \mathcal{E}_\mathcal{V}=\nabla\mathcal{E}^T\left(\mathcal{F}_\mathcal{Z}-\Gamma\left(\lambda\mathcal{G}\tanh(\hat{N}_1)-\frac{1}{2\gamma^2}\mathcal{K}\mathcal{K}^T\Gamma^T\nabla\Phi^T\hat{\mathcal{W}}\right)\right),$$

$\Gamma=I_n\otimes(\mathcal{M}+\mathcal{C})$, $\nabla\mathcal{E}=\left[\nabla\mathcal{E}_1^T,...,\nabla\mathcal{E}_N^T\right]^T$. Based on Assumption 1, it can be inferred that $\|\mathcal{F}_\mathcal{Z}\|\le b_\mathcal{F}\|\mathcal{Z}\|$ for a positive constant $b_\mathcal{F}$, $\|\mathcal{G}\|\le b_\mathcal{G}$, $\|\mathcal{K}\|\le b_\mathcal{K}$, $\|\nabla\mathcal{E}\|\le b_{\nabla\mathcal{E}}$, where $b_\mathcal{G}=\max_{1\le i\le N,h=q,v}g_{i,h}^{\max}$, $b_\mathcal{K}=\max_{1\le i\le N,h=q,v}k_{i,h}^{\max}$, $b_{\nabla\mathcal{E}}=\max_{1\le i\le N}\mathcal{E}_{\mathcal{Z}_i}^{\max}$. Furthermore, $\|\mathcal{W}\|<b_\mathcal{W}$, $\|\nabla\Phi\|<b_\Phi$. Since $\tilde{\mathcal{W}}=\mathcal{W}-\hat{\mathcal{W}}$, $\mathcal{E}_\mathcal{V}$ is bounded by

$$\|\mathcal{E}_\mathcal{V}\|\le b_1\|\mathcal{Z}\|+b_2\|\tilde{\mathcal{W}}\|+b_3 \tag{6.37}$$

where $b_1=b_{\nabla\mathcal{E}}b_\mathcal{F}$, $b_2=\frac{1}{2\gamma^2}\lambda_{\max}(\Gamma^T\Gamma)b_{\nabla\mathcal{E}}b_\mathcal{K}^2b_\Phi$, $\lambda_{\max}(.)$ is the largest eigenvalue, $b_3=\lambda_{\max}(\Gamma^T\Gamma)b_{\nabla\mathcal{E}}b_\mathcal{G}+\frac{1}{2\gamma^2}\lambda_{\max}(\Gamma^T\Gamma)b_{\nabla\mathcal{E}}b_\mathcal{K}^2b_\mathcal{W}b_\Phi$. The following overall saturated control policy is presented as

$$U(\mathcal{U}^*)=2\lambda\int_0^{u^*}\tanh^{-T}(\lambda^{-1}\upsilon)Rd\upsilon=\mathcal{W}^T\nabla\Phi\Gamma\mathcal{G}\tanh(N_1)+\lambda^T\lambda\bar{R}\ln(\bar{1}_N-\tanh^2(N_1)) \tag{6.38}$$

where $N_1=\frac{1}{2}\lambda^{-1}R^{-1}\mathcal{G}^T\Gamma^T\nabla\Phi^T\mathcal{W}$, $\bar{R}=[\bar{R}_{11},...,\bar{R}_{NN}]$. Using (6.34, 6.38) and (6.33), the overall HJI Equation (6.22) is changed to

$$\mathcal{W}^T\nabla\Phi\mathcal{F}_\mathcal{Z}+Q(\mathcal{Z})+\mathcal{G}-\mathcal{K}+\mathcal{E}_\mathcal{H}=0 \tag{6.39}$$

where $\mathcal{G}=\lambda^T\lambda\bar{R}\ln(\bar{1}_N-\tanh^2(N_1))$, $\mathcal{K}=\frac{1}{\gamma^2}\|\mathcal{P}\|^2$, $\mathcal{P}=\frac{1}{2}\Gamma\mathcal{K}\nabla\Phi^T\mathcal{W}$, $Q(\mathcal{Z})=\mathcal{Z}^T\bar{Q}\mathcal{Z}$, $\bar{Q}>0$. It is obvious that $\|\mathcal{G}\|\le c_\mathcal{G}$, for a positive constant $c_\mathcal{G}$. Employing $b_{i,\varepsilon}$ in Remark 3, $\dot{V}^*$ is satisfies the following inequality:

$$\dot{V}^*\le-\lambda_{\min}(\bar{Q})\|\mathcal{Z}\|^2+b_1\|\mathcal{Z}\|+b_4\|\tilde{\mathcal{W}}\|+b_5 \tag{6.40}$$

$b_4=b_2+\frac{1}{2\gamma^2}\lambda_{\max}(\Gamma^T\Gamma)b_\mathcal{K}^2b_\mathcal{W}b_\Phi^2$, $b_5=b_3+c_\mathcal{G}+\lambda_{\max}(\Gamma^T\Gamma)b_\mathcal{W}b_\Phi(\lambda b_\mathcal{G}+\frac{1}{4\gamma^2}\lambda_{\max}(\Gamma^T\Gamma)b_\mathcal{K}^2b_\mathcal{W}b_\Phi)$ $+\max_{1\le i\le N}b_{i,\varepsilon}$. Equation (6.30) can be rewritten as

$$\Psi_i=-\sigma_i^T(\tilde{\mathcal{W}}_i-\mathcal{W}_i)+\int_{t-T}^t\left(Q_{ii}+\hat{\mathcal{G}}_i-\hat{\mathcal{K}}_i\right)d\tau \tag{6.41}$$

where $\hat{\mathcal{G}}_i=\lambda^T\lambda\bar{R}_{ii}\ln(\bar{1}-\tanh^2(\hat{N}_{1i}))$, $\hat{N}_{1i}=\frac{1}{2}\lambda^{-1}(\beta_i+c_i)R_{ii}^{-1}\mathcal{G}_i^T\phi_i^T\hat{\mathcal{W}}_i$, $\hat{\mathcal{K}}_i=\frac{1}{\gamma^2}\|\hat{\mathcal{P}}_i\|^2$, $\hat{\mathcal{P}}_i=\frac{1}{2}\nabla\Phi_i^T\hat{\mathcal{W}}_i$. Substituting Q_{ii} from (6.22) into (6.41) yields

$$\Psi_i=-\sigma_i^T(\tilde{\mathcal{W}}_i+\mathcal{W}_i)+\int_{t-T}^t(\hat{\mathcal{G}}_i-\mathcal{G}_i+\mathcal{K}_i-\hat{\mathcal{K}}_i+\Pi_{0i})d\tau \tag{6.42}$$

where $\mathcal{G}_i = \lambda^T \lambda \bar{R}_{ii} \ln(\bar{1} - \tanh^2(N_{1i}))$, $N_{1i} = \frac{1}{2}\lambda(\beta_i + c_i)R_{ii}^{-1}\mathcal{G}_i^T\Phi_i^T\mathcal{W}_i$, $\mathcal{K}_t = \frac{1}{\gamma^2}\|\mathcal{P}_i\|^2$, $\mathcal{P}_i = \frac{1}{2}\nabla\Phi_i^T\mathcal{W}_i$, $\Pi_{0i} = -W_i^T\nabla\Phi_i\mathcal{F}_{z_i} - \mathcal{E}_{\mathcal{H}_i}$. Using the derivation in [30], $\hat{\mathcal{G}}_i - \mathcal{G}_i$ is rewritten as

$$
\begin{aligned}
\hat{\mathcal{G}}_i - \mathcal{G}_i &= \lambda^T\lambda\bar{R}_{ii}\left(N_{1i}\mathrm{sgn}(N_{1i}) - \hat{N}_{1i}\mathrm{sgn}(\hat{N}_{1i}) + \epsilon_{\hat{N}_{1i}} - \epsilon_{N_{1i}}\right) \\
&= \lambda^T\lambda\bar{R}_{ii}\left(N_{1i}\left(\mathrm{sgn}(N_{1i}) - \mathrm{sgn}(\hat{N}_{1i})\right) + \epsilon_{\hat{N}_{1i}} - \epsilon_{N_{1i}}\right) \\
&\quad + \lambda\bar{R}_{ii}R_{ii}^{-1}\mathcal{G}_i^T\nabla\Phi_i^T\tilde{\mathcal{W}}_i\mathrm{sgn}\left(\hat{N}_{1i}\right)
\end{aligned}
\tag{6.43}
$$

where $\epsilon_{\hat{N}_{1i}}$, $\epsilon_{N_{1i}}$ are bounded approximation errors. Then,

$$
\Psi_i = -\sigma_i^T\tilde{\mathcal{W}}_i + \sigma_i^T\mathcal{W}_i + \int_{t-T}^{t}\Omega_{1i}d\tau
\tag{6.44}
$$

where $\Omega_{1i} = \lambda\bar{R}_{ii}R_{ii}^{-1}\mathcal{G}_i^T\nabla\Phi_i^T\tilde{\mathcal{W}}_i\mathrm{sgn}(\hat{N}_{1i}) - \frac{1}{\gamma^2}\left(\|\hat{\mathcal{P}}_i\|^2 - \|\mathcal{P}_i\|^2\right) + \Pi_{1i}$, $\Pi_{1i} = \Pi_{0i} + \lambda^T\lambda\bar{R}_{11}$ $(N_{1i}(\mathrm{sgn}(N_{1i}) - \mathrm{sgn}(\hat{N}_{1i})) + \epsilon_{\hat{N}_{1i}} - \epsilon_{N_{1i}})$. According to the boundedness of the terms in Ω_{1i}, $\|\Omega_{1i}\| \leq \psi_{1i}\|\tilde{\mathcal{W}}_i\| + \psi_{2i}$, where ψ_{1i} and ψ_{2i} are some positive constants. Using the boundedness for (6.44), we have

$$
\Psi_i \leq -\sigma_i^T\tilde{\mathcal{W}}_i + \sigma_i^T\mathcal{W}_i + \frac{1}{T}\left(\psi_{1i}\|\tilde{\mathcal{W}}_i\| + \psi_{2i}\right)
\tag{6.45}
$$

Using (6.45), the second sum in (6.32) is rewritten as

$$
\begin{aligned}
\sum_{i=1}^{N}\tilde{\mathcal{W}}_i^T\dot{\tilde{\mathcal{W}}}_i &= \sum_{i=1}^{N}\tilde{\mathcal{W}}_i^T\chi_i\Theta_i\Psi_i \\
&\leq -\frac{\beta_1}{T}\|\tilde{\mathcal{W}}\|^2 + \frac{\beta_2}{T}\mathcal{W}\|\tilde{\mathcal{W}}\| + \frac{b_\Phi\mu\psi_1}{T}\|\tilde{\mathcal{W}}\|^2 + \frac{2b_\Phi\mu\psi_2}{T}\|\tilde{\mathcal{W}}\| \\
&= -\psi_3\|\tilde{\mathcal{W}}\|^2 + \psi_4\|\tilde{\mathcal{W}}\|
\end{aligned}
\tag{6.46}
$$

where $\mu = \sum_{i=1}^{N}1/\left(\sigma_i^T\sigma_i + 1\right)^2$, $\psi_3 = \frac{1}{T}(\beta_1 - \mu b_\Phi\psi_1)$, $\sum_{i=1}^{N}\|\sigma_i\| = \sum_{i=1}^{N}\|\Delta\Phi_i\| < 2b_\Phi$, $\psi_1 = \sum_{i=1}^{N}\psi_{1i}$, $\psi_4 = \frac{1}{T}(\beta_2 b_W + 2\mu b_\Phi\psi_2)$, $\psi_2 = \sum_{i=1}^{N}\psi_{2i}$. Substituting (6.40) and (6.46) into (6.32) yields

$$
\dot{L} \leq -\lambda_{\min}(\bar{Q})\left(\|\mathcal{Z}\| - \frac{b_1}{2\lambda_{\min}(\bar{Q})}\right)^2 - \psi_3\left(\|\tilde{\mathcal{W}}\| - \frac{a_1}{2\psi_3}\right)^2 + a_2
\tag{6.47}
$$

where $a_1 = \psi_4 + b_4$, $a_2 = b_5 + b_1^2/(4\lambda_{\min}(Q)) + a_1^2/(4\psi_3)$. Choosing $\beta_1 > \mu b_\Phi \psi_1$, then $\dot{L} < 0$ when

$$\|\mathcal{Z}\| > \sqrt{\frac{a_2}{\lambda_{\min}(\bar{Q})}} + \frac{b_1}{2\lambda_{\min}(\bar{Q})} = b_{\mathcal{Z}}, \|\tilde{\mathcal{W}}\| > \sqrt{\frac{a_2}{\psi_3}} + \frac{a_1}{4\psi_3} = b_{\tilde{\mathcal{W}}} \qquad (6.48)$$

According to [50], the overall consensus and approximation errors are UUB. The value function is followed by $\|V^* - \hat{V}\| \le \|\nabla\Phi\|\|\tilde{\mathcal{W}}\| + \|\nabla\mathcal{E}\| \le b_{\tilde{\mathcal{W}}} b_c + b_{\nabla\mathcal{E}} = b_V$, $\forall t > 0$. b_V can be smaller by setting adaptive rates χ_i, appropriately. Then, the value function approaches to the suboptimality. This completes the proof.

6.4 NUMERICAL SIMULATION STUDIES

This section conducts simulation studies on distributed cooperative control for Van der Pol oscillator agents. First, Algorithm 1 is compared with the existing one for control of SASS, and then applied to MASS.

6.4.1 COMPARISON

Comparison is made between the proposed algorithm and the two-NN algorithm using the knowledge of known dynamics [29]. Consider a Van der Pol oscillator system (6.3) with dynamics SASS [29]:

$$\begin{cases} \dot{x}_{0,1} = x_{0,2} \\ \dot{x}_{0,2} = -x_{0,2} - \dfrac{1}{2} x_{0,2}\left(1 - x_{0,1}^2\right) - x_{0,1}^2 x_{0,2} + x_{0,1} u_0 \\ y_1 = x_{0,1} \end{cases} \qquad (6.49)$$

The learning parameters are chosen as [29], i.e., $|u_0| < 0.1$, $\phi_0(x_0) = [x_{0,1}^2, x_{0,1}x_{0,2}, x_{0,2}^2]^T$, $x_0 = [x_{0,1}, x_{0,2}]^T$, $Q_{00} = \|x_0\|^2$, $R_{00} = 1$, $\alpha_0 = 10$, $\gamma = 1$. The probe noise $\chi_0 = \sin^2(2t) + \dfrac{1}{2}\cos(2t) - 0.5\sin^2(t)\cos(t) + \dfrac{1}{2}\sin^5(2t)$ excites the control signals. $\hat{\mathcal{W}}(0) = 0$. The states are initialized $x_{0,1}(0) = 1, x_{0,2}(0) = -1$.

Figure 6.1 shows the evolution of NN weights of two algorithms. Although the NN weights of two algorithms converge to the same values, i.e., $\hat{\mathcal{W}}_0 = [2.48\ 0.99\ 2.23]^T$, the convergence rate of the two-NN algorithm using the known dynamics is slower than Algorithm 1 using one NN. In addition, it is easy to see that the advantage of Algorithm 1 is outstanding as it does not require the known dynamics. The trajectories of the proposed algorithm will be combined with the results of other agents in MASS in the following subsection to reduce the space.

6.4.2 H_∞ OPTIMAL CONSENSUS CONTROL FOR MASS

After evaluating the control performance for a single agent in comparing the Algorithm 1 with the other in [29], we continue applying Algorithm 1 to MASS.

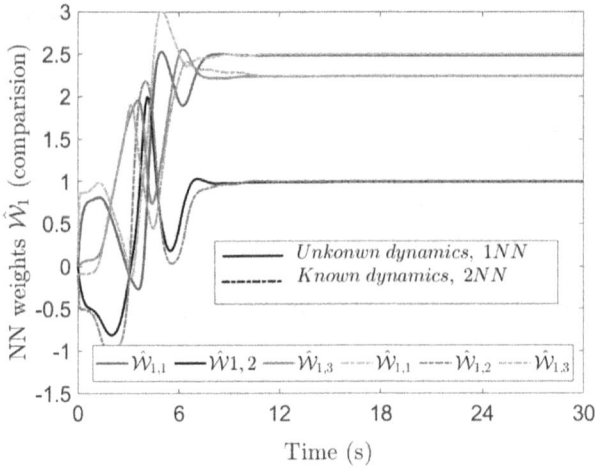

FIGURE 6.1   Evolution of NN weights of two algorithms.

The distributed consensus configuration of MASS is presented in Figure 6.2, where agent 1 follows leader 0, and its neighbor is agent 2. Note that agents 1 and 2 are disturbed by external disturbances d_1 and d_2. The dynamics of agent 0 is given in (6.49), and dynamics of agents 1 and 2 are presented as follows.

$$\begin{cases} \dot{x}_{1,1} = x_{1,2} + d_{1,1} \\ \dot{x}_{1,2} = -x_{1,2} - \dfrac{1}{2}x_{1,2}(1-x_{1,1}^2) - x_{1,1}^2 x_{1,2} + x_{1,1}u_2 + \sin(4x_{1,1}-1)x_{1,1}d_{1,2} \\ y_1 = x_{1,1} \end{cases} \quad (6.50)$$

$$\begin{cases} \dot{x}_{2,1} = x_{2,2} + d_{2,1} \\ \dot{x}_{2,2} = -x_{2,2} - \dfrac{1}{2}x_{2,2}(1-x_{2,1}^2) - x_{2,1}^2 x_{2,2} + x_{2,1}u_2 + \sin(4x2_{2,1}-1)\cos(x_{2,2})d_{2,2} \\ y_2 = x_{2,1} \end{cases} \quad (6.51)$$

where $|u_1| < 0.1$, $|u_2| < 0.05$. All learning parameters of agent 1 and 2 are selected as the agent 0. The agent states are initialized as $x_{1,1}(0) = 1$, $x_{1,2}(0) = -1$, $x_{2,1}(0) = 0.5$, $x_{2,2}(0) = -0.5$.

The evolution of NN parameters of agent 1 is presented in Figure 6.3, and the evolution of parameters of agent 2 is presented in Figure 6.4, where the convergence

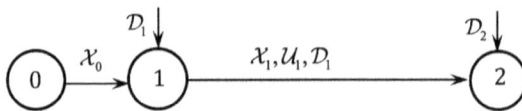

FIGURE 6.2   Distributed consensus configuration.

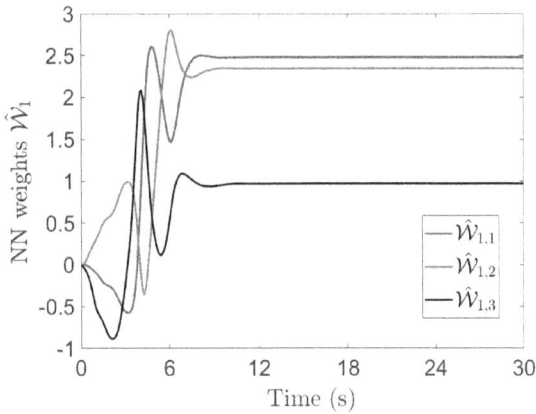

FIGURE 6.3   Evolution of NN of agent 1.

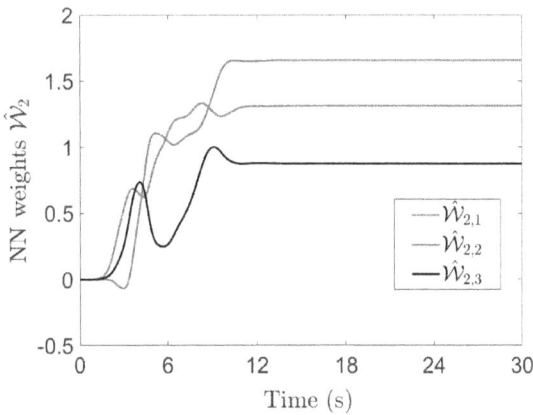

FIGURE 6.4   Evolution of NN of agent 2.

is obtained after 10(s), then the PE condition can be stopped. In Figure 6.5, the top subfigure shows phase plane trajectories of the closed-loop systems while the subplot at the bottom shows the output. It can be seen that the consensus is reached, where the output y_0 of agent 0 is controlled to return to the origin while the output y_1 of agent 1 is synchronized with y_0. Similarly, the output y_2 of agent 2 is synchronized with y_1. Figure 6.6 shows the consensus errors of the agents, where the tracking errors go to zero when $t \to \infty$.

In Figure 6.7, approximate optimal control inputs of the agents are saturated at the maximum and minimum limitations. The optimal control signals of agent 0 and 1 are bounded at ± 0.1, while that of agent 2 is bounded at ± 0.05, which are consistent with saturating actuators in agent dynamics.

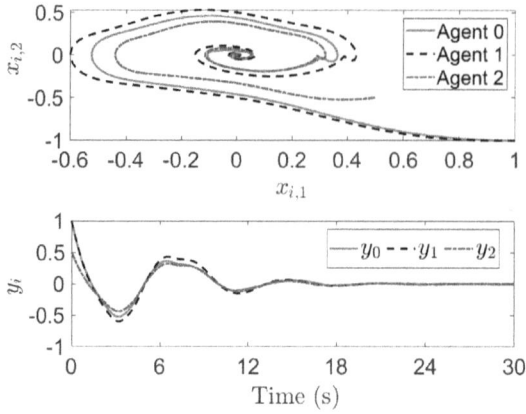

FIGURE 6.5   Trajectories. Top: phase plane trajectories. Bottom: output performance.

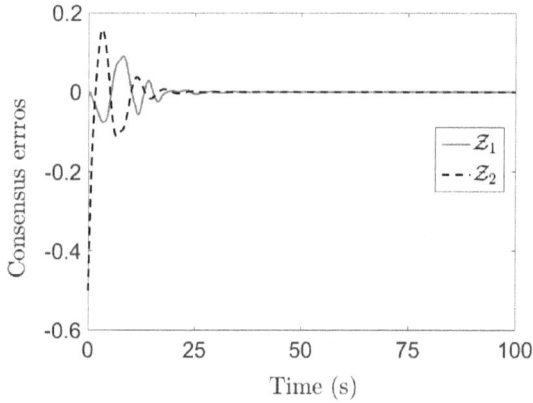

FIGURE 6.6   Consensus errors.

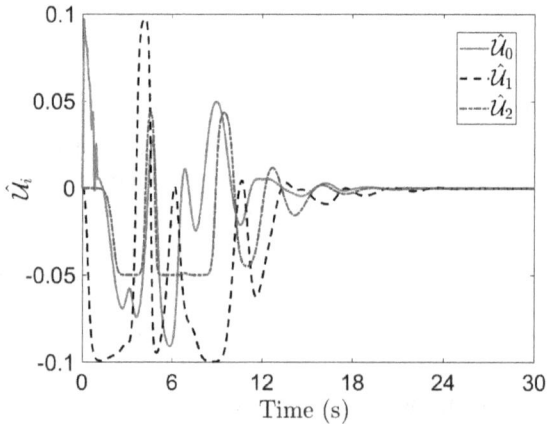

FIGURE 6.7   Approximate optimal control inputs of agents.

6.4.3 H_∞ OPTIMAL CONSENSUS CONTROL FOR MAS
UNDER CONNECTION UNCERTAINTIES

To further verify the robust ability under the effect of uncertain connection between agents, the distributed cooperative configuration with uncertainty is considered. Figure 6.8 shows the distributed consensus configuration under connection uncertainties, where Δ_1, Δ_2 are connection uncertainties, which are assumed to be added to dynamics of agents as follows.

$$\begin{cases} \dot{x}_{1,1} = x_{1,2} \\ \dot{x}_{1,2} = -x_{1,2} - \dfrac{1}{2}x_{1,2}(1-x_{1,1}^2) - x_{1,1}^2 x_{1,2} + \Delta_1 + x_{1,1}u_2 + \sin(4x_{1,1}-1)x_{1,1}d_{1,2} \\ y_1 = x_{1,1} \end{cases} \tag{6.52}$$

$$\begin{cases} \dot{x}_{2,1} = x_{2,2} + d_{2,1} \\ \dot{x}_{2,2} = -x_{2,2} - \dfrac{1}{2}x_{2,2}(1-x_{2,1}^2) - x_{2,1}^2 x_{2,2} + \Delta_2 + x_{2,1}u_2 + \sin(4x2_{2,1}-1)\cos(x_{2,2})d_{2,2} \\ y_2 = x_{2,1} \end{cases} \tag{6.53}$$

where $\Delta_1 = 0.1\left(y_0 + y_1^2\right) + 0.8y_0 y_1$, $\Delta_2 = 0.1\left(y_1^2 + y_2^3\right) + 0.5y_1 y_2$.

Figure 6.9 shows the consensus performance under connection uncertainties. At the early states, the formation quality is worse, and the convergence is slower than that in

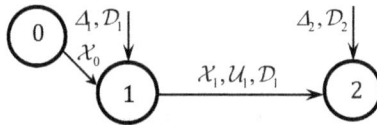

FIGURE 6.8   Distributed consensus configuration under connection uncertainties.

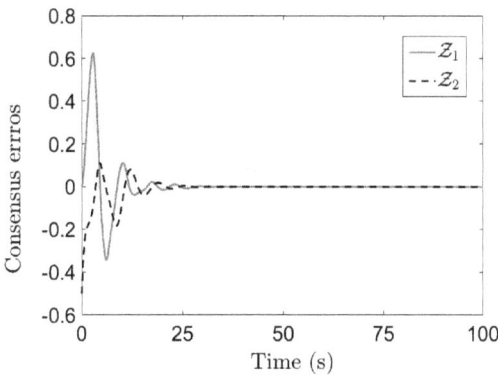

FIGURE 6.9   Consensus performance under connection uncertainties.

the absence of uncertainty (Figure 6.6). However, the formation still maintains robust stable and guarantees that the agent tracking errors are close to zero when $t \rightarrow \infty$.

6.5 CONCLUSION

This chapter has proposed an optimal consensus control method via machine learning with application to Van der Pol oscillators in MASS with input constraints and external disturbances. The algorithm is developed based on ADP and the differential games theory, in which the control structure for each agent only uses a single NN. In addition, the control and disturbance rejection policies are estimated and system identification for unknown dynamics are avoided. By using the appropriate Lyapunov theory, the closed-loop tracking and approximation errors are proven to be UUB. The value function is also approached to the suboptimality. The comparative simulation studies of single-agent and multi-agent of Van der Pol oscillators are performed to present the applicability of the designed algorithm. The problem of switching topologies is focused on the future work.

REFERENCES

1. Chen G., and Song Y. 2014. Cooperative tracking control of nonlinear multi-agent systems using self-structuring neural networks. *IEEE Transaction on Neural Networks and Learning Systems* 25 (8): 1496–1507.
2. Das A., and Lewis F. L. 2011. Cooperative adaptive control for synchronization of second-order systems with unknown nonlinearities. *International Journal of Robust Nonlinear Control* 21 (13): 1509–1524.
3. Hou Z. G., Cheng L., and Tan M. 2009. Decentralized robust adaptive control for the multiagent system consensus problem using neural networks. *IEEE Transaction on Systems Man and Cybernetics, Part B: Cybernetics* 39 (3): 636–647.
4. Peng Z., Wang D., Zhang H., and Sun G. 2014. Distributed neural network control for adaptive synchronization of uncertain dynamical multi-agent systems. *IEEE Transaction on Neural Networks and Learning Systems* 25 (8): 1508–1519.
5. Bertsekas D. P. 2017. *Dynamic Programming and Optimal Control*. 4th ed. Belmont, MA: Athena Scientific.
6. Vamvoudakis K. G., and Jagannathan S. 2016. *Control of Complex Systems: Theory and Applications*. Kidlington, Oxford : Butterworth-Heinemann.
7. Khalil H. K. 2002. *Nonlinear Systems*. 3rd ed. Englewood Cliffs, NJ, USA: Prentice-Hall.
8. Krstic M., Kokotovic P. V., and Kanellakopoulos I. 1995. *Nonlinear and Adaptive Control Design*. New York, NY: Wiley.
9. Li Z., and Zhisheng D. 2017. Cooperative control of multi-agent systems: A consensus region approach. In *Multi-Agent Systems*. Boca Raton, FL: CRC Press.
10. Basar T., and Bernhard P. 1995. *Optimal Control and Related Minimax Design Problems: A Dynamic Game Approach*. Boston, MA: Birkhuser.
11. Lin H., Wei Q., Liu D., and Ma H. 2016. Adaptive tracking control of leader-following linear multi-agent systems with external disturbances. *International Journal of Systems Science* 47: 3167–3179.
12. Jiao Q., Modares H., Xu S, and Lewis F. L. 2016. Vamvoudakis KG. Multiagent zero-sum differential graphical games for disturbance rejection in distributed control. *Automatica* 69: 24–34.

13. Vamvoudakis K. G., Modares H., Kiumarsi B., and Lewis FL. 2017. Game theory-based control system algorithms with real-time reinforcement learning: how to solve multiplayer games online. *IEEE Control Systems Magazine.* 37 (1): 33–52.
14. Sutton R. S., and Barto A. G. 2018. *Reinforcement Learning – An Introduction.* 2nd ed. Cambridge, MA: MIT Press.
15. Werbos P. J. 2009. Intelligence in the brain: A theory of how it works and how to build it. *Neural Networks.* 22 (3): 200–212.
16. Schwartz H. M. 2014. *Multi-Agent Machine Learning: A Reinforcement Approach.* John Wiley & Sons, Inc.
17. Zargarzadeh H., Dierks T., and Jagannathan S. 2014. Adaptive neural network-based optimal control of nonlinear continuous-time systems in strict-feedback form. *International Journal of Adaptive Control and Signal Processing* 28 (3–5): 305–324.
18. Zargarzadeh H., Dierks T., and Jagannathan S. 2015. Optimal control of nonlinear continuous-time systems in strict-feedback form. *IEEE Transaction on Neural Networks and Learning Systems* 26(10): 2535–2549.
19. Sun K., Sui S., and Tong S. 2018. Fuzzy adaptive decentralized optimal control for strict feedback nonlinear large-scale systems. *IEEE Transactions on Cybernetics* 48 (4): 1326–1339.
20. Tong S., Sun K., and Sui S. 2018. Observer-based adaptive fuzzy decentralized optimal control design for strict-feedback nonlinear large-scale systems. *IEEE Transactions on Fuzzy Systems* 26 (2): 569–584.
21. Tan L. N. 2018. Omnidirectional vision-based distributed optimal tracking control for mobile multi-robot systems with kinematic and dynamic disturbance rejection. *IEEE Transactions on Industrial Electronics* 65 (7): 5693–5703.
22. Tan L. N. 2020. Distributed H∞ optimal tracking control for strict-feedback nonlinear large-scale systems with disturbances and saturating actuators. *IEEE Transactions on Systems, Man, Cybernetics, Systems* 50 (11): 4719–4731.
23. Tan LN. 2021. Event-triggered distributed H∞ constrained control of physically interconnected large-scale partially unknown strict-feedback systems. *Transactions on Systems, Man, Cybernetics, Systems* 51 (4): 2444–2456.
24. Tran T. T., Ge S. S., and He W. 2018. Adaptive control of a quadrotor aerial vehicle with input constraints and uncertain parameters. *International Journal of Control* 91 (5): 1140–1160.
25. Liu Z., Liu J., and He W. 2017. Partial differential equation boundary control of a flexible manipulator with input saturation. *International Journal System Science* 48 (1): 53–62.
26. Wang B., Wang J., Zhang B., and Li X. 2017. Global cooperative control framework for multiagent systems subject to actuator saturation with industrial applications. *IEEE Transactions on Systems, Man, Cybernetics, Systems* 47 (7): 1270–1283.
27. Zhang H., Cui X., Luo Y., and Jiang H. 2018. Finite-horizon H∞ tracking control for unknown nonlinear systems with saturating actuators. *IEEE Transactions on Neural Network and Learning Systems* 29 (4): 1200–1212.
28. Abu-Khalaf M., and Lewis F. L. 2005. Nearly optimal control laws for nonlinear systems with saturating actuators using a neural network HJB approach. *Automatica* 41 (5): 779–791.
29. Vamvoudakis K. G., Miranda M. F., and Hespanha J. P. 2016. Asymptotically stable adaptive-optimal control algorithm with saturating actuators and relaxed persistence of excitation. *IEEE Transactions on Neural Network and Learning Systems* 27 (11): 2386–2398.
30. Modares H., Lewis F. L., and Naghibi-Sistani M. B. 2013. Adaptive optimal control of unknown constrained-input systems using policy iteration and neural networks. *IEEE Transactions on Neural Network and Learning Systems* 24 (10): 1513–1525.

31. Chen Z., Li Z., and Chen C. L. P. 2017. Adaptive neural control of uncertain MIMO nonlinear systems with state and input constraints,. *IEEE Transactions on Neural Network and Learning Systems* 28 (6): 1318–1330.
32. Van der Pol B. 1927. *Forced Oscillations in a Circuit with Non-Linear Resistance (Receptance with Reactive Triode).* London: Edingburg and Dublin Phil.
33. Haddad W. M., and Chellaboina V. 2008. *Nonlinear Dynamical Systems and Control: A Lyapunov-Based Approach.* Princeton, NJ: Princeton Univ.
34. Strogatz S. H. 2014. *Nonlinear Dynamics and Chaos: With Applications to Physics, Biology, Chemistry, and Engineering.* 1st ed. Boulder, CO:Westview Press, a member of the Perseus Books Group.
35. Veerman F., and Verhulst F. 2009. Quasiperiodic phenomena in the van der pol-Mathieu equation. *Journal of Sound and Vibration* 326 (1): 314–320
36. Van der Pol B. 1934. The nonlinear theory of electric oscillations. *Proceedings of the Institute of Radio Engineers* 22 (9):1051–1086.
37. Calise A. J., Hovakimyan N., and Idan M. 2001. Adaptive output feedback control of nonlinear systems using neural networks. *Automatica* 37 (8): 1201–1211.
38. Loria A., Panteley E., and Nijmeijer H. 2001. A remark on passivity-based and discontinuous control of uncertain nonlinear systems. *Automatica* 37(9): 1481–1487.
39. Hovakimyan N., Nardi F., Calise A., and Kim N. 2002. Adaptive output feedback control of uncertain nonlinear systems using singlehidden-layer neural networks. *IEEE Transactions on Neural Networks* 13 (6): 1420–1431.
40. Ye A., and Lewis F. L. 1995. Feedback linearization using neural networks. *Automatica* 31 (11): 1659–1664.
41. Ahmed H., Ríos H., Ayalew B., and Wang Y. 2018. Second-order sliding-mode differentiators: An experimental comparative analysis using van der pol oscillator. *International Journal of Control* 91 (9): 2100–2112.
42. Gao W., and Jiang Z. 2018. Learning-based adaptive optimal tracking control of strict-feedback nonlinear systems. *IEEE Transactions on Neural Networks and Learning Systems* 29 (6): 2614–2624.
43. Vrabie D., and Lewis F. 2009. Neural network approach to continuous-time direct adaptive optimal control for partially unknown nonlinear systems. *Neural Networks* 22 (3): 237–246.
44. Sun K., Sui S., and Tong S. 2018. Fuzzy adaptive decentralized optimal control for strict feedback nonlinear large-scale systems. *IEEE Transactions on Cybernetics* 56 (2): 569–584.
45. Nguyen T. L. 2018. Distributed optimal control for nonholonomic systems with input constraints and uncertain interconnections. *Nonlinear Dynamics* 93 (2): 801–817.
46. Das A., and Lewis F. L. 2010. Distributed adaptive control for synchronization of unknown nonlinear networked systems. *Automatica* 46 (12): 2014–2021.
47. Ioannou P., and Fidan B. 2006. *Advances in Design and Control, Adaptive Control Tutorial.* Philadelphia, PA: SIAM.
48. Basar T., and Bernhard P. 1995. *Optimal Control and Related Minimax Design Problems: A Dynamic Game Approach.* Boston, MA: Birkhuser.
49. Vamvoudakis K. G., and Lewis F. L. 2011. Online solution of nonlinear two-player zero-sum games using synchronous policy iteration. *International Journal Robust Nonlinear Control* 22 (13): 1460–1483.
50. Lewis F. L., Jagannathan S., and Yesildirek. 1999. *A. Neural Network Control of Robot Manipulators and Nonlinear Systems.* Philadelphia, PA: Taylor and Francis.

Part III

---

*Machine Interaction
and Applications in the
Healthcare Industry*

7 Smart Applications of Internet of Things (IoT) in Healthcare

Praveen Kumar Gupta, Shweta Sudam Kallapur,
Anusha Mysore Keerthi, Soujanya Ramapriya,
A. H. Manjunatha Reddy, and Sumathra Manokaran
Department of Biotechnology, RV College of Engineering,
Bengaluru, India

CONTENTS

7.1   INTRODUCTION

The Internet of Things (IoT) has become one of the most powerful communication paradigms and attracted several analysis interests in the twenty-first century. It connects various objects, such as sensors, vehicles, houses, and appliances, along with the internet, which permit users to share information, knowledge, and resources. The emergence of IoT has created a key element within the environmental observance and healthcare applications. Wireless sensors can be deployed in various locations to monitor environmental conditions, and wearable sensors can be attached to the patients' body to measure physiological status. This information is transmitted to cloud infrastructure and bestowed to the targeted users. However, the present works principally focus either on environmental or healthcare observance applications. IoT is the concept of connecting any device to the internet and other connected devices. The IoT is a giant network of connected things and people—all of which collect and share data about the way they are used and about the environment around them. Devices and objects with inbuilt sensors are connected to an IoT platform that integrates information from various devices and applies analytics to share the foremost valuable data with applications engineered to deal with specific needs. These powerful IoT platforms will pinpoint precisely what data is beneficial. This information can be used to detect patterns, make recommendations, and detect possible problems before they occur. This chapter seeks to explore the current applications and future scope of the IoT as a defining technology in healthcare management. It aims to provide a detailed account of the latest technological advancements of implementation of IoT in healthcare and the impact it has had in transforming the healthcare sector. It provides a comprehensive review of the technologies used in the implementation of IoT in healthcare and the benefits and challenges in their implementation.

7.1.1   Role of IoT in Healthcare

Advances in data, telecommunication, and network technologies have played a major role in healthcare systems and have very significant contributions to the development of medical data systems. However, healthcare represents one among the prominent social

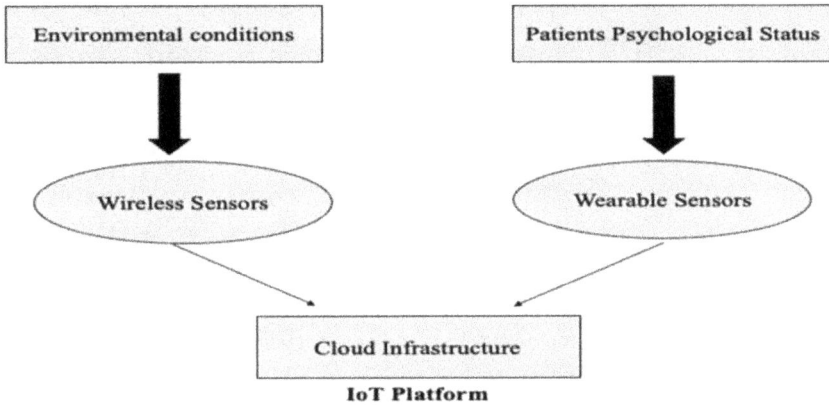

FIGURE 7.1   IoT platform.

and economic challenges that each country faces, and health care directors, clinicians, researchers, and other practitioners face increasing pressure to regulate to growing expectations from the general public as well as the industry sector. With the increase in the global aging population and the advent of new diseases and health risks, there is an immediate requirement to harness the power of information technology to improve the service efficiency of healthcare. Several connected devices have been developed to promote healthcare. Sensors have been developed that are designed to gather information and cloud-hosted analytics. Healthcare suppliers have increasingly become connected through the utilization of Wi-Fi, PCs, mobile computers, tablets, smartphones, and communication badges. These devices have enabled them to become much more proactive concerning delivery of health services. Management of big data, knowledge management, identification, authentication, and tracking of the current health conditions of patients and real-time sensing can all be carried out by implementing the IoT in healthcare [1] (Figure 7.1) shows the above information in a diagrammatical format.

7.2   METHODOLOGIES USED IN THE IMPLEMENTATION OF IoT IN HEALTHCARE

7.2.1   IoT-Based Approach for Specially Abled Persons and Palliative Care

Patients who require long-term care and/or suffer from physical impediments such as paralysis require special care to perform routine tasks. In most situations, the sole responsibility of the patient falls entirely upon the caregiver, resulting in a higher amount of stress for both parties. To overcome this issue, a model based on IoT has been developed that combines networking between physical devices and embedded sensors (Figure 7.2).

This system integrates mechanical devices, electronic components, and a digital computer to link the patient to the caregiver. For example, the speech recognition part of the system translates the patient's speech-to-text and transfers the message to the caregiver. Specific feelings such as hunger, pain, and illness can be conveyed through this feature. Bluetooth connection enables the display of the patient's heartbeat, blood pressure, and other vital signs at all times. The patient and the caregiver can also use

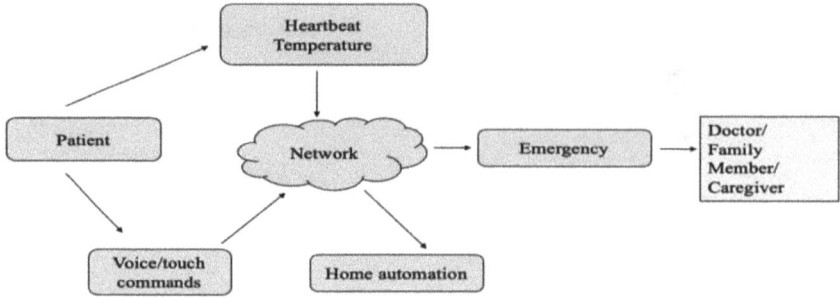

FIGURE 7.2 Working methodology of the system for specially abled patients using IoT [2].

the artificial intelligence segment of the system to view the prognosis and further courses of treatment available. Thus, this system can enable specially abled patients to increase their self-confidence and improve their capability to perform tasks [3].

7.2.2 Clinical Decision Support Systems (CDSSs) and Electronic Health Record (EHR)

Clinical decision support system (CDSS) is a computational system with a specialized algorithm that can be used to generate patient-specific treatment suggestions. CDSSs are integrated with electronic health records (EHRs), a database containing patients' medical records that can be queried with keywords to find and retrieve individual patient records. The CDSS algorithm can then be used to obtain recommendations on treatment decisions. Data mining and relevant clinical research are employed in CDSS algorithms (Figure 7.3). The methodologies used in the implementation of CDSS are outlined below.

CDSS METHODOLOGIES

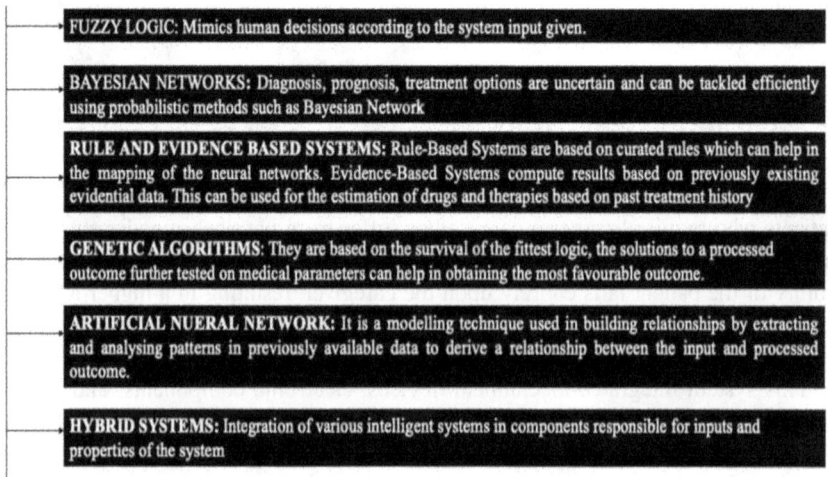

FUZZY LOGIC: Mimics human decisions according to the system input given.

BAYESIAN NETWORKS: Diagnosis, prognosis, treatment options are uncertain and can be tackled efficiently using probabilistic methods such as Bayesian Network

RULE AND EVIDENCE BASED SYSTEMS: Rule-Based Systems are based on curated rules which can help in the mapping of the neural networks. Evidence-Based Systems compute results based on previously existing evidential data. This can be used for the estimation of drugs and therapies based on past treatment history

GENETIC ALGORITHMS: They are based on the survival of the fittest logic, the solutions to a processed outcome further tested on medical parameters can help in obtaining the most favourable outcome.

ARTIFICIAL NUERAL NETWORK: It is a modelling technique used in building relationships by extracting and analysing patterns in previously available data to derive a relationship between the input and processed outcome.

HYBRID SYSTEMS: Integration of various intelligent systems in components responsible for inputs and properties of the system

FIGURE 7.3 CDSS methodologies.

7.2.2.1   Fuzzy Logic Rule-Based Approach

Fuzzy logic mimics human decisions according to system input given. Fuzzy logic allows data computation in the form of relative property and thus enhancing severity and decision logics like the working of a human brain.

7.2.2.2   Bayesian Networks

Diagnosis, prognosis, treatment options, etc. are uncertain and can be tackled effectively using probabilistic methods such as Bayesian network.

7.2.2.3   Rule and Evidence-Based Systems

Rule-based systems are based on curated rules that can help in the mapping of the neural networks. Evidence-based systems compute results based on previously existing evidential data. This can be used for the estimation of drugs and therapies based on past treatment history.

7.2.2.4   Genetic Algorithms

They are based on the survival of the fittest logic; the solutions to a processed outcome further tested on medical parameters can help in obtaining the most favorable outcome.

7.2.2.5   Artificial Neural Network

It is a modeling technique used in building relationships by extracting and analyzing patterns in previously available data to derive a relationship between the input and processed outcome.

7.2.3   Remote Medical Assistance

Patients present in areas that have poor connectivity to healthcare facilities can make use of smart mobile apps that notify doctors about the condition immediately. Distribution of drugs using IoT-based machines that work based on the app data can also be achieved. Figure 7.4 shows the summary of IoT technologies in remote medical assistance.

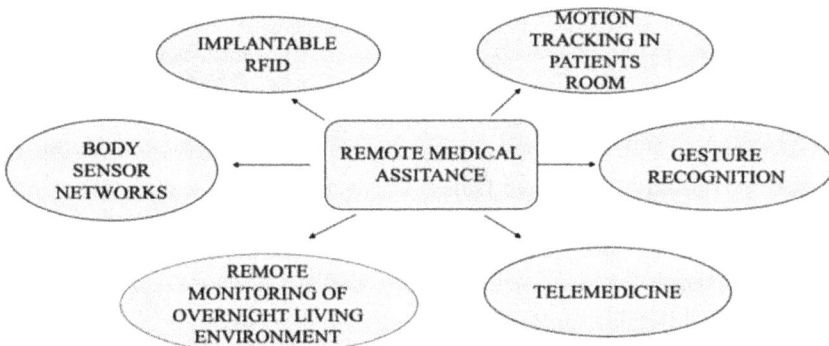

FIGURE 7.4   IoT technologies in remote medical assistance.

7.2.3.1   Radio Frequency Identification (RFID) Technology

By using wireless sensors inside homes and on personal items, it becomes possible to monitor a person's behavior and environment while protecting their privacy. This enables the compilation of statistics and allows the activation of prompts and alarms for remote assistance. Radio frequency identifications (RFIDs) are energetically autonomous being battery-less and can be used for this approach. A typical RFID system is composed of a digital device called tag that is embedded in an antenna. An IC chip with a unique identification code (ID) is used with a reader, radio scanner device. This enables continuous and reliable monitoring of critical parameters such as the temperature, humidity, and gaseous components. Sensing volatile molecules in a noninvasive manner can serve in breath analysis to recognize marker gases for distinguishing between healthy and sick people. Air monitoring can also be used to avoid inhalation of antiseptics in surgical rooms to monitor hazardous gases in hospital environments and prevent epidemics. A passive RFID tag becomes capable to detect changes of the chemical/physical parameters of the environment when it is integrated with specific chemical compounds or the microchip networks [4].

7.2.3.2   Implantable RFID

This can be utilized to evaluate patient health-state from within the body by labeling prosthetics, orthopedic fixtures, sutures, and stents. This enables monitoring of biophysical processes like tissue regrowth and prosthetics displacement in real time turning them in multifunctional devices.

7.2.3.3   Human Motion Tracking Inside Patient Rooms

RFID devices can be installed in inpatient rooms. The scattering and shadow effects caused by the human body when there is motion in the proximity of the tags can cause the communication links established with fixed readers to be disturbed. These changes in the signal can be measured and carry information about human activity. This technology can be used in the early detection of cardiovascular pathologies and progressive slowing of walking capabilities.

7.2.3.4   Gesture Recognition

Gesture recognition RFID wearable tags can be used for motion detection. Applications include monitoring of compulsory limb movements, sleep disorders by tags equipped with inertial switches.

7.2.3.5   Remote Monitoring of Overnight Living Environment

This technology can be used in the detection of patient's jerky movements during sleep, motion patterns, prolonged periods of inactivity, sleep posture, and interactions with nearby objects. Temperature and humidity sensors can be used to monitor fever progression and the patient's urine output. These devices can be connected to automatic alarms that issue alerts to family or first aid centers and can be useful when the patient is remotely located. This information can further be used for diagnosis and follow up therapies and for studying behavioral and clinical aspects of the patient. This feature can be particularly useful for patients suffering from a disorder such as sleep apnea and restless legs syndrome.

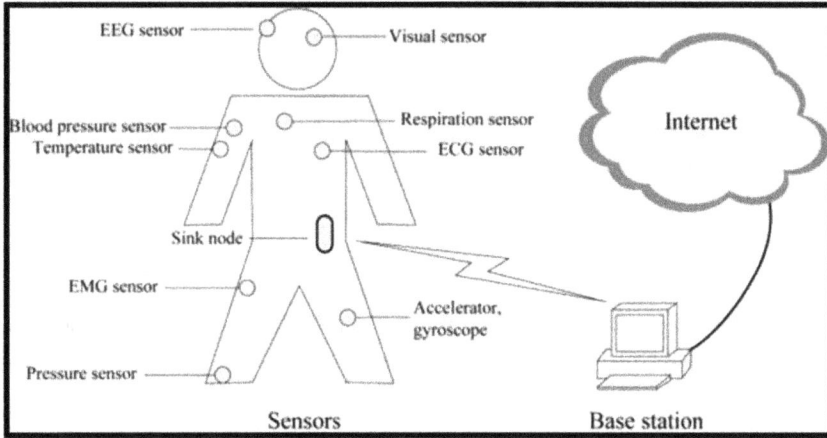

FIGURE 7.5 The architecture of a BSN [4].

7.2.3.6 Wireless Sensor Network (WSN)

Wireless sensor network (WSN) consists of a large number of sensor nodes distributed over a geographical region. These nodes are compact and battery-powered devices that can be used in any environment. It can be used to monitor factors like temperature, humidity, sound, pressure, object motion, gaseous particles, or characteristics of an object such as size and position. Each sensor node is made up of a power unit, transceiver unit, sensing unit, and processing unit. They can effectively be used in healthcare in smart nursing homes, in-home assistance, telemedicine, and wireless body area networks (BANs). Figure 7.5 shows the architecture of a body sensor network.

7.2.3.7 Health Monitoring

It involves monitoring of important health specifications such as oxygen saturation and blood pressure. Sensors and location tags can be used to track both healthcare

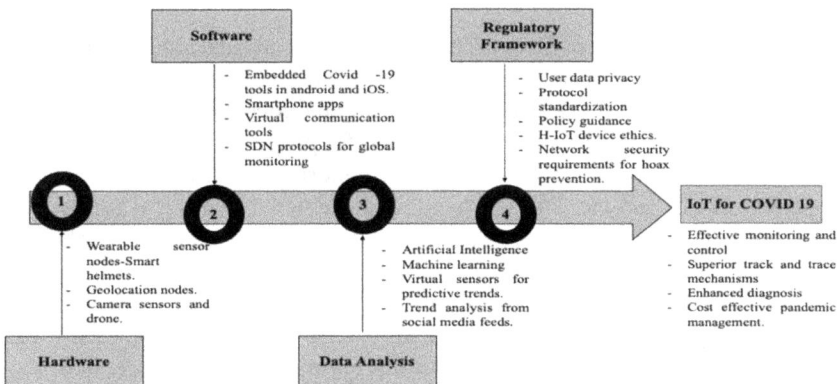

FIGURE 7.6 IoT for COVID-19 ecosystem.

personnel and patients. These devices can provide services like medical data access, asthma detection, and blood glucose monitoring. Figure 7.6 gives a complete overview of the IoT for COVID-19 ecosystem.

7.2.3.8 At-Home Healthcare

This technology can be used to provide affordable healthcare to the elderly while they live independently.

7.2.3.9 Telemedicine

Telemedicine is an approach where clinical work is performed using the internet and communication technology. This has applications like remote surgical procedures, patient-physician interaction, and medical evaluation reducing healthcare costs overall.

7.2.3.10 Body Sensor Networks

The body sensor network (BSN) technology is one of the core technologies of IoT developments in the healthcare system. This approach is where a patient can be monitored using a collection of low-powered and lightweight wireless sensor nodes to collect life-critical information and operate in hostile environments. BSNs are a kind of WSN formed by physiological parameter sensors placed in, on, or around the human body. The main techniques it covers are sensors, data fusion, and network communication. It is useful in disease monitoring and prevention. Its main purpose is to provide integrated ubiquitous computing hardware, software, and wireless communication technology platform for healthcare monitoring [4]. Commonly used sensors in BSNs include accelerometers used for human energy expenditure detection, artificial cochlea to generate auditory sensation, artificial retinas to stimulate visual sensation, blood pressure sensors to measure systolic and diastolic pressures, camera pill to detect gastrointestinal abnormalities by wireless endoscopy techniques, carbon dioxide sensors, ECG/EEG sensors, gyroscopes, humidity sensors, blood oxygen saturation sensors, respiration sensors, and pressure sensors [5].

7.2.4 IoT in Public Health Crisis Management: COVID-19 Pandemic Handling

The IoT as a network of sensors collecting data both locally and remotely has proved useful in the field of electronic-health (E-Health) management. A combination of BANs and field monitoring devices has allowed for the collection of patient vitals and the provision of track and trace services critical for pandemic management. Locally based E-Health mechanisms can collect health information such as blood pressure, temperature, heart rate, etc. This information can be stored locally and accessed by a healthcare professional. Local systems can also be used to alert the patient when they need to consult medical personnel and when they need to take medication. Remote-based e-health is essential for healthcare providers in enabling remote access of patients and patient data. Patient vitals and location can be transmitted at regular intervals to nearby or distant medical facilities for monitoring purposes. In times of a global pandemic such as the 2019 coronavirus (COVID-19), it is critical that social distance guidelines are adhered to and patients are effectively tracked and traced.

These two aspects help significantly in controlling the spread of the virus worldwide. The ability of IoT services in providing remote data collection and monitoring of patients in quarantine has made it a critical aspect in fighting the spread of virus pandemics. Health workers and authorities need data to manage a rapidly spreading respiratory pandemic. For COVID-19, data can be used to start the diagnosis of infection and also trace the direction of spread in the community. Primary essential data required includes body temperature, location, and travel history. These parameters can alarm officials on whether there is a need for further investigation and testing or not. Initially, health workers resorted to a manual method of measuring temperatures using infrared thermometers and verbal questioning of people on their history and locations. This posed a risk to health workers due to the increased contact with potentially infected subjects. It had also become an increasingly difficult approach as infection rates reached millions. Figure 7.7 gives a comprehensive overview of IoT linked technologies used during COVID-19.

A popular option for low-cost sensor hardware in fighting a pandemic is the use of already existing smartphone hardware. The smartphone accelerometer, microphone, camera, and temperature sensors are used in combination with machine learning (ML) algorithms to detect early COVID-19 symptoms. Geolocation sensors and drone technology are also being used in the detection and diagnosis of COVID-19. A thermal imaging IoT drone mechanism to detect suspected high body temperatures due to COVID-19 has also been developed. The smart helmet uses thermal imaging cameras and geolocation tags to detect and report suspected COVID-19 fevers. The first line of software solutions in tracking and tracing infections has been demonstrated by Android and iOS mobile phone operating systems. Both systems have integrated Application Programming Interfaces (APIs) to alert the user on a potential COVID-19 exposure

Technology	Description	Examples
Wearables	App enabled technology for receiving and processing data worn on body	
• Smart Thermometers	Temperature monitoring	Kinsa, Tempdrop, iFever
• Smart Helmet	Temperature monitoring, capturing location and face image, reducing human interaction	KC N901, China
• Smart Glasses	Temperature monitoring and capturing, reducing human interaction	Rokid, China Vuzix & Onsight
• IoT-Q-Band	Tracking quarantined cases	Electronic wristband, Ankle bracelet
• EasyBand	Monitoring social distancing, alerting to closeness by LED	Pact wristband
• Proximity Trace	Monitoring workers for social distancing, tracing contacts of contaminated employees	Hardhat TraceTag Instant Trace
Drones	Aircraft equipped with sensors, cameras, GPS and communications	
• Thermal imaging drone	Temperature capturing in crowd	Pandemic Drone
• Disinfectant drone	Sterilizing contaminated areas	DJI
• Medical delivery drone	Increase treatment accessibility	Delivery Drone Canada
• Surveillance drone	Crowd social distancing monitoring	MicroMultiCopter
• Announcement drone	Broadcast COVID information	Spain and Kuwait
Robots	Programmable machine that can handle complex actions	
• Autonomous robot	Detecting symptoms, controlling social distancing, disinfection and sterilization, collect swab tests, checking patient's respiratory signs	Intelligent Care Robot Spot Robot
• Telerobots	Reduce risk of infection for medical staff	DaVinci surgical robots
• Collaborative robots	Lowering health care workers' fatigue, disinfecting hard to reach areas	Asimov Robotics eXtremeDisinfection Robot
Smartphone Applications	Mobile device application software for monitoring and tracking	
• nCapp, China	Updates database, provides consulting, controls patient's long-term health	
• DetectaChem, USA	Taking low cost COVID 19 test using kit connected to smartphone application	
• StayHomeSafe, Hong Kong	Monitoring airport arrivals using wristband	
• Aarogya Setu, India	Case monitoring and linking with health services	
• TraceTogether, Singapore	Capture people close to user using encryption IDs. Government access of user information, notifying people in close contact	

FIGURE 7.7   IoT-enabled/linked technologies used during COVID-19.

Data parameter	Extraction Technology	Monitoring System
Temperature	Thermometers, thermal imaging, robots, drone, smart helmets,	AI based image processing. Big data, visual dashboards
Location	Smartphone location tracing, wearable trackers, smart city and ITS	Smartphone company AI algorithms, machine learning algorithms embedded in smartphone apps, ITS big data, visual dashboards
Imaging	Public CCTV, drone video surveillance, ITS based cameras	AI based image processing and identification
Social Media	Twitter, Instagram, Facebook	AI character recognition flags key words in posts, machine-based data filtering applications

FIGURE 7.8   Data collection and monitoring systems for IoT-based COVID-19 management.

risk. Using embedded smartphone proximity sensors, the API provides an opportunity to alert a user if someone near them recently tested positive. Governments in India and Singapore have promoted the development of smartphone applications to ensure effective tracking and tracing of people that have tested positive. As part of a campaign for global monitoring of the virus spread, the required increased computing power by IoT devices is being provided by SDN and cloud services. Furthermore, to limit interaction between patients, healthcare givers, and IoT managers, virtual communication tools are being developed leading to the implementation of virtual clinics. The use of ML techniques and artificial intelligence in general is becoming a popular route in the analysis and decision-making process of sensed data. ML algorithms have been used for training thermal images depicting both negative and positive COVID-19 tests. Deep learning a component of artificial intelligence can effectively be used to diagnose the COVID-19 infection based on CT scans and X-ray images with minimal error. Another important data analysis feature being used in tackling the spread of the coronavirus is trend analysis in combination with ML. CovidSens uses social media feeds and ML to track and trace virus propagation. Based on user experiences posted on social media, the application can filter useful new information for the benefit of the government and the general public [6]. Figure 7.8 gives a comprehensive overview about the various data collection and monitoring systems for IoT based COVID-19 management.

7.3   BENEFITS OF IoT IN HEALTHCARE

7.3.1 Simultaneous Reporting and Monitoring

With real-time monitoring of the condition in place by means of a smart medical device connected to a smartphone app, connected devices can collect medical and other required health data and use the data connection of the smartphone to transfer collected information to a physician. The IoT device collects and transfers health data such as blood pressure, oxygen, and blood sugar levels; weight; and ECGs. These data are stored in the cloud and can be shared with an authorized person, who could be a physician, your insurance company, a participating health firm, or an external consultant, to allow them to look at the collected data regardless of their place, time, or device.

7.3.2 End-to-End Connectivity and Affordability

IoT can automate patient care workflow with the help of healthcare mobility solutions and other new technologies, and next-gen healthcare facilities. IoT in healthcare enables interoperability, machine-to-machine communication, information exchange, and data movement that make healthcare service delivery effective. With connectivity protocols, such as Bluetooth LE, Wi-Fi, Z-wave, Zigbee, and other modern protocols, healthcare personnel can change the way they spot illness and ailments in patients and can also innovate revolutionary ways of treatment. Consequently, technology-driven setup brings down the cost, by cutting down unnecessary visits, utilizing better quality resources, and improving the allocation and planning.

7.3.3 DATA ASSORTMENT AND ANALYSIS

The vast amount of data that a healthcare device sends in a very short time owing to their real-time application is hard to store and manage if the access to the cloud is unavailable. Even for healthcare providers, acquiring data originating from multiple devices and sources and analyzing it manually is a tough task. IoT devices can collect, report, and analyze the data in real time and cut the need to store the raw data. This all can happen to overcloud with the providers only getting access to final reports with graphs. Moreover, healthcare operations allow organizations to get vital healthcare analytics and data-driven insights that speed up decision-making and are less prone to errors.

7.3.4 TRACKING AND ALERTS

An on-time alert is critical in case of life-threatening circumstances. Medical IoT devices gather vital data and transfer that data to doctors for real-time tracking while dropping notifications to people about critical parts via mobile apps and other linked devices. Reports and alerts give a firm opinion about a patient's condition, irrespective of place and time. It also helps make well-versed decisions and provide on-time treatment. Thus, IoT enables real-time alerting, tracking, and monitoring, which permits hands-on treatments, better accuracy, apt intervention by doctors, and improves complete patient care delivery results.

7.3.5 REMOTE MEDICAL ASSISTANCE

In case of an emergency, patients can contact a doctor who is many kilometers away with smart mobile apps. With mobility solutions in healthcare, the medics can instantly check the patients and identify the ailments on-the-go. Also, numerous healthcare delivery chains that are forecasting to build machines that can distribute drugs based on a patient's prescription and ailment-related data available via linked devices. IoT will improve the patient's care in the hospital. This, in turn, will cut on people's expenses on healthcare.

CASE STUDY: USAGE OF A SUSTAINABLE ELECTRONIC WEARABLE BAND TO CHECK THE STATUS OF QUARANTINED COVID-19 PATIENTS

A group of scientists in India have developed a wearable band, christened as IoT Q Band, to check the status of COVID patients who are under quarantine. The band works on the principles of IoT. An inbuilt global positioning system in the bands gives live updates on the location of the quarantined patients to the people monitoring them and generates an alarm if the quarantine zone is stepped out of. It can be viewed by the concerned people monitoring them in the form of an app that has an user friendly interface.

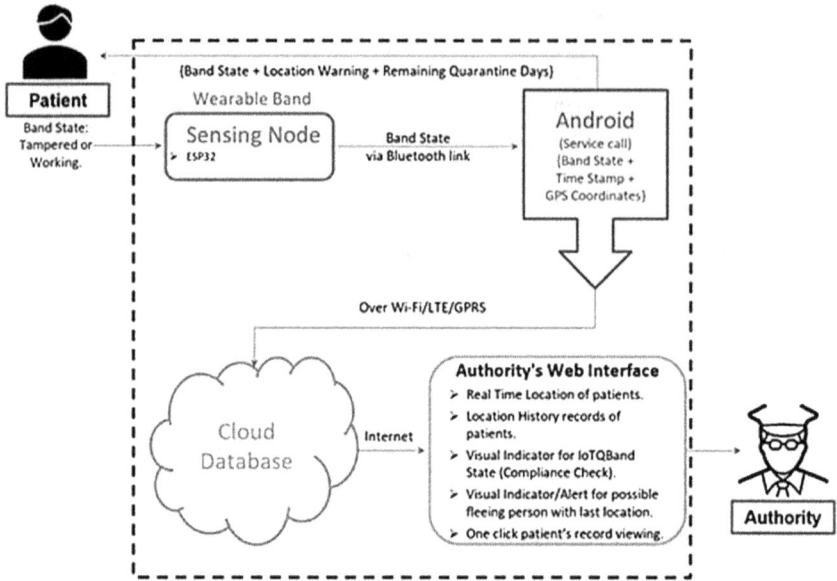

FIGURE 7.9 Architecture of the IoT Q band system.

Figure 7.9 shows the architecture of the IoT Q band system. The band has been designed in such a way that it can be worn on the arms or legs according to patient's convenience. It is connected to the monitoring server via Bluetooth. The data of the patient's location is transmitted to the server after pauses of two minutes duration each. For example, if the quarantine is violated, an alert is immediately sent to the overseer, who can track the status and catch hold of them immediately. This can be seen in the app interface as shown in the figures below.

The patients who have interfered with the working of the band or who have stepped out of the quarantine zone have been alerted to the people monitoring them via the red color in the app interface. The patients who are following the rules and are within the safe zone are represented by the green color in the app interface. Figure 7.10 shows the working of the IoT Q band.

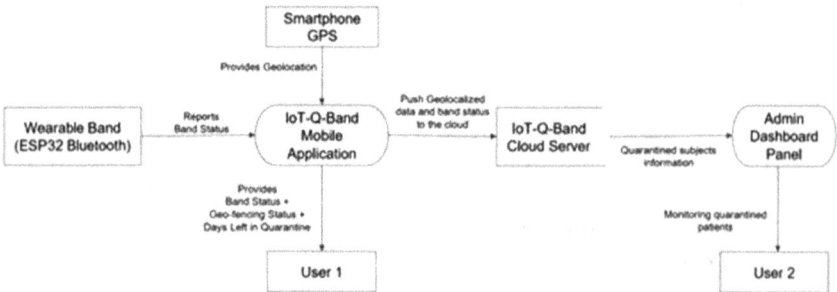

FIGURE 7.10 Working of the IoT Q band. User 1 is the patient and user 2 is the person overseeing the patient.

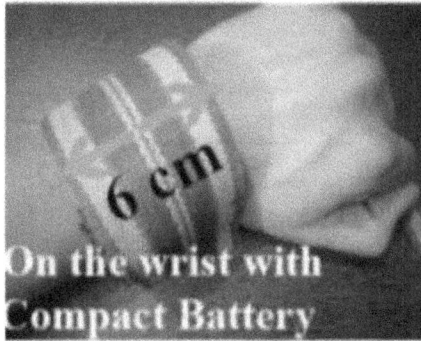

(a) Inner Side of the Band

(b) Outer Side of the Band

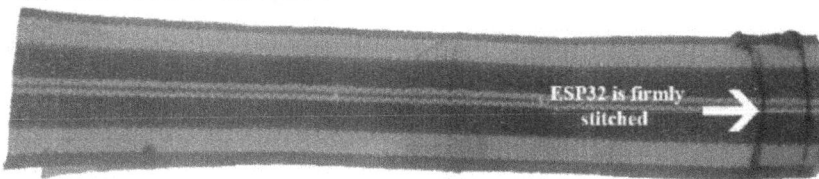

FIGURE 7.11   Appearance of the IoT band; (a) inner side of the band; (b) outer side of the band.

Figure 7.11 shows the actual appearance of the IoT Q band. The data can also be sorted by the defaulting patient's parameter. The status of the patient is shown in verbal format here, as well as the numerical distance of the patient from the quarantine zone. Thus, this provides a clear picture of the location of the quarantined patients and allows the concerned authorities to track and monitor them accordingly.

The simple design of the IoT Q band makes it a sustainable solution in order to solve the problem of COVID patients violating quarantine norms. Due to its low production cost, sustainability, and high effectiveness, this band can be used in low income and developing countries to gain control over the pandemic.

7.4   CHALLENGES OF IoT IN HEALTHCARE

7.4.1   DATA SECURITY AND PRIVACY

One of the most significant threats to IoT is data security and privacy. IoT devices store and transmit data in real time. Most of the devices used lack protocols for data stored and their standards. Data regulation is ambiguous. These factors make

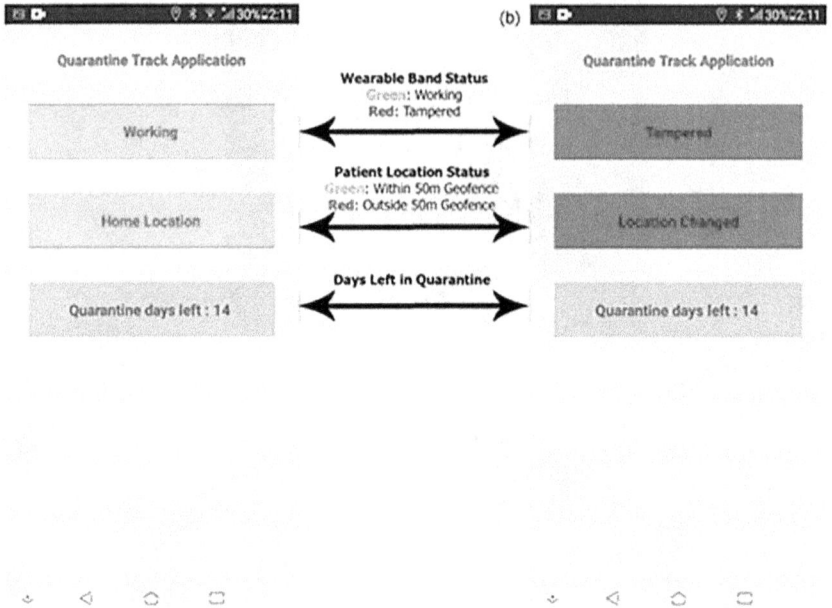

FIGURE 7.12 App interface; (a) app status when the quarantine is being followed; (b) app status when the quarantine is violated.

the information highly susceptible to cybercriminals who can hack into the system and take control of personal health information (PHI) of both patients as well as doctors. Cybercriminals can misuse patient's data to create fake identities to buy drugs that they can in turn sell later. Hackers can also file a fake insurance claim in the patient's name. Figure 7.12 shows the IoT Q band app interface and Figure 7.13 shows how the final patient data is displayed in the app.

All Patients Data

Name	Address	Days Left	Mobile	Distance from Home (in Kms)	BandStatus
Sample-name	Address	14	923xxxx123	14.438	Working
Test114	TestAdd114	14	981xxxx114	0.001	Working
Test116	addTest116	11	989xxxx116	0.019	Tampered
Test119	addTest119	13	981xxxx119	0.016	Working

Defaulted Patients Data

Name	Address	Days Left	Mobile	Distance from Home (in Kms)	BandStatus
Sample-name	Address	14	923xxxx123	14.438	Working
Test116	addTest116	11	989xxxx116	0.019	Tampered

FIGURE 7.13 Patient data displayed.

7.4.2 INTEGRATION OF MULTIPLE DEVICES AND PROTOCOLS

The integration of multiple devices is one of the major challenges faced in the implementation of IoT in the healthcare platform. The reason is that the devices do not have a standard protocol regarding their working. Each device works in its manner, and hence, the integration of all these together causes ambiguity in data aggregation and communication between these devices.

7.4.3 DATA OVERLOAD AND ACCURACY

IoT devices collect a huge amount of data. The information gathered is extremely vast and hence causes a threat to the decision-making ability of doctors. Moreover, the integration of multiple devices is an addition to this problem.

7.4.4 COST

IoT is not an affordable option for the common man yet. It is increasing its cost day by day as the devices are becoming sophisticated simultaneously. Due to this reason, it is not being implemented in most of the developing countries.

7.5 DISCUSSION

In this chapter, we have attempted to delineate the different advantages and methodologies that are used in the IoT in healthcare. We feel that these methodologies have several abilities to transform the healthcare system in a real-world scenario. IoT for ICU patients can prove particularly useful during a pandemic, wherein large-scale monitoring of several patients is required. Monitoring of public health parameters in the event of such a crisis is crucial, such as the number of infections, recovery tracing, and the number of deaths. These can be achieved using the described methodologies, thereby reducing the burden on healthcare workers. In the case of patients who have long-term issues such as diabetes and movement disorder and disabilities, when they do not have access to constant care, remote monitoring can be done using apps and wearable devices that provide recurring reports to doctors about their present health status. In countries where the number of healthcare centers per unit of population is low, IoT in healthcare can be used to implement telemedicine, remote diagnosis, and treatment. IoT gateway can be used to store patient records obtained from clinical trials in a cloud and this data can be analyzed using various genetic algorithms. Mobile applications that use IoT can be used for tracing disease spread, data about the nature of symptoms, blood sugar levels, weight, blood pressure, the general health status of the population, and lifestyle habits. A notable example in this context would be the Aarogya Setu app developed by the Ministry of Electronics and Information Technology to aid the government in tracking the status of COVID-19 infections in India. CDSS in tandem with EHR allows healthcare providers to store patient history. It gives suggestions about the further course of treatment based on the available records. RFID technology can be used for pharmaceutical tracking to optimize drug inventory, manufacturing, distribution, and addiction control.

IoT as a technology in healthcare management is integral in ensuring high-quality healthcare access to all sections of the society, particularly the disadvantaged and patients with disabilities. It can also aid in enabling healthcare access to a geographically dispersed or isolated population. IoT can bring down the cost of healthcare making it more affordable and bridge the interaction gap between doctors and patients. IoT can also help balance the doctor to patient ratio, allowing physicians to cater to more patients. IoT is also essential in public health crisis management by ensuring simultaneous data reporting and monitoring. It is critical in effectively sorting and analyzing real-time data, tracking, and alerts. IoT is also helpful in research purposes like clinical studies for drug and therapeutics development where massive amounts of data on public health are monitored and analyzed. IoT is a technological advancement that has the potential to revolutionize healthcare and is integral to the future of healthcare management.

7.6   CONCLUSION

The implementation of the IoT is clearly transforming the face of healthcare for many decades to come. IoT is enabling healthcare to provide increasingly accessible medical facilities to many patients. The main concerns plaguing the complete integration of IoT in healthcare are costs and privacy issues. Once these issues have been solved, a finalized system that fully integrates digital and electronic devices along with the healthcare system can be developed and implemented in the real world.

REFERENCES

1. Gope, Prosanta, and Tzonelih Hwang. "BSN-Care: A secure IoT-based modern healthcare system using body sensor network." *IEEE Sensors Journal* 16, no. 5 (2015): 1368–1376.
2. Boyanapalli, Arathi, and Rohini Patil. "Assistive technology using IoT for physically disabled people." *International Journal of Innovative Technology and Exploring Engineering* 13 (2019): 13–17. ISSN: 2278-3075.
3. Amendola, Sara, Rossella Lodato, Sabina Manzari, Cecilia Occhiuzzi, and Gaetano Marrocco. "RFID technology for IoT-based personal healthcare in smart spaces." *IEEE Internet of Things Journal* 1, no. 2 (2014): 144–152.
4. Lai, Xiaochen, Quanli Liu, Xin Wei, Wei Wang, Guoqiao Zhou, and Guangyi Han. "A survey of body sensor networks." *Sensors* 13, no. 5 (2013): 5406–5447.
5. Gupta, Rajan, Manan Bedi, Prashi Goyal, Srishti Wadhera, and Vaishnavi Verma. "Analysis of COVID-19 tracking tool in India: Case study of Aarogya Setu mobile application." *Digital Government: Research and Practice* 1, no. 4 (2020): 1–8.
6. Schaefer, Guenter. "Identification pill with integrated microchip: smartpill, smartpill with integrated microchip and microprocessor for medical analyses and a smartpill, smartbox, smartplague, smartbadge or smartplate for luggage control on commercial airliners." U.S. Patent 5,792,048, issued August 11, 1998.

8 Machine Learning and IoT Augmented Classification and Prediction of Obstructive Sleep Apnea (OSA)

Nishita Anand and Reeth Nalamitha
Department of Medical Electronics,
Dayananda Sagar College of Engineering,
Bangalore, India

Vidya M. J.
Giritra Solutions, Nagarabhavi II Stage,
Bangalore, India

Padmaja K. V. and Rajasree P. M.
Department of Electronics & Instrumentation Engineering,
R V College of Engineering,
Bangalore, India

CONTENTS

DOI: 10.1201/9781003268796-11

8.1   INTRODUCTION

Sleep apnea is an ailment that occurs during night sleep as a result of suspensions in breathing, causing interruptions of sleep. The main types of sleep apnea are obstructive sleep apnea (OSA), central sleep apnea, and mixed sleep apnea. OSA is the most frequently occurring of all and results from complete or incomplete closure of the upper airway gap that is the pathway for oxygen supply to the heart. This is because of the relaxation caused by muscles around the airway gap during sleeping. The occurrence of CSA is due to the absence of respiratory effort initiated by the brain. MSA happens because of combined episodes of OSA and CSA. In all these cases, intermittent cessation of breathing causes an insufficient supply of oxygen to the heart, which leads to problems in its functioning. Initially, it causes the development of low heart rate, i.e., bradycardia condition and reduced oxygen level in the blood. The brain sends a signal to the arteries to pump blood at a higher rate, which in turn raises the heart rate, i.e., tachycardia condition. The episodes of bradycardia and tachycardia that result due to apnea effect are harmful. In the aggravated conditions, there can be continuous episodes of tachycardia even during the daytime causing unfavorable cardiac-related ailments. Polysomnography (PSG) is used to study sleep disorders. The PSG study parameters are breath airflow, respiratory movement, blood desaturation oxygen, ECG, EEG, etc. The ECG signal gets changed in amplitude and frequency due to breathing complications caused by sleep apnea condition, resulting in insufficient oxygen supply to the heart. Hence, ECG signal can be used to study sleep apnea that creates fluctuations in the ECG signal parameters by diverging from their absolute readings. Therefore, a validated method for early detection, classification, and estimation of sleep apnea, on the basis of ECG signal analysis, could prove to be useful as a noninvasive sleep apnea screening method. The ECG signal in each of its beat cycle consists of wave components, particularly P wave, which is a result of atrial depolarization, QRS complex caused by ventricular polarization, S wave and T wave caused by ventricular polarization and ventricular repolarization, respectively, as shown in Figure 8.1. The ECG parameters are depicted by wave amplitudes, time intervals, specifically PQ, QRS, and TU intervals, along with PR segment, ST segment, and TP segments as indicated in Figure 8.1 [2] and Table 8.1 [3].

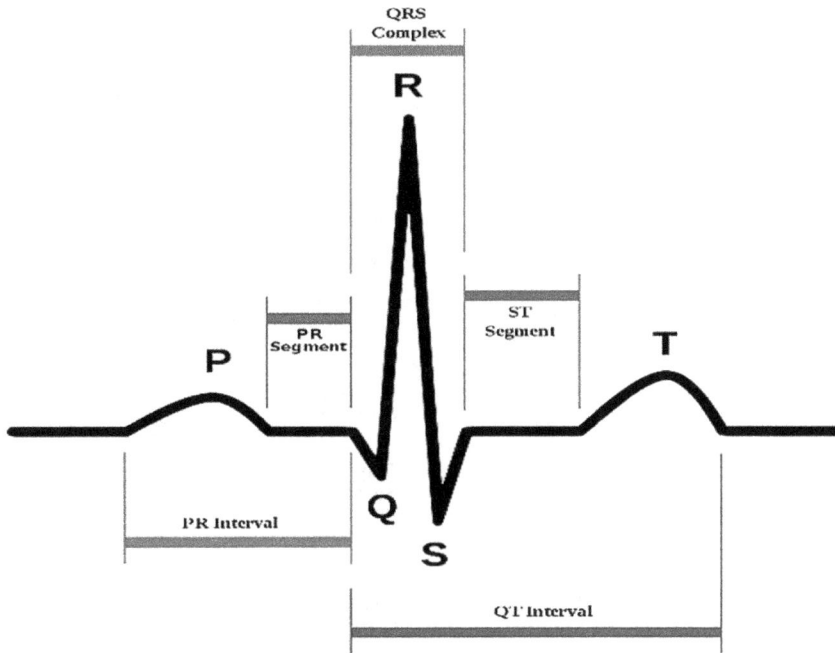

FIGURE 8.1   Ideal or normal ECG waveform.

8.1.1 MOTIVATION

It has been reported in the United States that 4% of men and 2% of women satisfy the standards for OSA [4]. Lately, the occurrence has been recorded to be as high as 14% in men and 5% in women. Occurrence also increases with age. At least 90% anterior to a posterior collapse of the airway for more than 10 seconds results in the

TABLE 8.1
ECG Parameters

ECG Parameter	Typical Amplitude (mV) and Wave Duration (Seconds)
P wave	0.25 mV
R wave	1.60 mV
Q wave	25% of R wave
T wave	0.1–0.5 mV
PR interval	0.12–0.20 seconds
QT interval	0.35–0.44 seconds
ST segment	0.05–0.15 seconds
QRS interval	0.09–0.10 seconds
Heart rate	60–100 bpm

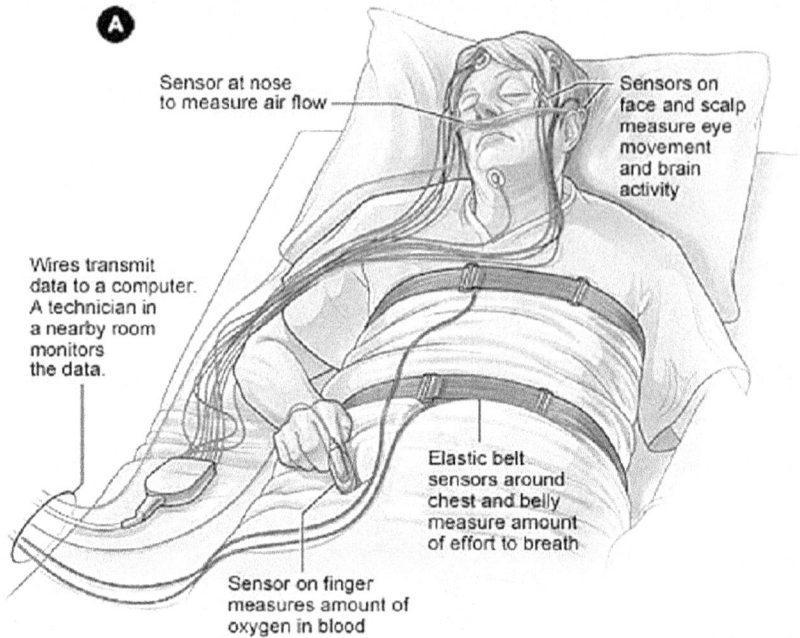

FIGURE 8.2 PSG set up in the lab [7].

occurrence of an event of OSA. Prevalence of hypopnea makes the subject 4× more likely to have a stroke and 3× more likely to experience heart disease [6].

The conventional method to detect sleep apnea is through PSG, which, as shown in Figure 8.2, is an advanced lab set up managed under expert personnel where the subject is observed through the night, with various sensors strapped on to assess different parameters. As many as 80% of the cases go undiagnosed [8], partly due to the high cost (Rs. 15,000–20,000 [9]) of monitoring and disturbance as well as low availability and invasiveness. Therefore, there is a need to develop home monitoring systems for easy analysis and prediction.

8.1.2 OBJECTIVES

The objectives of the chapter are:

1. Preprocessing using digital filters;
2. Peak or QRS complex detection using Pan-Tompkins algorithm;
3. Feature extraction from detected QRS complex;
4. Classification using support vector machine (SVM).

8.1.3 LITERATURE REVIEW

Pan Jiapu and Tompkins [10] proposed an algorithm for QRS complex detection of ECG signals by examining the slope, amplitude, and breadth of the QRS complex followed by the adaptive dual threshold technique, wherein the detection accuracy of

99.3% with improved sensitivity is attained by testing the algorithm using the MIT/BIH ECG arrhythmia database records.

Mihaela Lascu et al. [11] proposed a filtering artifact noise and interference from ECG signal effectively using time domain filtering and signal averaging process implemented in LabVIEW. Its performance reports are better than linear filters.

Mohamed Elgendi et al. [12] proposed a QRS detector based on adaptive quantized threshold and the second dynamic threshold while working with ECG signals of lower SNR, lower QRS amplitude, and nonstationary characteristics. The testing was done on the CSE and MIT-BIH database, achieving sensitivity and a positive prediction rate of 97.55% and 99.55%, respectively.

Mohammad Pooyan et al. [13] combined two algorithms: the Pan-Tompkins algorithm and the state logic machine, to find QRS complexes of ECG signals belonging to the category of normal sinus and arrhythmia gathered from the MIT-SCD database. An accuracy value of 95% has been attained using morphological features of the QRS complex. De Chazal et al. [14] used the wavelet transform-based QRS detection process along with the Pan-Tompkins algorithm and derivative-based method and attained accuracy and specificity of 93.4% and 90%, respectively.

Laila Almazaydeh et al. [15] discusses the use of an automatic classification method based on support vector machine (SVM) for sleep apnea detection by the analysis of a set of fundamental and derived features from 15-second epochs of ECG signal records from MIT's ECG Apnea database, attaining accuracy of 96.5%.

8.1.4   BRIEF METHODOLOGY FOR THE IMPLEMENTATION OF THE ACQUISITION, TRAINING, AND TESTING OF OSA

The methodology adopted in the proposed work is as follows:

1. Acquisition of raw ECG signal from "Apnea-ECG database from MIT's PhysioNet".
2. Load and plot the ECG signal into MATLAB using inbuilt MATLAB commands for signal analysis.
3. To remove 60-Hz powerline interference, use a notch filter with inbuilt command "iirnotch ()" and "filter ()".
4. Next, to further remove "baseline wander" and "detect QRS peak" for subsequent feature extraction, the peaks are marked using the "Pan-Tompkins algorithm".
5. Once the R peaks are detected, we calculate the distinguishing ECG features most effective for apnea detection.
6. The dataset is then split into training and test datasets, and an SVM is applied to classify if the dataset is apneic or not.

8.2   FUNDAMENTALS

8.2.1   DATA ACQUISITION

8.2.1.1   ECG Data

Sleep apnea can be detected with the help of ECG, which is considered one of the most efficient features. Sleep apnea episodes have been linked to changes in the periodic RR interval of the ECG. This comprises bradycardia during an apneic event, followed by tachycardia on its suspension, to compensate for the lowered heart rate [6]. RR interval defines the time interval between two consecutive R peaks. The RR interval [18] time series for each ECG beat can be formulated as shown below [17]:

$$rr(i) = r(i+1) - r(i), \quad i = 1, 2, \ldots, n-1 \tag{8.1}$$

8.2.1.2 PhysioNet Data

There are 35 recordings in the PhysioNet Apnea-ECG database that contain one ECG signal each. The sampling rate is 100 Hz. The standard sleep laboratory ECG electrode positions were used (modified lead V2).

8.2.2 SIGNAL PREPROCESSING

8.2.2.1 Pan-Tompkins Algorithm

The Pan-Tompkins algorithm implements the following four stages

Stage I: The band-pass filter (BPF) removes different types of noise affecting the ECG signal quality, measured with signal-to-noise ratio (SNR). Powerline interference is the noise at 50/60 Hz due to power sources in the vicinity of the ECG signal acquisition system and can be filtered using a notch filter. Low-pass filter (LPF) that filters the baseline wander noise is of low frequency caused by the patient's respiration activity or body movement. Motion artifacts result from improper contact between the electrode and the skin resulting in significant artifacts and can be filtered out using a high-pass filter (HPF). First, the ECG signal is transformed to the frequency domain using fast Fourier transform (FFT) and then applied with a digital BPF, 5–15 Hz. An LPF followed by HPF implements a BPF.

The transfer function (TF) of LPF is as given in Eq. (8.2) [10].

$$H_{lp}(z) = \frac{(1 - z^6)^2}{(1 - z^{-1})^2} \tag{8.2}$$

The difference equation of LPF is given in Eq. (8.3) [10]

$$y(nT) = 2y(nT - T) - y(nT - 2T) + 1/32(x(nT) - 2x(nT - 6T) + 2x(nT - 12T) \tag{8.3}$$

where y is the output of discrete filter, x is the discrete input signal, T is the sample period, and n is an integer. Here, the cut-off frequency is about 11 Hz with filter processing delay of six samples.

HPF is resulted as an all pass filter minus a LPF and its TF is represented in Eq. (8.2) [10].

$$H_{hp}(z) = \frac{\pm 1 + 32\, z^{-16} + z^{-32}}{(1 - z^{-1})} \tag{8.4}$$

and difference equation is as shown in Eq. (8.5) [10].

$$y(nT) = 32x(nT-16T)-\left[y(nT-T)+x(nT)-x(nT-32T)\right] \quad (8.5)$$

HPF uses the cut-off frequency fc = 5 Hz and processing delay of 16 samples, i.e., 80 milliseconds. Thus, the BPF is used to filter out the noise and retain only the signal information keeping QRS complex, P wave peak, and T wave peak. The goal here is to improve SNR and the overall detection sensitivity. A sample raw ECG signal, X05, and the corresponding preprocessed signal after BPF are obtained, that is passed on to the Stage II for suppressing of P and T waves of ECG.

   Stage II: Derivative operation is used to detect the QRS complex with largest slope suppressing P and T waves.

Derivative TF and the difference equation are shown in Eqs. (8.6) and (8.7), respectively [10].

$$H(z) = \left(\frac{1}{8}T\right)\left(-z^{-2}-2z^{-1}+2z^{1}+z^{2}\right) \quad (8.6)$$

The derivative output of the preprocessed signal is shown in Figure 8.5.

$$y(nT) = \left(\frac{1}{8}T\right)\left[-x(nT-2T)-2x(nT-T)+2x(nT+T)+x(nT+2T)\right] \quad (8.7)$$

   Stage III: The slope information got from differentiator in Stage II is enhanced by squaring process as shown in Eq. (8.8) [7]. It provides selective amplification of large differences arising from QRS peaks and attenuation of small differences resulting from P and T waves.

$$y(nT) = \left[x(nT)^{2}\right] \quad (8.8)$$

Output of the squaring operation stage III is shown in Figure 8.8(iv).

   Stage IV: Squared output of Stage III is passed through moving window integrator (MWI) that produces a large amplitude pulse for every QRS, lower amplitude pulses for noise spikes. The smoothened output from the moving window integration filter is as shown in Eq. (8.9) [10].

$$y(nT) = \left(\frac{1}{N}\right)[x(nT-(N-1)T+x(nT-(N-2)T+\cdots+x(nT))] \quad (8.9)$$

where N, the window width, is taken as 30 which is adequate for sampling frequency fs = 100 Hz. The output of MWI stage IV indicates the position and amplitude of all the detected peaks.

8.2.2.2 Thresholding

The threshold that is applied during the QRS complex detection is given by:

$$Th = 0.45 \text{ of max Amplitude} \tag{8.10}$$

8.2.3 Feature Extraction

Feature extraction is the process of gathering ECG parameters keeping the detected QRS complex in each beat cycle as a reference as discussed. Once the QRS complex or R peak is located, RR interval, the time interval between consecutive QRS complexes, heart rate, number of QRS complexes detected per minute, PR interval, time difference of onset of the P wave and the onset of QRS complex, etc. can be computed in the time domain, and their statistical distribution of parameters are obtained for the minute-wise duration.

On acquiring the RR interval of individual signals, the subsequent ECG features that are most useful for apnea detection can be calculated:

1. Mean;
2. Standard deviation;
3. NN50 measure: count of neighboring RR intervals in which the first one surpasses the second by a value more than 50 ms;
4. pNN50 measures: the proportion of NN50 over the total count of RR intervals;
5. The SDSD measures: standard deviation of the differences between neighboring RR intervals;
6. The RMSSD measures: the square root of the mean squares of differences between adjacent RR intervals;
7. Median;
8. Interquartile range: the difference between 75th and 25th percentiles of the RR interval value distribution;
9. Mean absolute deviation values: achieved by subtracting the mean RR interval values from all the RR interval values;
10. Beats per minute.

8.2.4 Classification using SVM

We use SVM as a classification (also known as supervised learning) method in order to investigate apneaic epoch detection. In our implementation, we use a linear kernel function to map the training data into kernel space. In the optimization process, we use a method called sequential minimal optimization to find the separating hyperplane. For data randomization, we separate the apnea and non-apnea data. We then separate training data and testing data, with 80% for the training and 20% for the testing. After the signals are separated, we perform the training for SVM.

8.3 DESIGN AND IMPLEMENTATION OF PAN-TOMPKINS ALGORITHM AND FEATURE EXTRACTION

8.3.1 BLOCK DIAGRAM OF PROPOSED SYSTEM

The data acquired from the Apnea-ECG database by MIT PhysioNet undergoes preprocessing for noise removal as well as QRS detection. The Pan-Tompkins algorithm was used to detect the QRS peaks. It consists of four main stages: the BPF, derivative filter, squarer, and integrator. Once the peaks are detected, feature extraction is performed to extract the distinguishing feature for further analysis. Subsequently the dataset is split into two: training and test sets and classification is done using SVM in MATLAB Classification Learner. The block diagram for the detection and prediction of OSA is depicted in the Figure 8.3.

8.3.2 IMPLEMENTATION OF NOTCH FILTER

A notch filter attenuates signal over a narrow range of frequencies while leaving the signal and other frequencies unaltered. We use a notch filter to remove a powerline interference of 60 Hz and is depicted in the Figure 8.4.

8.3.3 IMPLEMENTATION OF PAN-TOMPKINS ALGORITHM

The implementation of Pan-Tompkins Algorithm involves the adoption of bandpass filter, a combination of low pass filter with cutoff frequency of 11 Hz and a high pass

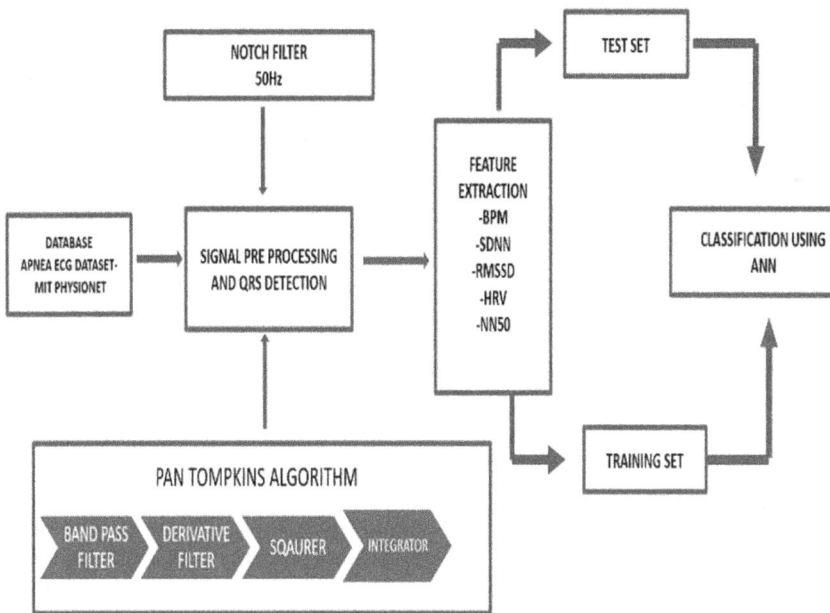

FIGURE 8.3 Block diagram for the detection and prediction of OSA.

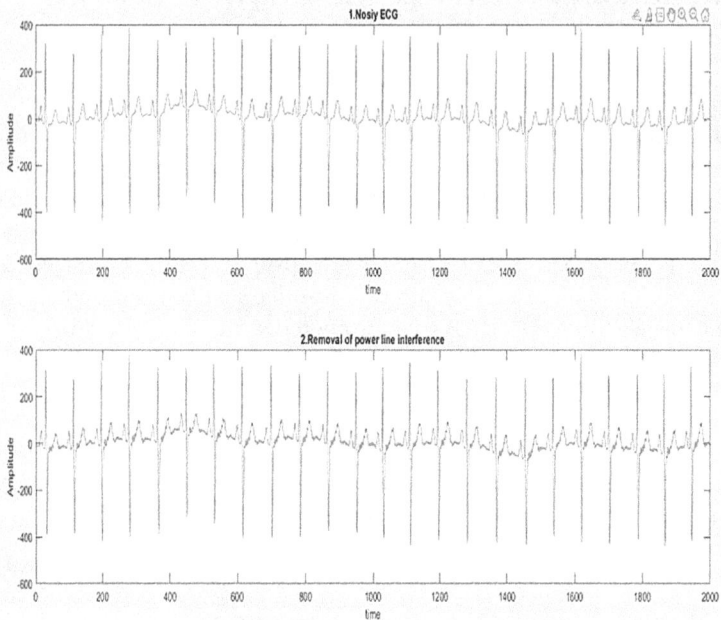

FIGURE 8.4 Plot of noisy ECG signal, implementation of notch filter.

filter with a cutoff frequency of 5 Hz. The obtained results of the band pass filter are shown in the Figure 8.5 and Figure 8.6. Then, subsequently the QRS complex with largest slope is detected using the derivative filter and its output is depicted in Figure 8.7. To amplify the large differences arising from the QRS peaks with attenuation of the lower amplitudes of P and T, a squarer is deployed and its results are shown in Figure 8.8. Later, the large amplitude of every QRS complex for noise spikes are removed using the Integrator and its results are shown in Figure 8.9. The results are discussed in detail in section 8.4.

8.4 RESULTS AND DISCUSSIONS

Preprocessing of the ECG signal along with feature extraction and classification is carried out on Matlab 2017a.

Database: The Apnea-ECG database of MIT's PhysioNet.org consisting of the ECG signal recordings are gathered [17]. Around 70 nighttime ECG recordings are sampled at 100 Hz.

The QRS complex is the most characteristic band in the ECG, with a high slope and apparent wave crest, according to the principle of the threshold method for QRS detection. After that, using a threshold to detect QRS complex becomes possible after concentrating this information. However, the first issue in the actual process is that the ECG data collected from the acquisition device is loud, which makes detection difficult. Frequency interference, baseline drift, and EMG noise are the three

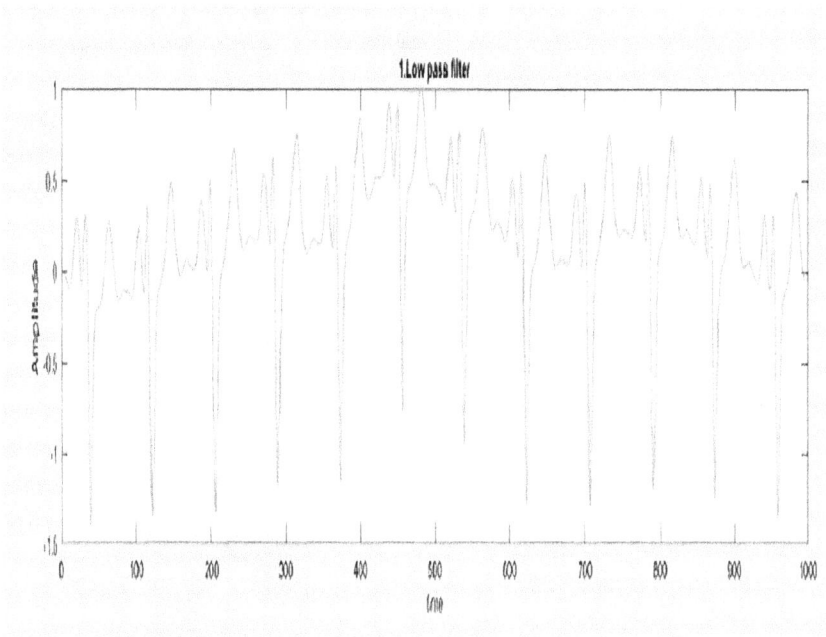

FIGURE 8.5 Low-pass filter with fc = 11 Hz.

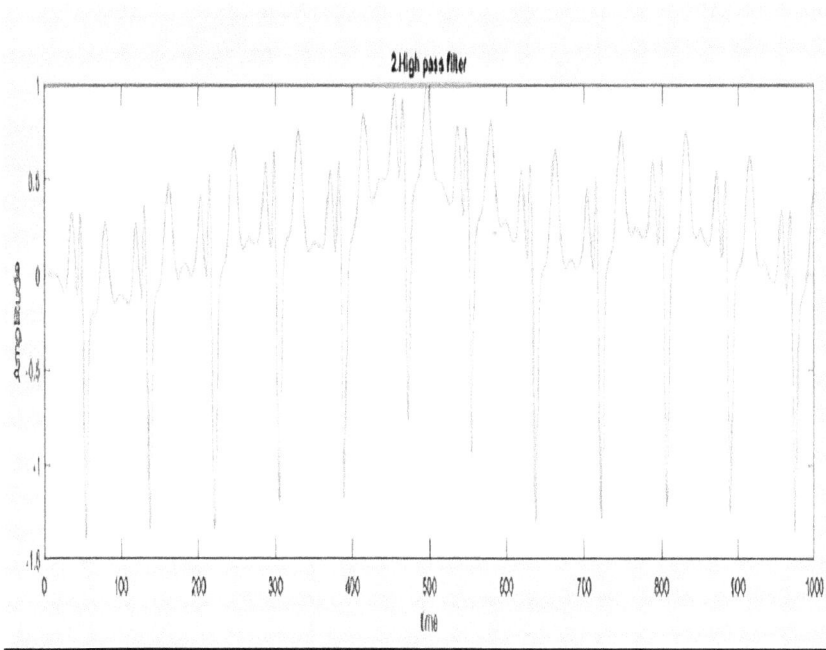

FIGURE 8.6 High-pass filter with fc = 5 Hz.

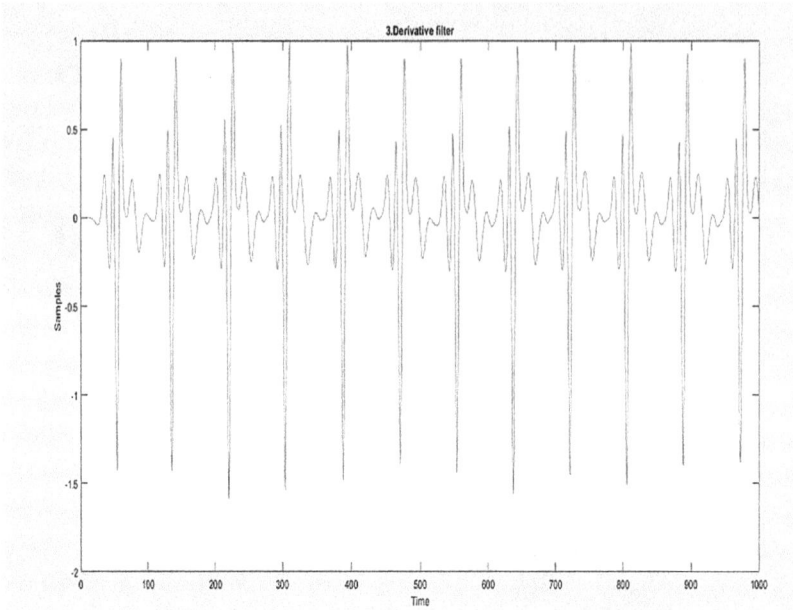

FIGURE 8.7   Derivative filter to detect the QRS complex with largest slope suppressing P and T waves.

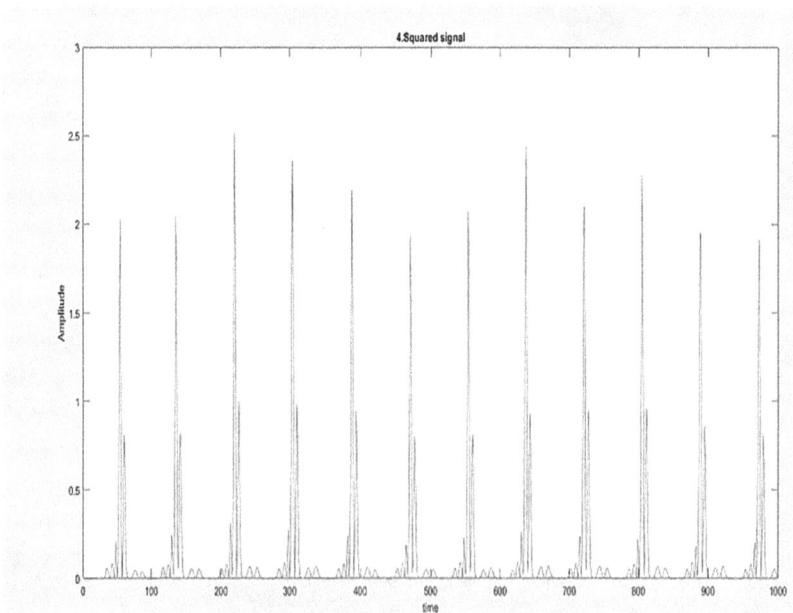

FIGURE 8.8   Squarer provides selective amplification of large differences arising from QRS peaks and attenuation of small differences resulting from P and T waves.

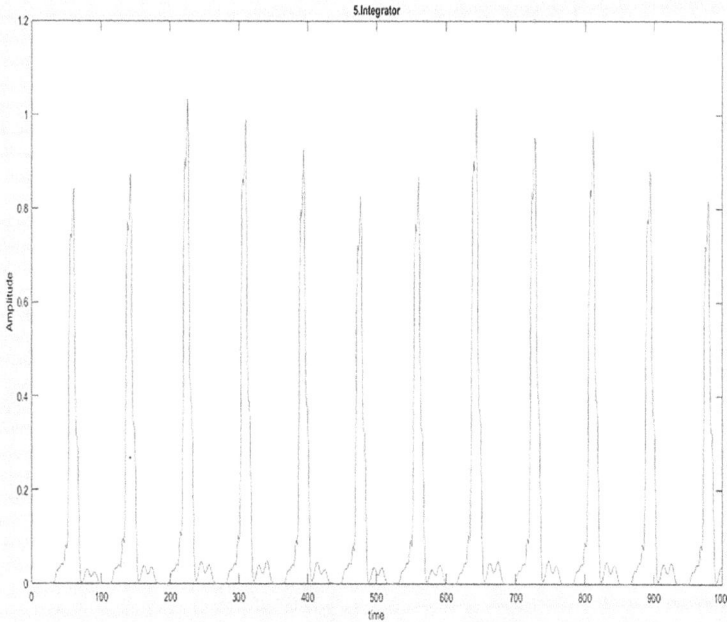

FIGURE 8.9 Integrator that produces a large amplitude pulse for every QRS, lower amplitude pulses for noise spikes

basic types of noise. As a result, a data preprocessing phase is required to reduce noise while focusing and amplifying the required signals. Preprocessing is critical to the success of this approach. Because the major energy of QRS lies in the range of 5–15 Hz, most of the noises indicated above can be filtered out using a band-pass filter. The data is then differentiated to determine the slope's highlight. The T wave, on the other hand, has a slope. The characteristic is increased by squaring to divide T and QRS waves, and then this information is captured using window integration. The preprocessing stage is then done.

After the preprocessing step, the signals are combined to make detection easier, and then the key step of detecting peaks begins. Peaks are considered candidates for the QRS complex in this method. When peaks are discovered, the samples in the vicinity are examined and results are shown in Section 8.4.1. The Section 8.4.2 deals with feature extraction from ECG data, while Section 8.4.3 deals with the results for OSA prediction. Section 8.4.3 is a distinguishing result for the prediction of OSA. Section 8.4.4 is a measuring scale and comparison scaling for the prediction of Obstructive Sleep Apnea.

8.4.1 RESULTS OF PREPROCESSING AND QRS DETECTION THRESHOLD

The R peaks of the ECG waveform are identified, and thresholding is conducted at 45 percent of the maximum amplitude of the ECG waveform. As seen in Figure 8.10, the QRS peaks are indicated.

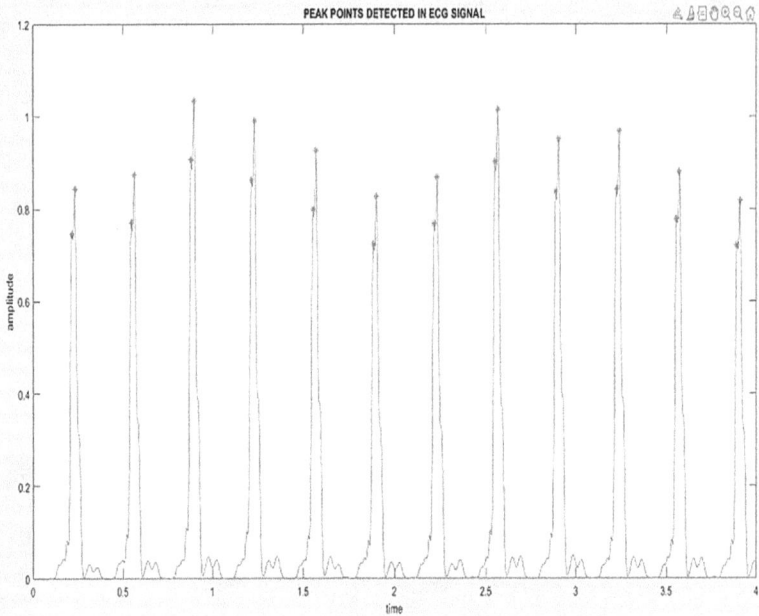

FIGURE 8.10   R peak detection using threshold = 0.45(Max (ECG)).

8.4.2   RESULTS OF FEATURE EXTRACTION

Heart Rate Variability (HRV), or the variation in heart rate from one beat to the next, is one of the most important aspects for future patient diagnosis. This is a noninvasive way to check how well the autonomic nervous system is working. Low HRV has been linked to a range of ailments, including diabetes, sleep problems, and emotional regulation concerns. Figure 8.11 shows a staircase representation of HRV.

8.4.3   DISTINGUISHING RESULTS FOR PREDICTION OF OSA

Table 8.2 compares the heart rate (beats per minute) to the standard deviation of the RR interval to assess the prognosis of myotonic dystrophy patients. The percentage of apnea is used to determine whether or not the patient is apneic.

8.4.4   MEASURING SCALE

Table 8.3 describes the standard value for normal and apneic patients on the basis of the features extracted. Based on these reference criterions, the patient is classified as an apneic or non-apneic. The degree of sleep apnea can also be deduced from the given range. The section 8.5 provides the conclusions on the performance characteristics of QRS complex.

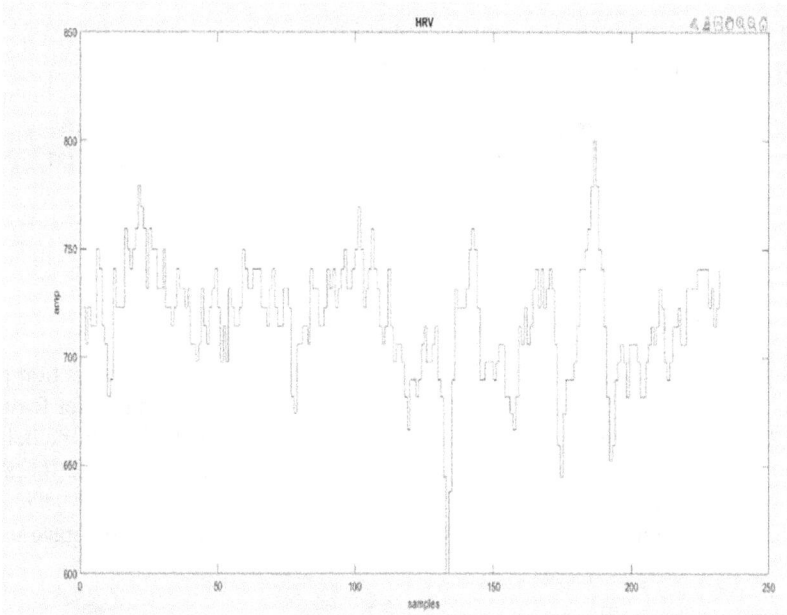

FIGURE 8.11 Staircase representation of HRV.

TABLE 8.2
Distinguishing Results for the Prediction of OSA

Subject No.	BPM (Beats/ minute)	SDNN (Standard Deviation of RR Intervals)	RMSSD (Root Mean Square of Successive Differences)	pNN50 (%)	% Apnea (minutes)	Decision (1/0)
X01	128.3563	10.328	11.314	15	78.36	1
X02	140.2332	12.22	12.649	7	50.45	1
X03	65.4165	58.653	52.608	27	3.724	0
X04	98.3815	30.692	67.717	22	0.023	0
X05	73.4105	51.54	30.165	19	65.99	1
X06	74.8122	52.25	54.053	25*	1.224	0
X07	162.5757	14.376	14.787	12	43.586	1
X08	133.7034	39.331	31.191	6	70.2133	1
X09	120.2959	32.068	26.363	10	36.75	1
X10	113.1773	13.856	16.971	17	19.004	1
X11	82.8454	63.908	58.595	35	1.5554	0
X12	119.9403	33.32	21.107	9	13.05	1
X13	130.7446	39.653	38.696	8	54.098	1
X14	104.1688	36.383	26.873	8	90.667	1
X15	112.2583	32.654	24.646	8	30.298	1

Note: Decision: "1"—apneic patients; "0"—non-apneic patients.
* *Outlier.*

TABLE 8.3
Reference Criteria for Prediction of OSA

SDNN (ms)	RMSSD (ms)	BPM	pNN50 (%)
Normal: 50–100	Normal: 45–103	Normal: 60–100	Normal: 24–137
Apneic: <50	Apneic: <40	Apneic: >100	Apneic: <22

8.5 CONCLUSION AND FUTURE SCOPE

QRS peaks with higher amplitudes were observed and made QRS detection possible. The efficiency of the classifier can be improved by extracting better features of the ECG signal. The signal-to-noise ratio has improved to 12% with the help of preprocessing. The performance characteristics of QRS complex detection after the implementation of the Pan-Tompkins algorithm have an accuracy of 94%, sensitivity of 95%, and precision of 92%. The classifier with ten features gave us an accuracy of 90%.

8.5.1 FUTURE SCOPE

More features are to be extracted from the above dataset, and it needs to be divided into test and training sets. An ANN with LM algorithm applied to classify whether the subject has sleep apnea or not [2]. Better feature extraction methodology, improving the classification result of cardiac arrhythmias in ECG signal can be a part of future development. Beat arrhythmias are arranged in a recognized way to analyze the classification exactness using various classifiers. To achieve more reliable classification correctness than the existing ECG beat classifier, the network configuration is altered according to the cost function of the multilayer neural network. Cardiac arrhythmias are recognized with the help of a real-time process as the methodology uses the automated detection of R peaks and feature extraction procedures.

REFERENCES

1. Bali, Jyoti; Nandi, Anilkumar; Hiremath, P. S.; Patil, Poornima G. "Detection of Sleep Apnea in ECG Signal using Pan Tompkins Algorithm and ANN Classifiers" COMPUSOFT, An International Journal of Advanced Computer Technology, Vol. 7, No. 11, November-2018 (Volume-VII, Issue-XI).
2. Sivaranjni, V.; Rammohan, T. "Detection of Sleep Apnea Through ECG Signal Features", 2016 2nd International Conference on Advances in Electrical, Electronics, Information, Communication and Bio-Informatics (AEEICB), Chennai, 2016
3. Reddy, K. Giridhara; Vijaya, P. A. "ECG Signal Characterization and Correlation To Heart Abnormalities." (2017). Semantic Scholar.
4. Marshall, Nathaniel S.; Wong, Keith K. H.; Liu, Peter Y.; Cullen, Stewart R. J.; Knuiman, Matthew W.; Grunstein, Ronald R. "Sleep Apnea as an Independent Risk Factor for All-Cause Mortality: The Busselton Health Study", SLEEP, Vol. 31, No. 8, 2008.

5. Jennifer M. Slowik; Jacob F. Collen. Obstructive Sleep Apnea. [Updated 2020 Jun 7]. In: StatPearls [Internet]. Treasure Island (FL): StatPearls Publishing; 2020 Jan Publication- NCBI.nlm.nhi.

6. https://cpap.1800cpap.com/blog/sleep-apnea-and-depression/health/

7. https://www.disabled-world.com/health/neurology/sleepdisorders/sleepapnea/

8. Lee, Won; Nagubadi, Swamy; Kryger, Meir H.; Mokhlesi, Babak. "Epidemiology of Obstructive Sleep Apnea: A Population-based Perspective", Expert Review of Respiratory Medicine, 2008 June.

9. https://www.sleepblizz.com/sleep-study-test-cost-in

10. Pan, Jiapu; Tompkins, Willis J. "A Real-Time QRS Detection Algorithm", IEEETransactions On Biomedical Engineering, Vol. BME-32, No. 3, 1985.

11. Lascu, Mihaela; Lascu, Dan. "Labview Based Event Detection using Pan Tompkins Algorithm", Proceedings of 7th WSEAS International Conference on Signal Processing, computational Geometry and Artificial Vision, Athens Greece, 2007.

12. Elgendi, Mohamed; Jonkma, Mirjam; De Boer, Friso. "Improved QRS Detection Algorithm using Dynamic Thresholds", International Journal of Hybrid Information Technology, Vol. 2, No. 1, 2009.

13. Pooyan, Mohammad; Akhoondi, Fateme. "Providing An Efficient Algorithm For finding R Peaks In ECG Signals And Detecting Ventricular Abnormalities With Morphological Features", Journal of Medical Signals & Sensors, 2016.

14. De Chazal, Philip; Reilly, Richard B. "A Patient Adapting Heart Beat Classifier Using ECG morphology and Heart Beat Interval Features", IEEE Transactions on Biomedical Engineering, Vol. 53, No. 12, 2006.

15. Almazaydeh, Laiali; Elleithy, Khaled; Faezipour, Miad. "Detection of Obstructive Sleep Apnea Through ECG Signal Features" 2012 IEEE International Conference on Electro/ Information Technology, 2012.

16. Rodríguez, Ricardo; Mexicano, Adriana; Bila, Jiri; Cervantes, Salvador; Ponce, Rafael. "Feature Extraction of Electrocardiogram Signals by applying Adaptive Threshold and Principal Component Analysis", Journal of Applied Research and Technology, Vol. 13, 2015, 261–269.

17. https://physionet.org/content/apnea-ecg/1.0.0/- MIT's Physionet ECG apnea Database.

18. Vidya Madapur Jagadish, Padmaja Karanam VenkobRao. "Affirmation of Electronic Patient Record through Bio-electric Signal for Medical Data Encryption Authenticity", 2017 IEEE Region 10 Conference (TENCON), Malaysia, 2017, pp 1332–1336, 978-1-5090-1134-6/17/$31.00 © 2017, IEEE, doi:10.1109/TENCON.2017.8228064.

9 Machine Learning Methods for Electroencephalogram (EEG) Big Data in the Context of MIOT Smart Systems

Aileni Raluca Maria and Pasca Sever
Politehnica University of Bucharest,
Bucharest, Romania

CONTENTS

9.1 LEARNING MACHINE TECHNIQUES FOR EEG BIG DATA

Electroencephalogram (EEG) has a primary indication of epilepsy, as this system allows the diagnosis of epilepsy, as well as the monitoring of the treatment and evolution. Statistically, epilepsy is a disease that affects about 1% of the world's population.

The EEG represents the recording of signals from neurons [1–3] of the cerebral cortex [4], a set of fluctuating field potentials, produced by the simultaneous activity of a large number of neurons and captured by electrodes located on the scalp. EEG is used in the diagnosis of epilepsy, encephalopathy, in monitoring brain activity during anesthesia, in patients with coma, and in determining brain death. On the entire surface of the skin of the head are arranged 10–20 metal electrodes connected by wires to the recording device. It measures the electrical potential detected by each electrode and compares the electrodes two by two, each

DOI: 10.1201/9781003268796-12

comparison translating through a path called a bypass. Electroencephalographic reactivity is evaluated using simple tests such as eye-opening, hyperpnea (slow and full breathing), and intermittent light stimulation obtained with short and intense light discharges whose frequency is gradually increased. The EEG assessment takes approximately 20 minutes and does not require hospitalization.

In the case of an EEG, the risks are minimal. Still, intermittent light stimulation or hyperventilation can produce epileptic seizures. Therefore, the examination is performed under the supervision of a physician who can recognize the crisis and immediately establish appropriate safety and therapeutic measures.

It can also be used in:

- Evaluation of drugs;
- Assessment of brain damage in cases of stroke (tumors), tumors, or trauma;
- Evaluation of memory disorders, dementias, or psychoses;
- Monitoring of cerebral blood flow during a surgical procedure and control of anesthesia;
- Detection of inflammation in the brain (encephalitis, meningoencephalitis);
- The study of sleep pathology and sleep disorders;

EEG diagnosis is useful in the study of disease pathology:

- Parkinson's disease is a progressive neurological disorder characterized by limb tremors, muscle rigidity, and bradykinesia. EEG can be used for early-stage monitoring of Parkinson's disease to study abnormalities that may occur in the brain and to determine the evolution of the disease [5, 6].
- Alzheimer's disease is a disease that involves progressive nerve degeneration due to a diminished number of neurons, brain atrophy, and the presence of "senile plaques," manifesting through memory loss and disorientation in time and space. EEG can be used for early diagnosis [7, 8] and the dynamics of the disease and abnormalities that occur in the brain [9, 10] compared to other neurological disorders [11].
- Epilepsy is a chronic disease of the brain that manifests through partial (focal) or generalized seizures, due to spontaneous electrical discharges that occur in the brain.

Manifestations consist of involuntary movements of different body segments and abnormal neurovegetative sensations in the body. EEG analysis can be used to diagnose and monitor the patient in various stages of the disease (focal or generalized seizures, sleep, etc.) [12–15].

- Central motor neuron syndrome [16] occurs due to strokes, brain and medullary tumors, myelitis, spinal fractures, amyotrophic lateral sclerosis, and manifests as paraplegic hemiplegia or tetraplegia.
- Cerebral palsy [17] is caused by abnormal development or deterioration of the parts of the brain that control movement, balance, and posture, and is manifested by spastic cerebral palsy (muscular rigidity), ataxic cerebral palsy (poor coordination), or athetoid cerebral palsy (patients present

TABLE 9.1
CNNs Types

CNNs	Single Channel	Multichannel
1D	Audio waveforms	3D model animation by skeleton
2D	Audio data	Color image
3D	Volumetric data (voxels)-CT scans	Color video data

writing problems). EEG analysis is not used in paralysis detection but can be used to detect paralysis degree [18, 19].

- Muscular dystrophy (Duchenne muscular dystrophy or Becker muscular dystrophy) is a hereditary disease characterized by the progressive deterioration of the muscles of the body, generating muscular weakness and disability. EEG analysis can be used to investigate potential muscular dystrophies [20, 21].

EEG obtained by wearable scalp EEG devices can be used in monitoring of several neurological disorders even if they are not equipped with so many channels and electrodes. The classification methods for investigating offset and onset seizures are usually based on learning machine methods and artificial neural network (ANN) in EEG-based recognition [22–24], and more and more studies indicate the use of deep learning (DL) techniques [25], such as convolutional neural networks (CNNs) [26]. The CNNs are 1D, 2D, and 3D type (Table 9.1) and can be used for signal processing and classification (ECG, EEG). However, the 1D CNNs can be used in training a limited data set of 1D raw signals (voltage, current), while the 2D CNNs need data sets with massive data size (big data) and scale in order to avoid the over fitting.

9.2 MIOT PERSPECTIVE. REAL-TIME MONITORING OF EEG SIGNALS USING SMART SYSTEMS

The usage of the medical Internet of Things (MIOT) in telemedicine/healthcare (personal health-care monitoring and health-care payment applications) has an intensive use in specific internet of things (IoT) EEG use cases. MIOT allow medical devices connection through IoT [27–29]. Smart health diagnosis can be enabled by remote and telemonitoring using invasive or noninvasive (implantable sensors) wearable devices and body area networks in star configuration. Continuous monitoring generates a huge amount of data and big data in the case of the video-EEG monitoring (6–8 hours), which needs to be stored and analyzed.

In order to analyze a massive data volume, artificial intelligence (AI) techniques can be used, such as machine learning tools like ANN or DL, to handle this amount of medical data.

A second approach is tracking, monitoring, and maintenance of assets, using IoT and radio frequency identification (RFID) [30]. This can be achieved in hospital on the level of medical professional devices and personal monitoring by semiprofessional

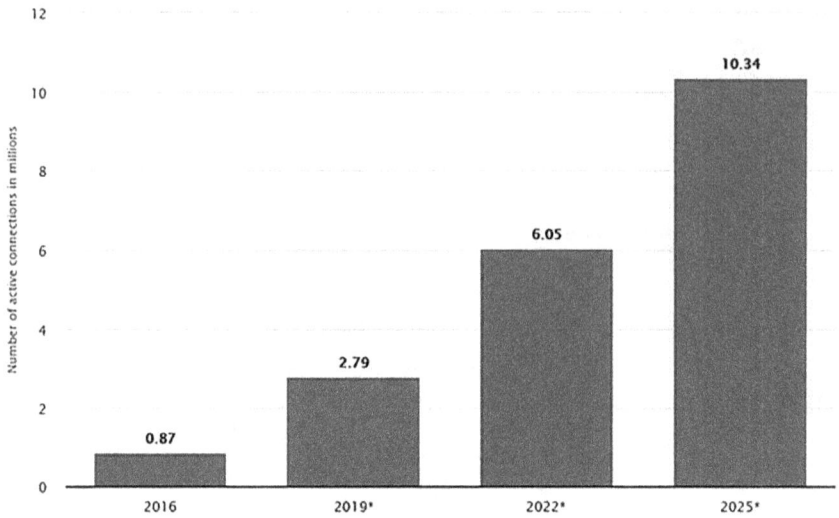

FIGURE 9.1   IoT active connections 2016–2025.

devices [31–34]. According to statistical studies, a huge number of IoT active connections will be active in health care in the European Union (EU) in 2016, 2019, 2022, and 2025. However, the number of IoT health-care active connections is expected to increase through the years. If it was at 0.87 million connections in 2016, and it is expected to reach at 10.34 million connections by 2025 (Figure 9.1). IoT revenues were amounted to 24 billion US dollars worldwide, with forecasts predicting that this number will increase to over 135 billion by 2025.

From 2017 until 2022, growth in IoT health-care applications is expected to accelerate as IoT is a key in the digital transformation of the health care [35, 36]. According to the 2017 Thales Data Threat Report, health-care edition [37], 30%of health-care organizations used IoT for sensitive data, and by 2019, over 40% of health-care organizations were expected to use IoT-enabled biosensors as IDC predicts.

Moreover, health-care ecosystem includes personal health care, the pharma industry, health-care insurance, smart wearable, real-time health systems (RTHS) [38–40], health-care building facilities, robotics, biomedical sensors, smart beds/sofa, smart pills, remote, and IoT applications based on AI (ANN, DL) which help in the treatment of specific diseases/disorders [41] or in studies or predictive modeling of the disease/patient behavior [42].

In the case of telemonitoring, patient' data is collected and is available in real time for medical staff, being integrated with electronic health-care records (EHR) [43, 44].

RTHS are part of IoT in health care as big data analytics tools, and processes are utilized to evaluate, model, and predict data for predictive analytics as part of medical support decision systems in continuous improvement [45–47] (Mind Commerce, end 2016).

In context of the paradigm "smart hospitals" [48], EHR systems [49] have been developed to boost health-care improvement and to be more patient-centric. RTHS used for remote monitoring-based EHR should provide real-time data capabilities (storage—cloud [50], signal processing, and data anonymization—edge computing [51], big data analysis [52]) in context of the IoT [53] and connected/wearable devices for remote clinical decision support (RCDS) [54].

On a wearable level, we also see an increasing use of specialized wearables across all parts of the body (next-generation hearing aids, implantable wearables, skin patches, smart contact lenses, etc.) [55–57].

9.3 DEEP NEURAL NETWORKS ARCHITECTURES

Deep neural networks (DNN) are based on four major network architectures:

1. Unsupervised pretrained networks (UPNs) with specific architectures:
   - Autoencoders;
   - Deep belief networks (DBNs);
   - Generative adversarial networks (GANs);
2. CNNs with specific architectures:
   - LeNet18;
   - AlexNet19;
   - ZF Net20;
   - GoogLeNet21;
   - VGGNet22;
   - ResNet23;
3. Recurrent neural networks with specific architectures:
4. Recursive neural networks;

By supervised learning of CNNs, the epileptic seizure of 23-channel scalp EEG data can be classified in MATLAB, obtained from CHB-MIT Scalp EEG Database.

The time-frequency signal analysis is presented in Figure 9.2, and the scalogram, the absolute value of the continuous wavelet transform (CWT) of a signal plotted as a function of time and frequency, in Figure 9.3. The CNN architecture [58, 59] for classification of the EEG signals in order to investigate the seizures is presented in Figure 9.4.

The proposed CNN model (Figure 9.4) is trained using the scalogram images (Figure 9.3), and the block image input has as input the scalogram images. The convolution layer (conv) performs convolution operations such as scanning the input images and the resulting output is the feature map. The pooling layer (maxpool) selects the maximum value of the current view. The input to the fully connected layer is the output from the final maxpool layer, a flattened vector. The softmax layer will normalize the output of the network to a probability distribution over predicted output classes and will return the vector of output probabilities $p \in \mathbb{R}^n$ of the inputs to in a particular class. The final layer classification will compute the cross-entropy loss for classification and weighted classification with mutually exclusive classes.

9.4 CLASSIFICATION METHODS FOR EEG DATA

EEG classification includes signal processing techniques such as artifact removing (noise reduction), feature extraction, and classification.

Overall the methods to remove artifacts from EEG are principal component analysis (PCA), independent component analysis (ICA), blind source separation (BSS),

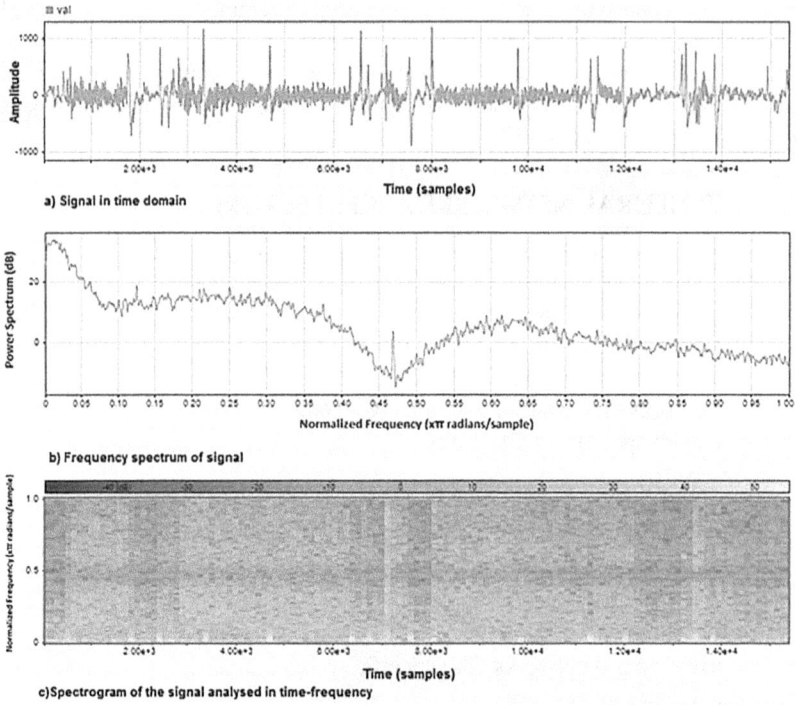

a) Signal in time domain

b) Frequency spectrum of signal

c)Spectrogram of the signal analysed in time-frequency

FIGURE 9.2   Time-frequency signal analysis ➔ FP1-F7.

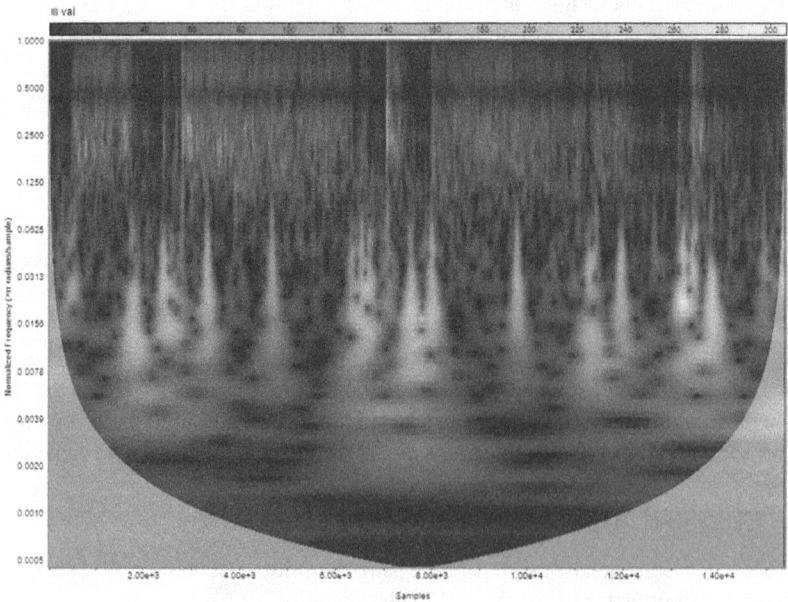

FIGURE 9.3   Scalogram time-frequency analyses ➔ FP1-F7.

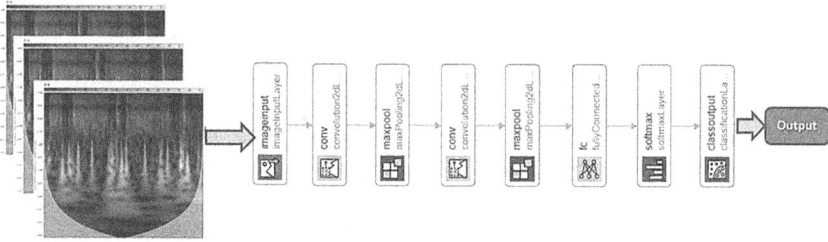

FIGURE 9.4   Deep learning—CNN architecture.

parallel factor analysis (PFA), and discrete wavelet transform (DWT) and independent component analysis (ICA).

Common methods to extract the features (decomposition) from EEG signals are time-frequency distributions (TFD), fast Fourier transform (FFT), DWT, eigenvector methods (EM), and autoregressive method (ARM).

The classification methods for EEG signals are based on EEG pattern frequency recognition (alpha, beta, delta, theta) using wavelet transformation, PCA, ICA, linear discriminant analysis (LDA), support vector machine (SVM), multilayer perceptron (MLP), naive Bayes (NB), k-nearest neighbor (KNN), and k-fold cross classification [59–61]. The common classification methods include simple linear methods (LDA for classification and multiple linear regression), SVM-like kernel methods, random forests, neural networks, or combination of methods.

To classify EEG, signals can be used, supervised, or unsupervised for learning methods [57]. The supervised learning approach involves labeled training data set used to train, test, and validate the method using a number of neurons or cross-validation. The analysis of the EEG signals can be used for the diagnosis and seizure detection in real time. EEG signals are difficult to interpret and are necessary preliminary actions (noise reduction, feature extraction, classification), present high variability even for the same patient.

In Figures 9.5 and 9.6, the 2D-3D EEG signals of an epileptic patient are presented. The signal was preprocessed using a filter in EEG lab (Figure 9.5).

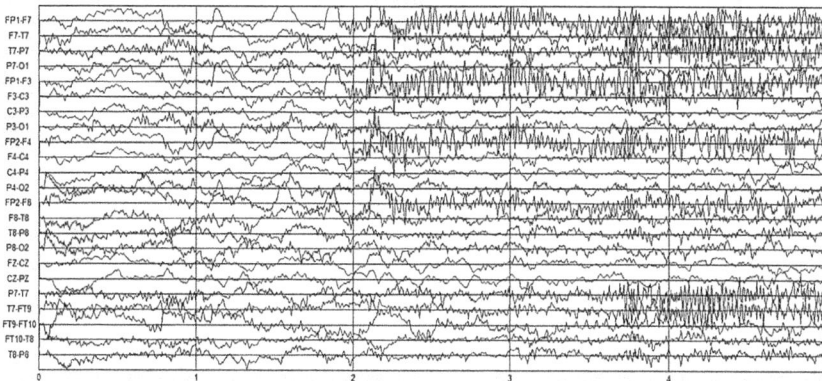

FIGURE 9.5   EEG signals 2D—23 channels.

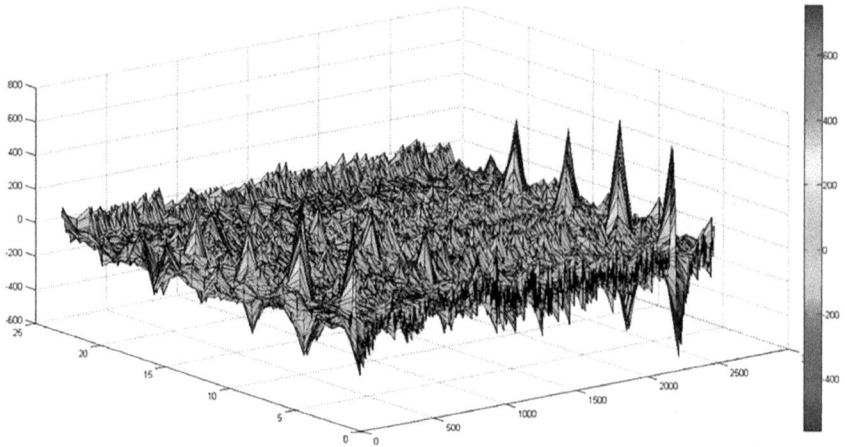

FIGURE 9.6   3D spectrogram signals EEG from 23 channels for patient with epileptic seizures.

By supervised classifier learner, support vector machine (SVM)-based linear kernel in MATLAB was classified as the epileptic seizure of 23-channel scalp EEG data, obtained from CHB-MIT scalp EEG database [62]. EEG data were collected from the Children's Hospital Boston and contain EEG recordings from pediatric subjects with intractable seizures. The goal was to obtain the classifications that could be implemented in support decision systems in order to easily obtain a decision to continue the medication or to plan a surgical intervention. The sampling rate was 256 Hz with a 16-bit resolution. Subjects were monitored several days to analyze their seizures and to see if it is necessary surgical intervention. In Figure 9.7, the initial distribution of data is presented; similarly in Figure 9.8, the final data distribution for the predicted model is presented. In Figure 9.9, parallel channels are represented based on holdout validation (50% training and 50% testing). In addition, using the multiclass method, on-vs-all in linear SVM used in the classification of the epileptic seizure was obtained from the predicted model (Figure 9.10).

In order to obtain the predicted model, a function was defined such as function [trainedClassifier, validationAccuracy] = trainClassifier (trainingData). For predictors, the functions validationPredictors = predictors (x.test,:) and correct Predictions = (validationPredictions == ValidationResponse) were used.

In Figures 9.7 and 9.8, data used and final data distribution obtained using as classifier linear SVM one-against-one are presented. This method represents parallel channels based on holdout validation (50% training and 50% testing) (Figure 9.9). The accuracy of the method was 67%. In Figure 9.10, the final data distribution obtained using as classifier SVM multiclass method one-against-all with accuracy 81% is presented.

9.5   FUTURE PERSPECTIVES OF THE EEG SMART SYSTEMS BASED ON DEEP LEARNING TECHNIQUE

The perspective of smart health-care system involves real-time support decision, data and big data analysis, smart sensors, and IoT devices. The IoT devices or smart sensors can provide remote continuous monitoring of the patients and generate real-time medical

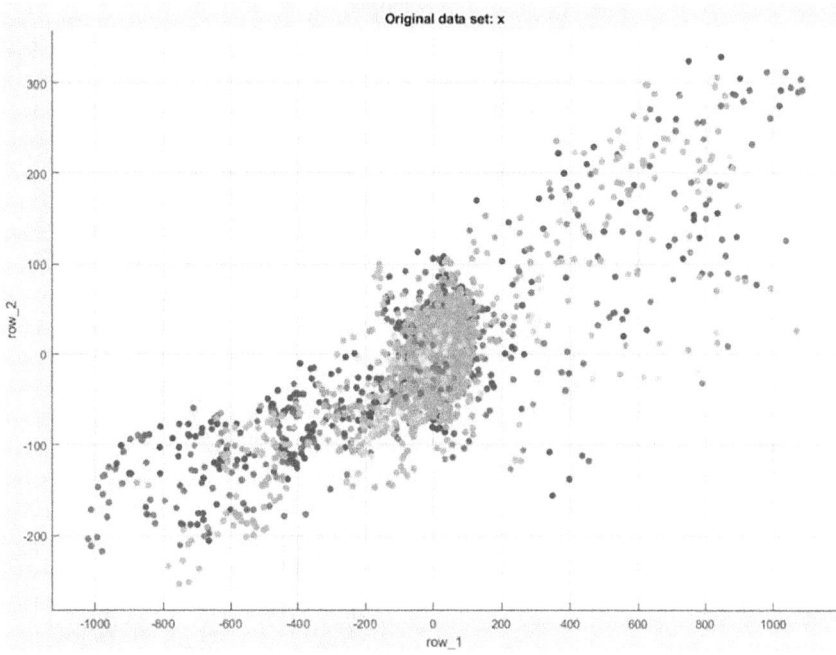

FIGURE 9.7 Initial EEG data distribution [63].

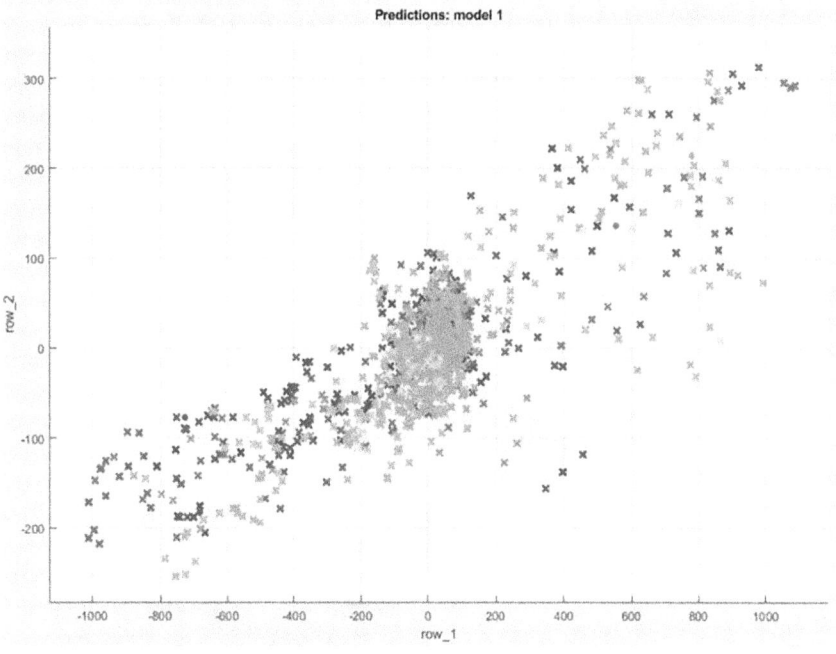

FIGURE 9.8 Final data distribution of the predicted model using SVM one-against-one [63].

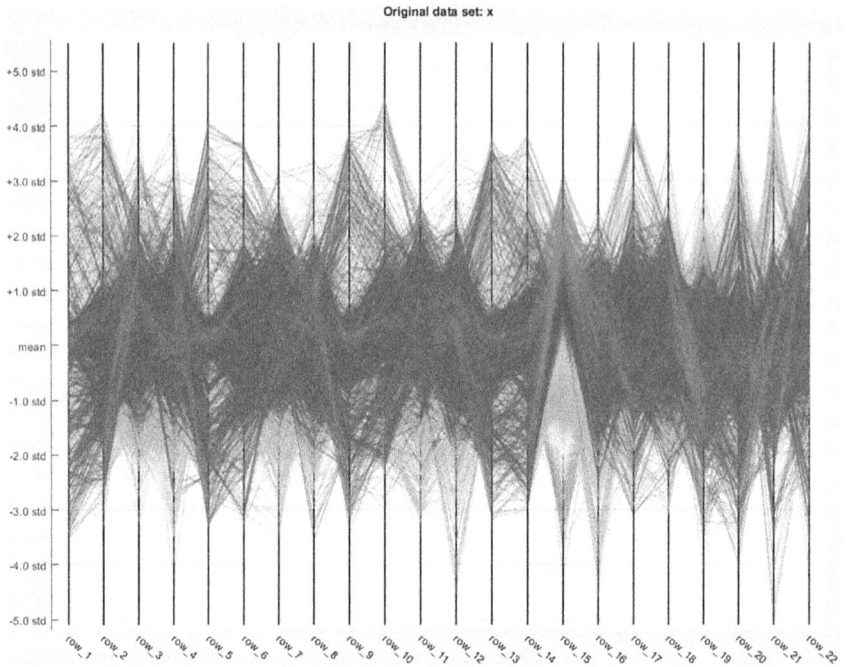

FIGURE 9.9   Parallel channels based on holdout validation (50% training and 50% testing).

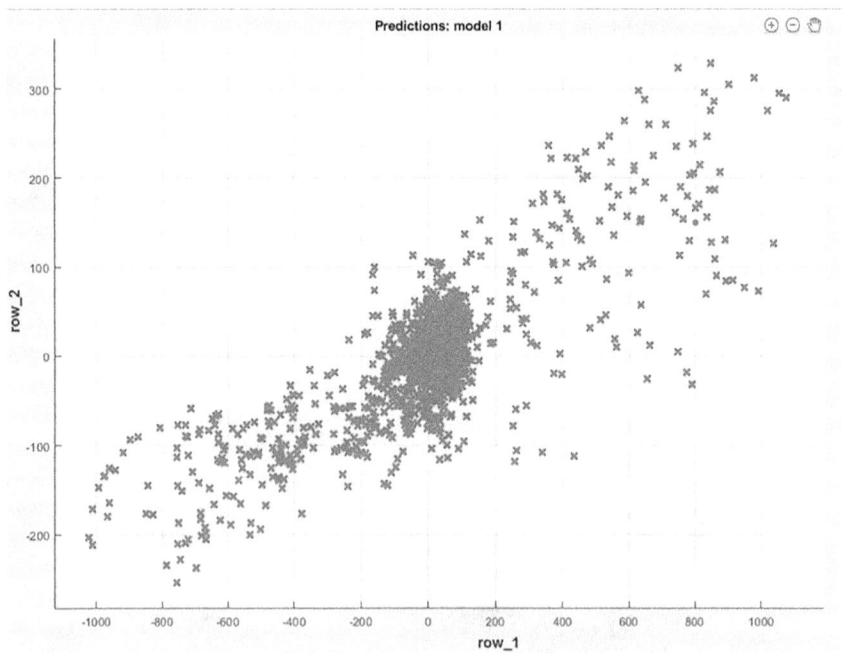

FIGURE 9.10   Final data distribution of the predicted model using SVM one-against-all [62].

data by recording patient responses such as ECG, EEG, heartbeat, blood pressure, blood glucose, temperature, and voice. The medical data collected are must be processed to determine the health status of the patient and if the patient requires emergency care.

EEG signals are used to predict epileptic seizure, to study other diseases' behavior (Parkinson's disease, Alzheimer's disease, stroke), or to evaluate if it is necessary to conduct surgical intervention. These studies offer significant information for medical actions and represent a challenge for medical staff and researchers. Classification of the EEG signals allows highlighting the seizure in order to be used by medical staff in support decision systems for easily obtaining the correct recommendation, drugs treatments, or surgical intervention. The predictive modeling can achieve the accuracy of EEG signals to around 99.9%.

REFERENCES

1. Ismail, W.W., Hanif, M., Mohamed, S.B., Hamzah, N. and Rizman, Z.I., 2016. Human emotion detection via brain waves study by using electroencephalogram (EEG). International Journal on Advanced Science, Engineering and Information Technology, 6(6), pp.1005–1011.
2. Paranjape, R.B., Mahovsky, J., Benedicenti, L. and Koles, Z., 2001, May. The electroencephalogram as a biometric. In Canadian Conference on Electrical and Computer Engineering 2001. Conference Proceedings (Cat. No. 01TH8555) (Vol. 2, pp. 1363–1366). IEEE.
3. Aileni, R.M., Pasca, S. and Florescu, A., 2020. EEG-brain activity monitoring and predictive analysis of signals using artificial neural networks. Sensors, 20(12), p.3346.
4. Sullivan, L.R., Beverwyk, A.J. and Davis, S.F., 2020. Electroencephalography. In Principles of Neurophysiological Assessment, Mapping, and Monitoring (pp. 155–169). Springer, Cham.
5. Brown, D.R., Richardson, S.P. and Cavanagh, J.F., 2020. An EEG marker of reward processing is diminished in Parkinson's disease. Brain Research, 1727, p.146541.
6. de Oliveira, A.P.S., de Santana, M.A., Andrade, M.K.S., Gomes, J.C., Rodrigues, M.C. and dos Santos, W.P., 2020. Early diagnosis of Parkinson's disease using EEG, machine learning and partial directed coherence. Research on Biomedical Engineering, 36, pp.1–21.
7. Dauwels, J., Vialatte, F., Musha, T. and Cichocki, A., 2010. A comparative study of synchrony measures for the early diagnosis of Alzheimer's disease based on EEG. NeuroImage, 49(1), pp.668–693.
8. Albert, B., Zhang, J., Noyvirt, A., Setchi, R., Sjaaheim, H., Velikova, S. and Strisland, F., 2016, July. Automatic EEG processing for the early diagnosis of traumatic brain injury. In 2016 World Automation Congress (WAC) (pp. 1–6). IEEE.
9. Jeong, J., 2004. EEG dynamics in patients with Alzheimer's disease. Clinical Neurophysiology, 115(7), pp.1490–1505.
10. Asadzadeh, S., Rezaii, T.Y., Beheshti, S., Delpak, A. and Meshgini, S., 2020. A systematic review of EEG source localization techniques and their applications on diagnosis of brain abnormalities. Journal of Neuroscience Methods, 6, p.108740.
11. Alturki, F.A., AlSharabi, K., Abdurraqeeb, A.M. and Aljalal, M., 2020. EEG signal analysis for diagnosing neurological disorders using discrete wavelet transform and intelligent techniques. Sensors, 20(9), p.2505.
12. Yakovleva, T.V., Kutepov, I.E., Karas, A.Y., Yakovlev, N.M., Dobriyan, V.V., Papkova, I.V., Zhigalov, M.V., Saltykova, O.A., Krysko, A.V., Yaroshenko, T.Y. and Erofeev, N.P., 2020. EEG analysis in structural focal epilepsy using the methods of nonlinear

dynamics (Lyapunov exponents, lempel–ziv complexity, and multiscale entropy). The Scientific World Journal, 2020, 1–13.

13. Hrachovy, R.A. and Frost Jr, J.D., 2006. The EEG in selected generalized seizures. Journal of Clinical Neurophysiology, 23(4), pp.312–332.

14. Tartara, A., Moglia, A., Manni, R. and Corbellini, C., 1980. EEG findings and sleep deprivation. European Neurology, 19(5), pp.330–334.

15. Gigli, G.L., Calia, E., Marciani, M.G., Mazza, S., Mennuni, G., Diomedi, M., Terzano, M.G. and Janz, D., 1992. Sleep microstructure and EEG epileptiform activity in patients with juvenile myoclonic epilepsy. Epilepsia, 33(5), pp.799–804.

16. Lazarou, I., Nikolopoulos, S., Petrantonakis, P.C., Kompatsiaris, I. and Tsolaki, M., 2018. EEG-based brain–computer interfaces for communication and rehabilitation of people with motor impairment: A novel approach of the 21st century. Frontiers in Human Neuroscience, 12, p.14.

17. Perales, F.J. and Amengual, E., 2017. Combining EEG and serious games for attention assessment of children with Cerebral Palsy. In Converging Clinical and Engineering Research on Neurorehabilitation II (pp. 395–399). Springer, Cham.

18. Spüler, M., López-Larraz, E. and Ramos-Murguialday, A., 2018. On the design of EEG-based movement decoders for completely paralyzed stroke patients. Journal of Neuroengineering and Rehabilitation, 15(1), p.110.

19. Whitham, E.M., Pope, K.J., Fitzgibbon, S.P., Lewis, T., Clark, C.R., Loveless, S., Broberg, M., Wallace, A., DeLosAngeles, D., Lillie, P. and Hardy, A., 2007. Scalp electrical recording during paralysis: Quantitative evidence that EEG frequencies above 20 Hz are contaminated by EMG. Clinical Neurophysiology, 118(8), pp.1877–1888.

20. Biere, J., Okkersen, K., van Alfen, N., Kessels, R.P., Gouw, A.A., van Dorst, M., van Engelen, B., Stam, C.J. and Raaphorst, J., 2020. Characterization of EEG-based functional brain networks in myotonic dystrophy type 1. Clinical Neurophysiology, 131, pp.1886–1895.

21. Subramony, S.H., Wymer, J.P., Pinto, B.S. and Wang, E.T., 2020. Sleep disorders in myotonic dystrophies. Muscle & Nerve, 62, pp.309–320.

22. Khosla, A., Khandnor, P. and Chand, T., 2020. A comparative analysis of signal processing and classification methods for different applications based on EEG signals. Biocybernetics and Biomedical Engineering, 40, pp.649–690.

23. Ko, W., Jeon, E., Jeong, S. and Suk, H.I., 2020. Multi-Scale Neural network for EEG Representation Learning in BCI. arXiv preprint arXiv:2003.02657.

24. Ray, P.P., Dash, D. and Kumar, N., 2020. Sensors for internet of medical things: State-of-the-art, security and privacy issues, challenges and future directions. Computer Communications, 160, pp.111–131.

25. Merlin Praveena, D., Angelin Sarah, D. and Thomas George, S., 2020. Deep learning techniques for EEG signal applications – A review. IETE Journal of Research, pp.1–8.

26. Raghu, S., Sriraam, N., Temel, Y., Rao, S.V. and Kubben, P.L., 2020. EEG based multiclass seizure type classification using convolutional neural network and transfer learning. Neural Networks, 124, pp.202–212.

27. Darwish, S., Nouretdinov, I. and Wolthusen, S.D., 2017. Towards composable threat assessment for medical IoT (MIoT). Procedia Computer Science, 113, pp.627–632.

28. Dimitrov, D.V., 2016. Medical internet of things and big data in healthcare. Healthcare Informatics Research, 22(3), pp.156–163.

29. Haghi, M., Thurow, K. and Stoll, R., 2017. Wearable devices in medical internet of things: Scientific research and commercially available devices. Healthcare Informatics Research, 23(1), pp.4–15.

30. Khan, S.F., 2017, March. Health care monitoring system in Internet of Things (IoT) by using RFID. In 2017 6th International Conference on Industrial Technology and Management (ICITM) (pp. 198–204). IEEE.

31. Milenković, A., Otto, C. and Jovanov, E., 2006. Wireless sensor networks for personal health monitoring: Issues and an implementation. Computer Communications, 29(13–14), pp.2521–2533.

32. Shin, M., 2012. Secure remote health monitoring with unreliable mobile devices. Journal of Biomedicine and Biotechnology, 2012, pp.1–5.

33. Dias, D. and Paulo Silva Cunha, J., 2018. Wearable health devices—Vital sign monitoring, systems and technologies. Sensors, 18(8), p.2414.

34. Kim, I., Lai, P.H., Lobo, R. and Gluckman, B.J., 2014, August. Challenges in wearable personal health monitoring systems. In 2014 36th Annual International Conference of the IEEE Engineering in Medicine and Biology Society (pp. 5264–5267). IEEE.

35. Dunbrack, L., Ellis, S., Hand, L., Knickle, K. and Turner, V., 2016. IoT and Digital Transformation: A Tale of Four Industries. IDC White Paper.

36. Sodhro, A.H., Pirbhulal, S. and Sangaiah, A.K., 2018. Convergence of IoT and product lifecycle management in medical health care. Future Generation Computer Systems, 86, pp.380–391.

37. Xu, S., Li, Y., Deng, R., Zhang, Y., Luo, X. and Liu, X., 2019. Lightweight and expressive fine-grained access control for healthcare Internet-of-Things. IEEE Transactions on Cloud Computing, 10, pp.474–490.

38. Imani, S., Bandodkar, A.J., Mohan, A.V., Kumar, R., Yu, S., Wang, J. and Mercier, P.P., 2016. A wearable chemical–electrophysiological hybrid biosensing system for real-time health and fitness monitoring. Nature Communications, 7(1), pp.1–7.

39. Kakria, P., Tripathi, N.K. and Kitipawang, P., 2015. A real-time health monitoring system for remote cardiac patients using smartphone and wearable sensors. International Journal of Telemedicine and Applications, 2015, pp.8.

40. Wan, J., Al-awlaqi, M.A., Li, M., O'Grady, M., Gu, X., Wang, J. and Cao, N., 2018. Wearable IoT enabled real-time health monitoring system. EURASIP Journal on Wireless Communications and Networking, 2018(1), p.298.

41. Sabra, S., 2018. Prediction of Venous Thromboembolism Using a Hybrid Semantic Based and Machine Learning Approach (Doctoral dissertation, Oakland University).

42. Basu, M., Sharmin, M., Das, A., Nair, N.U., Wang, K., Lee, J.S., Chang, Y.P.C., Ruppin, E. and Hannenhalli, S., 2017. Prediction and subtyping of hypertension from pan-tissue transcriptomic and genetic analyses. Genetics, 207(3), pp.1121–1134.

43. Crameri, K.A., Maher, L., Van Dam, P. and Prior, S., 2020. Personal electronic healthcare records: What influences consumers to engage with their clinical data online? A literature review. Health Information Management Journal, 51.

44. Tanwar, S., Parekh, K. and Evans, R., 2020. Blockchain-based electronic healthcare record system for healthcare 4.0 applications. Journal of Information Security and Applications, 50, p.102407.

45. Sadineni, P.K., 2020, October. Developing a model to enhance the quality of health informatics using Big Data. In 2020 Fourth International Conference on I-SMAC (IoT in Social, Mobile, Analytics and Cloud)(I-SMAC) (pp. 1267–1272). IEEE.

46. Raj, P., Chatterjee, J.M., Kumar, A. and Balamurugan, B., 2020. Internet of Things Use Cases for the Healthcare Industry. Springer Nature Switzerland AG.

47. Blumenthal, S., Potter, D., Simon, K. and Garcia, B., Patient Oncology Portal Inc, 2019. Methods and System for Real Time, Cognitive Integration with Clinical Decision Support Systems featuring Interoperable Data Exchange on Cloud-Based and Blockchain Networks. U.S. Patent Application 16/374,683.

48. Sebastian, M.P., 2019. Smart Hospitals: Challenges and Opportunities (No. 315).

49. Spatar, D., Kok, O., Basoglu, N. and Daim, T., 2019. Adoption factors of electronic health record systems. Technology in Society, 58, p.101144.

50. Mehmood, A., Mehmood, F. and Song, W.C., 2019, October. Cloud based E-Prescription management system for healthcare services using IoT devices. In 2019 International

Conference on Information and Communication Technology Convergence (ICTC) (pp. 1380–1386). IEEE.

51. Ray, P.P., Dash, D. and De, D., 2019. Edge computing for Internet of Things: A survey, e-healthcare case study and future direction. Journal of Network and Computer Applications, 140, pp.1–22.

52. Md, I.P., Lau, R.Y., Md, A.K.A., Md, S.H., Md, K.H. and Karmaker, B.K., 2020. Healthcare informatics and analytics in big data. Expert Systems with Applications, 152, p.113388.

53. Adhikary, T., Jana, A.D., Chakrabarty, A. and Jana, S.K., 2019, January. The internet of things (IoT) augmentation in healthcare: An application analytics. In International Conference on Intelligent Computing and Communication Technologies (pp. 576–583). Springer, Singapore.

54. Olmedo-Aguirre, J.O., Reyes-Campos, J., Alor-Hernández, G., Machorro-Cano, I., Rodríguez-Mazahua, L. and Sánchez-Cervantes, J.L., 2022. Remote healthcare for elderly people using wearables: A review. Biosensors, 12(2), p.73.

55. Martínez-Caro, E., Cegarra-Navarro, J.G., García-Pérez, A. and Fait, M., 2018. Healthcare service evolution towards the Internet of Things: An end-user perspective. Technological Forecasting and Social Change, 136, pp.268–276.

56. Internet of Things (IoT) in healthcare: Benefits, use cases and evolutions, online available: https://www.i-scoop.eu/internet-of-things-guide/internet-things-healthcare.

57. Yan, L.C., Yoshua, B. and Geoffrey, H., 2015. Deep learning. Nature, 521(7553), pp.436–444.

58. Krajca, V. and Petránek, S., 2014. 4. Classification of EEG graphoelements with supervised and unsupervised learning algorithms. Clinical Neurophysiology, 125(5), p.e26.

59. Amin, H.U., Mumtaz, W., Subhani, A.R., Saad, M.N.M. and Malik, A.S., 2017. Classification of EEG signals based on pattern recognition approach. Frontiers in Computational Neuroscience, 11, p.103.

60. Zhang, Y., Zhang, Y., Wang, J. and Zheng, X., 2015. Comparison of classification methods on EEG signals based on wavelet packet decomposition. Neural Computing and Applications, 26(5), pp.1217–1225.

61. A.H. Shoeb, 2009. Application of Machine Learning to Epileptic Seizure Onset Detection and treatment (Doctoral dissertation, Massachusetts Institute of Technology).

62. R.M. Aileni, S. Pasca and A. Florescu, 2019. Epileptic Seizure Classification based on Supervised Learning Models, 2019 11th International Symposium on Advanced Topics in Electrical Engineering (ATEE), Bucharest, Romania, pp. 1–4, doi: 10.1109/ATEE.2019.8725004.

10 Machine Interaction-Based Computational Tools in Cancer Imaging

Praveen Kumar Gupta, Anushree Vinayak Lokur, Shweta Sudam Kallapur, Ryna Shireen Sheriff, and A. H. Manjunatha Reddy
Department of Biotechnology, R.V College of Engineering, Bangalore, India

V. Chayapathy
Department of Electrical & Electronics Engineering, R.V College of Engineering, Bangalore, India

Sindhu Rajendran
Department of Electronics and Communications Engineering, R.V College of Engineering, Bangalore, India

Keshamma E.
Department of Biochemistry, Maharani's Science College for Women, Bangalore, India

CONTENTS

DOI: 10.1201/9781003268796-13

167

10.1   INTRODUCTION

Health sector using computational tools involves the usage of the tools of artificial intelligence (AI), advanced robotics, machine learning, and natural language processing, for the substitution of human expertise. AI in healthcare is a growing field due to the availability of advanced predictive techniques and randomized methods. Discriminative Gaussian processed models are tested on the human face recognition data from various sources in the future; they may be used to detect the image-based pattern recognition. Trunk branch holistic ensemble convolutional neural network has data with both face recognition and video processing. Convolution neural network (CNN) is trained with large amounts of clinical data images that aid in the detection of skin lesions. Using this information for the detection of skin cancer, computer-aided diagnostics (CAD) has rapidly entered radiology mainstream and is used as a second opinion for breast cancer identification. The algorithm has steps for image processing and image feature analysis. This algorithm helps in reducing the improper representation of temporary images that are used as nonconventional images and they are compared for matching the reference image to the current one that is developed. Figure 10.1 shows how the imaging of the bone cancer happens by temporal subtraction method [1].

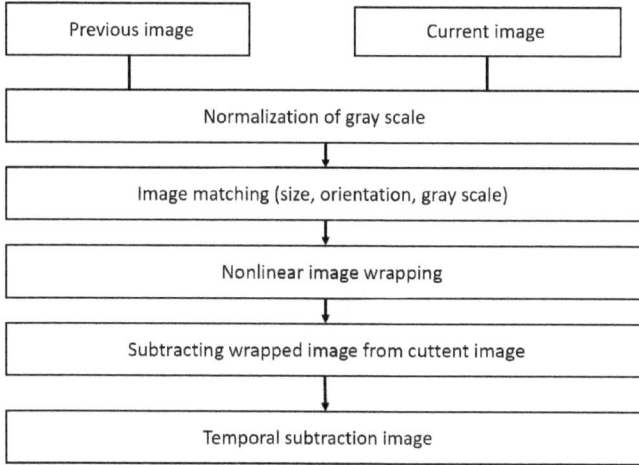

```
┌─────────────────────┐   ┌─────────────────────┐
│   Previous image    │   │   Current image     │
└─────────┬───────────┘   └─────────┬───────────┘
          │                         │
┌─────────┴─────────────────────────┴───────────┐
│           Normalization of gray scale          │
└────────────────────┬───────────────────────────┘
                     │
┌────────────────────┴───────────────────────────┐
│  Image matching (size, orientation, gray scale) │
└────────────────────┬───────────────────────────┘
                     │
┌────────────────────┴───────────────────────────┐
│           Nonlinear image wrapping              │
└────────────────────┬───────────────────────────┘
                     │
┌────────────────────┴───────────────────────────┐
│   Subtracting wrapped image from cuttent image  │
└────────────────────┬───────────────────────────┘
                     │
┌────────────────────┴───────────────────────────┐
│          Temporal subtraction image             │
└─────────────────────────────────────────────────┘
```

FIGURE 10.1   Imaging of bone cancer by temporal subtraction method.

AI is useful in the field of imaging of cancer using three methods: **identification, designation, and as a tumor tracker** [2].

- **Step 1:** Identification means delimitation of regions of interest obtained in radiography by using CADs. They are formulated with a pattern-recognition context or regions. In low-dose computed tomography (CT) (LDCT) scanning, sometimes cancers are missed. CAD is a tool that helps to discern this. It also detects the metastasis in brain magnetic resonance imaging (MRI), which improves the radiology interpretation and sensitivity and locates microcalcification clusters in the renal and early breast carcinoma.
- **Step 2:** Segmentation and cancer staging are diagnosed and then identified by characterization. The expanse of an abnormality can be defined by segmentation. This can variate from basic two-dimensional examination of the maximum tumor diameter to more covered or used volumetric surroundings in which the entire tumor and cells are assessed.
- **Step 3:** Radiologists interpret visually the data obtained to diagnose dubious lesions and classify them to be one of the two: benign or malignant.

AI can be used in diagnostic medical imaging for classification and identification of abnormalities. For example, picture archiving and communication systems (PACSs) are used in digital mammography interpretation that converts the single whole digitally obtained images of the breast into variables like pixels, and computer clusters this information and develops new image that gives the information of features associated with breast cancer. This information also helps in prediction studies and breast cancer risk factor analysis.

Machine learning is one of the growing fields of AI, which uses computer programming and statistics for the automatic extraction and analysis of complex data and helps to discover new materials. Cancer is a procedure that interacts with its microenvironment, which challenges the detection at early stages accurately, distinction between neoplastic and pre-neoplastic lesions, termination of the infiltrative tumor margins, tracking of tumor growth, and events that occur, which makes cells resistance to potential anticancer drugs, metastasis, and recurrence when initial detection of a neoplastic lesion results in requirement of separation from non-neoplastic mimics to optimize the type of the treatments. CTs and MRI help in the detection of clots when the patient is under clinical importance which results in initiation of the cascade observation, advanced testing followed by empirical intervention. Early detection of cancer plays a key role in cancer diagnosis [3].

Traditional radiographic imaging depends upon the qualitative features of the tumor such as tumor density, pattern of study, cellular matrix, tumor size, and relationship with surrounding tissue; these qualitative features are called semantic features. Size-based and shaped measurements will be done in one-, two-, and three-dimensional analysis. Radiomics is based on the virtual coding of the radiographic image of semantic features. Recent advances in AI have automatic radiographic patterns in medical imaging and artificial learning automatically learns the simple features from complex medical imaging data. AI approaches can be applied to the infamous types of cancer: breast, lung, prostate, and brain cancer. AI quantifies the information undetected by humans and complements clinical protocol designing; it also helps aggregation of the multiple data into integrated systems.

There are many updating and internet-available sources that make use of computing systems to collect, classify as well as process insights from different types of clinician-based data, dimensions, as well as distributions, mainly for oncology studies. There is a lot of data getting generated in each clinical trial studies that can be used for better patient enrollment for the next clinical trial studies. Cognitive computing in oncology clinical trials can reduce the cost and time required by the clinical trials and increase efficiency. Systems usually use a large set of data inflow that can classify and differentiate through available medical records and data management systems used by many clinical trial supporting companies. There are different online platforms that can classify data that is linear and nonlinear at a large scale recognizing tumors as well as data associations that were otherwise not voluntarily visible via manual or traditional based analytics [4].

AI in healthcare sector has been used and possibly helps improve doctors' treatment accuracy and reframe their first set of treatment predictions. For example, Watson for Oncology is a clinical decision start-up (CCDS) tool that was framed by experts at Memorial Sloan Kettering Cancer Centre. It has shown elevations in the field of coordination in giving doctors proper suggestions. Another example listed is Manipal Hospitals, a hospital system in Bangalore, India, which uses Watson therapy to treat and develop appropriate treatment consideration factor for many of their patients with cancer.

In cancer imaging, AI is applied to three main tasks: identification, designation, and tumor tracking [5]:

1. **Identification**: It indicates object localization in the radiographs and they are familiarized as computer-aided detection (CADe). This step reduces the error in the initial stage detection and it contains the pattern generation

and pattern recognition steps. CADe is used as an auxiliary tool along with LDCT screening to detect brain metastasis in MRI. This step detects the microcalcification clusters in the screening of mammography that indicates early breast carcinoma.

2. **Designation**: It captures tumor staging, segmentation, prognostication, and outcome-based specific modalities. The level of abnormality can be indicated by segmentation. It ranges from two-dimensional measurement of tumor diameter to the volumetric segmentation of tumor and surrounding tissues assessed. This information is used for the dosage administration calculation. The limitations and inconsistent reproducibility of cells are what define tumors in current practice. Then it detects whether a tumor is benign or malignant.

3. **Tumor tracking**: Temporal monitoring of tumors is limited to tumor predefined matrices, including the longest diameter of the tumor. In AI, the geometry of tumors is expressed via the revolutionary imaging instruments and biomarkers are developed for the longitudinal tracking of tumors. Liquid biopsy and circulating tumor DNA (ctDNA) released from tumor cell are used for monitoring and provide information about resistance-associated cancer mutations in real time. So liquid biopsies combined with radiography increase the efficiency of cancer treatment.

Medical imaging by using AI has a great role in quantifying intratumor characteristics. Sequencing studies conducted on the tumors indicate that the heterogeneity of the intratumor is the common feature of solid tumor cancers. A tumor contains billions of independent tumor cancer cells. Subclones of cancer can be caused by clonal expansion, where the cancer cells share a recent, single, or common ancestor.

Designing of the deep learning models was on the basis of CNNs, autoencoders (AEs), FCN (fully convolutional network), and DBN (deep belief network).

10.1.1 Basic Concepts of CNN, FCN, AE, DBNs

10.1.1.1 Convolutional Neural Networks (CNNs)

They are subgroups of feed forward networks where there is a signal flow in one dimension all over the network and no formation of loops [6].

$$F(x) = f_N \left(f_{N-1} \left(\ldots \left(f_1(x) \right) \right) \right)$$

where
N = numerical value of hidden layers
F_i = function of the corresponding layer i

Each convolutional layer f contains many convolution kernels. If kernel size is 1*1, then a fully connected layer is taken as a convolutional layer.

10.1.1.2　Fully Convolutional Networks (FCNs)

The main classification factor between FCNs and CNNs is that FCNs exchange the fully connected layer with upsampling layers and de-convolutional layers. Upsampling layer and de-convolutional layer represented as backward stretching of layers and convolutional layer and both of them are learning capacities based on training matrix, but they won't be made of probability to classify the images FCN uses score map, which contains input images of the exact size for each class [7].

10.1.1.3　Autoencoders (AEs)

This is a neural network that aids in supervised learning. Generally, there will be three kinds of layers: input, between, and output layers. Learning of feature representation having lower dimensionality obtained from input data is its goal. Training of AE contains two stages of tanning: encoding and decoding stages. The input x is encoded to a representation h using weight matrix $W_{x,h}$ and bias be $b_{x,h}$

$$h = \sigma(W_{x,h}x + b_{x,h})$$

σ = activation function is given by the below equation:

$$\sigma(x) = \frac{1}{1 + exp(-x)}$$

In the decoding step, decoding of the representation reconstructs the output by a fresh weight matrix as well as bias.

$$\hat{x} = \sigma'(W_{h,\hat{x}} h + b_{h,\hat{x}})$$

Sparse autoencoders (SAEs) are encoders in between sparsity that will consider into the hidden layers and makes input node higher than the input nodes. SSAE is a block of SAE containing only encoding layers and is mostly trained by greedy probabilistic theory. They train hidden layer separately and outcoming data of the current hidden layer is reused as input for subsequent hidden layer [8].

10.1.1.4　Deep Belief Network (DBN)

It is a probability model and uses the restricted Boltzmann machines (RBMs) instead of AEs; it has two layers: output and hidden layers.

An energy function E is initiated into the equation

$$E(v,h) = -a^T v - b^T h - v^T W h$$

where a, b = bias vectors of the visible and hidden layers.

RBM is trained to maximize the probability of the energy

$$\text{argmax}_{W,a}\ P(v) = \frac{1}{z}\sum_h \exp(-E(v,h))$$

where Z = partition function.

10.2 LUNG CANCER DETECTION AND MEDICAL IMAGING BY AI

One of the top reasons for death in men and women is lung cancer. Computer-aided treatment has become a part of breast cancer clinical diagnosis. National lung screening trial demonstrated screening with LDCT has reduced 20% mortality. Identification of lung cancer via LDCT screening is highly compliant with surgical cure. Some of the factors that promote as well as result in lung cancer are history of cardiovascular disorders, male gender, age, smoking, poor pulmonary functionality, etc. Application of AI in analysis of medical images started in the 1960s but the advances in the systematic investigation started in the 1980s. The computer output of the CAD is the second opinion in assisting radiologist interpretation of images. Detection of the cancer in lung, breast, and colon is done by screening examinations [9].

Convolutional neural network are used to differentiate, detect, and diagnose the lesions. Image-based biomarkers, conversely and longitudinally, can capture the radiographic images and define the pathophysiology of the tumor. The longest diameter of the tumor is mainly used for response assessment and staging of tumors. These features help to identify the semantic and radiomic features CANARY (computer-aided nodule assessment and risk yield) provides identification of risk stratification on the basis of semantic analysis to identify a set of lung adenocarcinomas. Zhu et al. developed an AI model to predict the survival time of the patients from different lung cancer pathological images and used CNN for the classification of the pulmonary nodules in 2D CT images.

10.2.1 TEMPORAL SUBTRACTION METHOD

Development of this method is based on the chest radiographs. This method of clinical imaging is useful when radiological examination is performed frequently on the same person so the same patient can have more numbers, greater than two, of sequential images for the classification, so that nonrigid image matching can be applied. In this 2-dimensional polynomial warping surface, shift values of x and y coordinates with cross-correlation value of each pair were observed. This correlation value indicates a small region of interest (ROI) (Figure 10.2) [10].

After this first attempt, the temporal subtraction methods are incorporated into CT scan images (Figure 10.3).

Lung cancer testing discovers different numbers of pulmonary nodules. A total of 96.4% are identified in the LDCT but there is no approach to classify the nodules as benign or malignant, so the temporal subtraction method is applied. Image-based biomarkers are developed to capture the radiographic phenotype. CANARY gives the risk-based randomization to identify the subset of lung adenocarcinoma.

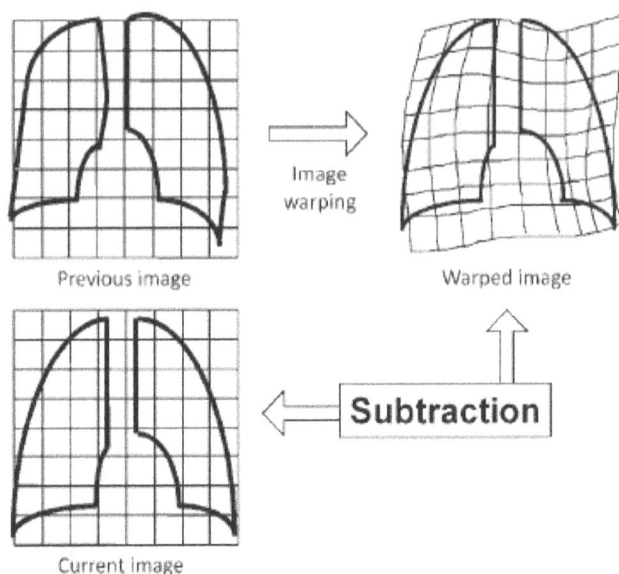

FIGURE 10.2 Principle of temporal subtraction via technique using warping of nonrigid image in successive chest radiography.

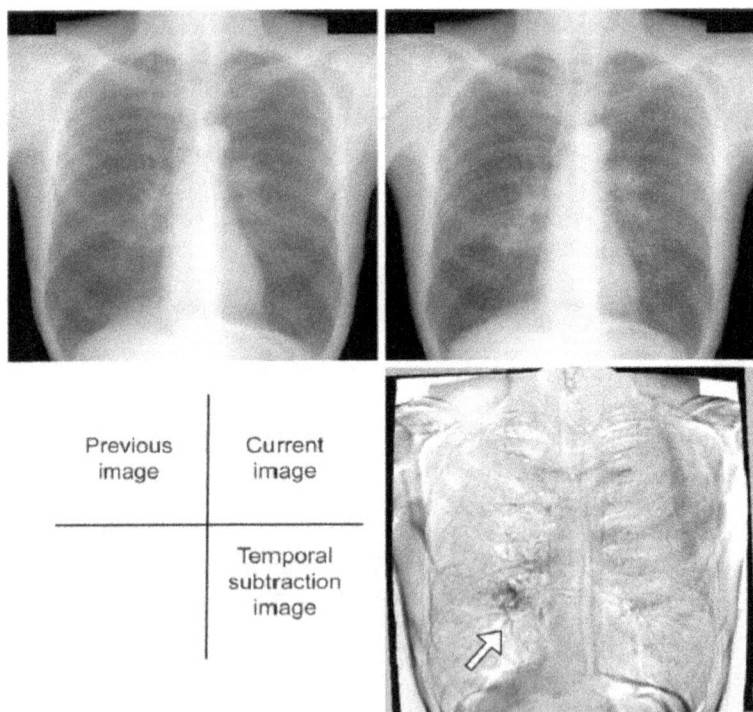

FIGURE 10.3 Example of temporal subtraction image in two sequential chest radiographs. Progression of a lung mass in right lower lung was clearly identified in the temporal subtraction image.

Primarily AI predicts the probability of lung cancer, distant metastasis in lung adenocarcinoma, and heterogeneity [11–13].

10.2.2 SUMMARY OF THE ROLE OF AI IN LUNG CANCER DETECTION AND IMAGING

Type of Tumor	Application	Imaging Modality	AI Algorithm	Type of Feature Imaging	Validation
Non-small cell lung carcinoma	• Prediction of cellular as well as molecular pathways • Tumor phenotype prediction	CT	• Small vector machine • Regression	Radiomic features that are predefined	Independent validation with multicenter data
Benign and malignant	Predict lung cancer in screening	CT	Small vector machine	Deep learning radiomic	Independent validation with multicenter data
Non-small cell lung carcinoma	Discrimination of non-ALK and ALK tumors	CT	Classifier of random forests	Semantic	Independent validation with multicenter data
Lung adenocarcinoma	• Differentiation between indolent and aggressive adenocarcinoma • Distant metastasis prediction	CT	• Previously built CANARY model • Regression	• Semantic (CANARY) • Radiomic features that are predefined	Independent validation with • Multicenter data • Single-center data
Non-small cell lung carcinoma	• Prediction of status of mutation, premature identification • Prediction of cancerous cells in the pulmonary nodules of the lung	CT	• Classifier of random forests • Multiple supervised technique	• Radiomic features that are predefined • Semantic	Independent validation with single-center data

10.3 ARTIFICIAL INTELLIGENCE IN BREAST CANCER SCREENING

AI is used in breast cancer because of available capabilities of tumor identification and automated image feature analysis. Picture archiving communication systems (PACSs) have a large space capacity, for which they are linked to electronic medical records. Search strategy is performed by medical literature such as MEDLINE, IEEE [8]. After this, biopsy and Pap smear are done, then images are processed. The importance of processing of images is done in two ways:

1. Improvement of images for human data interpretation;
2. Processing of images for classification understanding and insights by machine;

CNN-based support vector machine (SVM) is used to detect the cell division specifically mitosis in the breast cancer pathological images. AlexNet was used for training of CNN to classify tumors that are malignant or benign. Chen et al. proposed an AI model to detect mitosis of breast cancer. First, they trained FCN model to differentiate tumor types, then trained CaffeNet model on the large-scale image dataset. For improving the robustness, three networks having varied configurations of completely randomized layers underwent training for the generation of multiple scores/probabilities, as well as the average of these scores being selected for the final result [14].

10.3.1 IMAGE PROCESSING STEPS

10.3.1.1 Image Acquisition

Capturing the image and data is kept as virtual images. The image format is normally gray map because it will not erase the data while compressing images (Figure 10.4).

10.3.1.2 Image Preprocessing

In this step, quality of the image is improved by eliminating noise. Preprocessing is done through middle filter. Some important techniques used are FPN, bad-pixels, vignetting, temperature calibration, and noise smoothing.

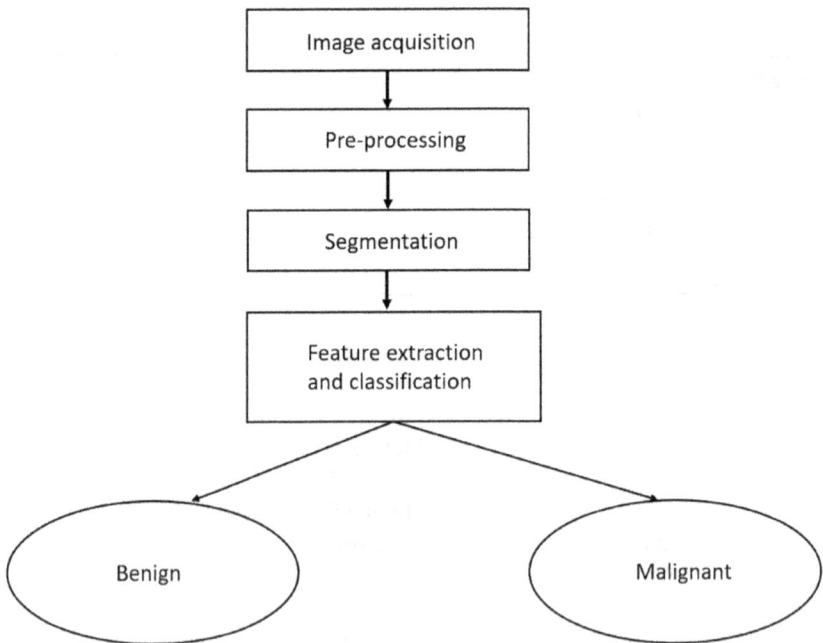

FIGURE 10.4   Image processing of breast cancer.

10.3.2 PREPROCESSING TECHNIQUES

FPN	FPN is the outcome of excluding receptiveness of markers to approaching irradiation
Bad-pixels	Do not give any specific information and only specifies about detectable image contrast
Vignetting	Vignetting is another background noise source on thermograms; given the image center, there will be a darkening of image corners
Temperature calibration	Transformation classification is utilized for transferring the infrared camera gray-scale values g to a linear increment ion having a quantity physical in nature
Noise smoothing	More elaborate noise removal techniques

10.3.2.1 Image Segmentation

This step involves cutting of an image and classifying it into constructive parts.

10.3.2.2 Feature Extraction

Converting the input images into a set of extracted features is called feature extraction. These are the common ways to extract features: (1) transform features, (2) spatial features, (3) corner and boundary characteristics, (4) shape features, and (5) color and textile features [15].

10.3.3 ADVANTAGES AND DISADVANTAGES OF THE IMAGING TECHNIQUES FOR CANCER

Imaging Method	Application	Advantages	Disadvantages
Mammography	Breast cancer detection in early stages	1. Utilizes low-level X-ray for imaging 2. This method is ideal for classification	1. Radiation risks 2. Double reading of mammography increases the cost 3. False results 4. Difficulty in result interpretation
Ultrasound	Used for soft and denser textured tissues	1. Usable for women with dense breast 2. Widely available 3. Quick and highly sensitive	1. Quality and image processing requires highly skilled person during scan
Thermography	Suitable for muscle tissue	1. Noninvasive	1. Low-quality images

10.3.4 CNS TUMOR IMAGING

There are three main challenges to existing tumors arising from nonneural tissues, including meningiomas, pituitary tumors, schwannomas, and lesions of skulls: (1) accurate diagnostics, (2) tracking of neoplastic diseases, and (3) capability of extraction of genotypic features from the phenotypic manifestation of tumor imaging. Machine learning techniques trained MRI for automatic imaging that helps in the individualized treatment of brain tumors. Glioblastomas are most common in adults and more than 4500 people are diagnosed with glioblastomas. Neural networks are developed neural networks to validate the therapeutic response of brain tumors [16].

It uses a reference database with MRI scans of almost 500 brain tumor patients. The algorithms were able to localize tumors, volumetrically measure the individual areas and precisely assess the response to therapy.

Main goal of using AI in the tumor imaging is locating the ROI, i.e., complete tumor, tumor core, and expansion of tumor core to achieve this

1. The group of features called circular concurrent sensitive (CCS) was developed. CCS characters fully utilize the histogram information of rays along with various length;
2. Gradient information is used to extract features from two-dimensional or three-dimensional images;
3. NMR feature selection algorithm is used to detect characters that have the least redundancy and more relevance [17].

10.3.4.1 Artificial Intelligence Model

Deep learning is used to model is used to visualize to study brain membrane's cortex features such as lines, shapes, dimensions and entire objects. It uses different units of graphical processing and improved mathematical randomization methods. Usually back propagation and stratification are used and CNNs are used for the image-based problems (Figure 10.5) [18].

- **Step 1: Image acquisition and preprocessing**
  Most of the real-time brain images are obtained from the web of Harvard Medical School. This step is used to reduce the noise and give high-quality resolution of image.
- **Step 2: Segmentation techniques for medical imaging**
  The most common techniques used for segmentation were C-means as well as fuzzy sets used in combination with a variety of techniques for better execution. These are the following techniques used for segmentation:
  1. Feedback pulse-coupled neural network (FPCNN)—uses feedback loop characteristic where stratification experience feedback optimization that is variable for the entire input.

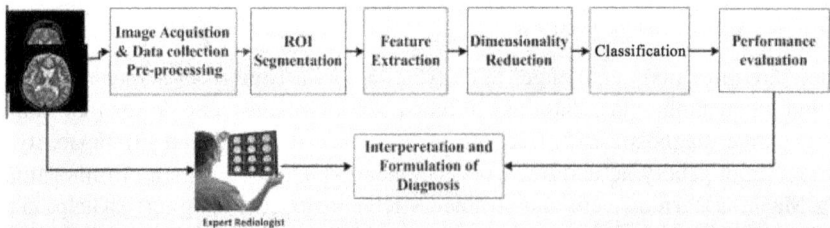

FIGURE 10.5 Processing of CNS tumor prediction.

2. Seeded region growing segmentation (SRGS) along with connected component labeling (CCL)—begins with detection of the brain from skin-neck-bone followed by ventricles and then distinguishing regions of the brain from the scalp and tumor tissues from normal tissues via MRI of the brain (based on pathological testing).
3. Possibilistic neuro fuzzy C-means algorithm (PNFCM)—Segmentation of tumor, white matter, edema, and gray matter on brain MRI (T1-weighted, T2-weighted, proton density) [19].

10.3.4.2 Advantages and Disadvantages of Segmentation Techniques

Technique Name	Advantages	Disadvantages
Level set	Important method for segmentation of medical image and handles curves, convolution, concavities, merging, or splitting	Approach requires the identification of inceptive noise as well as only provides important answers given that the noise is allocated almost symmetrically with respect to the object boundary
Region growing	Could correctly differentiate the regions that have the same properties	The user needs to manually choose the main point
SVM	Given a condition of high-feature space dimension and high generalization performance, this proves to be a good choice	The training time is very high
ANN	Noisy fields improve performance	Gradient-type learning method is utilized but is very time-consuming
PCNN	Differentiates between TF and TN	Satisfactory result
SOM	The training algorithm that was simpler as well as quicker	Options, including a variety of tuneable criteria, that can stop prospective users from tracking further SOM applications for new users

- **Step 3: Brain MRI classification post feature extraction**

The below techniques are used for feature extraction.

10.3.4.3 Feature Extraction Techniques

Feature Extraction	Classification Technique	Data
Pattern analysis	Hierarchical descending classification	The hospital's database provided images
Texture-based feature extraction + nonparametric Wilcoxon rank sum test method	Nonlinear least squares feature transformation with probabilistic neural network	General Hellenic Air-force Hospital, MRI unit
Manual feature extraction	Based on SVM	Collected data from 98 patients
Wavelet transform	Self-organizing map, support vector machine, support vector machine possessing kernel based on radial basis function	DSC-MRI data

Fractional anisotropy (FA) as well as mean diffusivity (MD) voxels having subject class indicative variable	Support vector machine classifier	DSC-MRI data

---

10.3.4.4   Feature Extraction by Artificial Neural Networks

This network identifies two kinds of signals: function signal and error signal. Function signals get propagated forward via the network's hidden layers to produce the output signal. Error signals are produced at the output neuron of the network, following which they get propagated backward through the same network (Figure 10.6) [20].

$$y = F_0\left(\sum_{j=0}^{M} W_{0j}\left(F_h\left(\sum_{i=0}^{N} W_{ji} X_i\right)\right)\right) \tag{10.1}$$

where
W_{0j} = synaptic weights;
X_j = input vector;
F_0 and F_h = activation functions of the neuron hidden layers;
W_{ji} = connection weights from hidden layer to input layer;

The learning phase is shown below. Network proceeds by adjusting free parameters. The system output is based on mean square error (E) described by Equation (10.2).

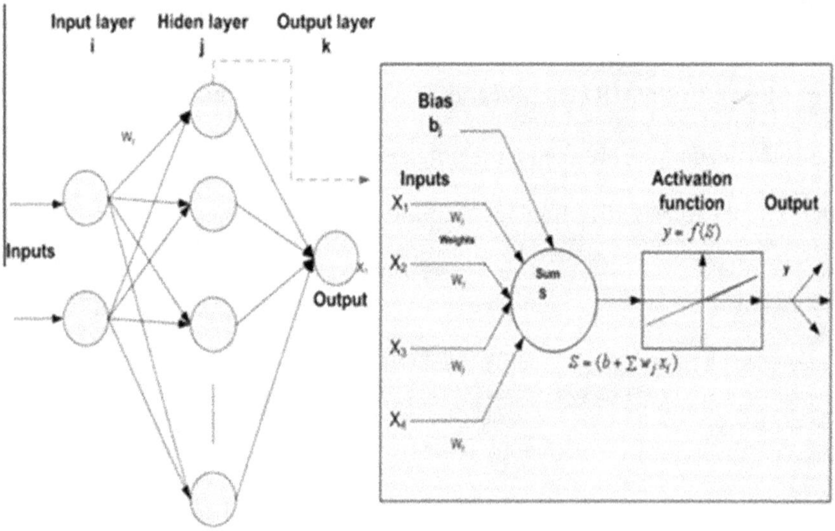

FIGURE 10.6   Architecture of artificial neural networks.

$$E = \frac{1}{2} \sum_{i=1}^{m} (y_i - d_i)^2 \tag{10.2}$$

y_i = output value;
d_i = anticipated output;

Testing the network with vectors will terminate the learning process, given that the error between subsequent errors and output values are minimized.

10.3.4.5 Performance Evaluation

Calculating the stability and functionality of the CAD system was done by utilizing conclusion matrix. Conclusion matrix possesses the details of actual, predicted, and existing classifications. Algorithm is required for calculation of matrix sensitivity, specificity, and accuracy:

$$\textbf{Sensitivity} \left(\text{true positives} \right) = \frac{TP}{(TP + FN)} \tag{10.3}$$

$$\textbf{Specificity} \left(\text{false positives} \right) = \frac{TN}{(FP + TN)} \tag{10.4}$$

$$\textbf{Accuracy} \left(\text{percent of accurately classified samples} \right) = \frac{(TP + TN)}{(TP + FN + TN + FP)}$$

where
  TP: (true positives) is the correctly classified positive cases;
  TN: (true negative) is the correctly classified negative cases;
  FP: (false positives) is the incorrectly classified negative cases;
  FN: (false negative) is the incorrectly classified positive cases;

The receiver operating characteristic (ROC) curve displays the trade-off between the sensitivity as well as (1-specificity) across a set of cut-off points. The accurate effective measure of inherent validity of a diagnostic test is by the area under the ROC [21].

10.3.4.6 Network Implementation

The model was trained using MATLAB version 10.0 using an amalgamation of PCNN wavelet as well as neural networks toolboxes.

10.4 ARTIFICIAL INTELLIGENCE APPROACHES TO LYMPHOMA AND MYELOMA

Lymphoma and myeloma are cancers of the immune systems. Histogram-oriented gradient (HOG)-based SVM is used for accurate detection of a number of leukocytes and specific subsets (T cell and B cell); HOG-based SVM helps in the separation of cells from background. CNNs are trained for separation of T cells from B cells with a specificity of 99%, accuracy of 98%, and sensitivity of 97%. HOG-based SVM classifier is used for initial selection of the detection windows. HOG functions are given illuminations and fluctuations in the cell size and shape. CNN detects T and B cells from the fluorescence microscopy images. Samples were collected from healthy tumor cells, but a conventional dataset is unavailable for the fluorescent images [22].

- **Step 1: Sample collection**
  Healthy donors were the sources of human blood samples. Isolation of leukocytes from peripheral blood is done with high accuracy as well as efficiency preventing clotting using a specially designed microfluid chip. 1 µL of the peripheral blood has to be poured on the chip. Once the sample flows through the chip sheath lead is added. Sheath liquid removes the non-trapped cells. A CMOS camera captured the images.
- **Step 2: Dataset preparation**
  A conventional dataset is unavailable, so 6300 images of cells and 2500 background images are collected.
- **Step 3: training of network by using artificial intelligence**
  Training CNN with miniature dataset is known as transfer learning. Transfer learning uses preexisting CNN and refines it with the latest dataset in a very short time. Transfer learning is done by eliminating the last layer of the network and using rest of the parameters. Last layer is exchanged with a fully connected network. It uses the stochastic gradient net algorithm.

$$Q(w) = \frac{1}{n} \sum_{i=1}^{n} Q_i(w), \tag{10.5}$$

- $Q(w)$ = parameter to be estimated;
- $Q_i(w)$ = ith observation in the data set;
  Iterations are done for minimizing error to calculate the weight of network after each iteration they use (Equation 10.6)

$$w := w - \eta \nabla Q(w) = w - \eta \sum_{i=1}^{n} \nabla Q_i(w)/n,$$

- **Step 4: Classification**
  There are two layers of classifications; first layer of classification HOG-SVM is used to separate cells from background noise in the window in which positive detection is shown. All others that neglected the output of

the first classifier are the inputs of the secondary classifier, i.e., AlexCAN. It is a function provided by MATLAB. In the case where a cell is absent in the pre-chosen window, it is considered a miss.

- **Step 5: Obtaining results by ROC**
  ROC is plot of true-positive rate versus false-positive rate; using this information test set is calculated by sensitivity accuracy and specificity.

10.4.1 PREDICTION MODELS FOR THE LYMPHOMA

Protein expression of FOXP1, molecular risk models, and gene expression prediction models are used to detect cancer prediction and prognosis [23].

10.4.2 SKIN CANCER DETECTION BY USING AI

Skin cancer is mainly found in fair-skinned individuals. It is the most common malignancy and is mainly diagnosed using initial clinical screening prior to dermoscopic analysis, biopsy, and histopathological examinations. AI is used for automatic classification of skin lesions. They took input as pixels and disease labels and trained single CNN using a dataset of 129,450 clinical images containing 2032 different diseases. To classify the clinical images, two different binary classes are there [24, 25]:

1. Malignant carcinoma versus benign keratosis;
2. Malignant melanomas in comparison to benign nevi;

The initial case shows the most habitual cancer and the next one is the deadliest skin cancer.

CNN is used to classify the skin lesions by using photographic and dermoscopic images. CNN can be delivered through mobile networks allowing patient remote access to dermatologist-level diagnosis. This is used to increase the primary care

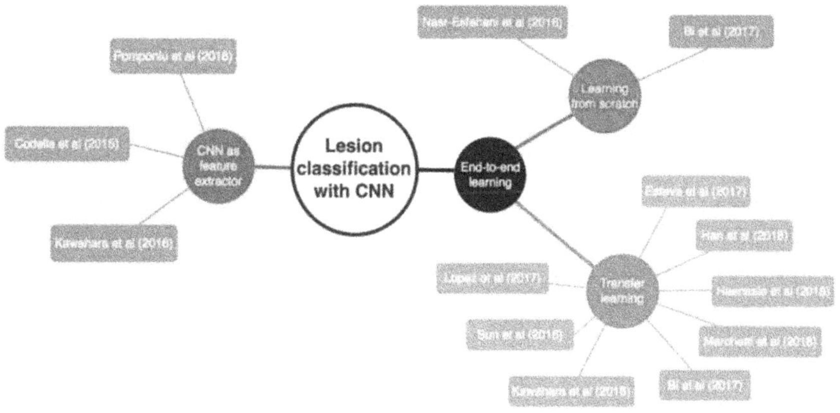

FIGURE 10.7   CNN model for skin lesions.

beyond the clinic and expands the scope of decision-making. The application of CNN is nonpigmented and non-melanocytic. The model is described in Figure 10.7.

Taking out the CNN layers that are completely connected and are pretrained with big datasets is a way to include a randomized CNN. ImageNet was used for grouping skin lesion, which used 399 images obtained through a standard camera for grouping of melanomas in comparison to benign nevi as well as data augmentation, including processing. The extraction of representational features was performed using a pertained AlexNet. Distance metrics were used alongside the k-nearest-neighbor classifier for the lesions. About 91.1% was the sensitivity attained by the algorithm as well as a specificity of 96.18%, including an accuracy of 95.64%. Manually annotated ROI for every skin lesion is major for nonexistent independent test dataset. Kawahara et al. utilized a linear classifier for the classification of ten unique skin lesions, and then the performance of feature extraction took place with the help of an AlexNet with a convolutional layer instead of a last fully connected layer [26, 27].

10.4.3   CNN Model Trained with Transfer Learning

CNN model was refined with transfer learning for the classification of the skin lesions. A tree-structured taxonomy is used where individual diseases form the leaves of the tree. Visually as well as clinically similar inner nodes are grouped together. The CNN does not have two-dimensional vector as output instead it gives probability distribution over each training class. Many CNNs were not trained for the same classification but three Res-Nets for various issues were trained: the original three class problems as well as another for two binary classifiers.

CNN is composed of multiple layers and each layer takes into consideration the identical image at a variety of resolutions. The final layer amalgamates the output obtained at each layer. The CNN identifies interactions of different image resolutions and the weighting parameters are completely optimized by end-to-end learning. DermQuest is used for training and testing and the performance of the CNN models CaffeNet and VGGNet is used for this classification problem [28, 29].

10.5   CONCLUSION

AI is successfully used for oncology but there are several disadvantages as well. With the large applications of imaging in the health sector, the machine of huge quantities of data is being consistently produced by providers. Standards, including the PACS as well as the digital imaging and communications in medicine (DICOM), prove that data is structured to fulfill simple visualization as well as retrieval. Clinical trials are mainly divided into four different phases to estimate efficiency and dosage, and to test the side effects of the drug. First, they will start with the preclinical studies at laboratory and a drug is generally tested first in healthy human beings for its efficiency testing. In phase 1 of clinical trials, mainly four things are focused on, which will evaluate key components required for the phase 2 clinical trials and which will take into consideration points such as evaluating short-term safety and generating pharmacodynamics and pharmacokinetic properties data collections identifying the optional dose for larger and durable clinical applications [30].

Using software tools, complex problems of biological connections may have a proven effect on the examination of response and evaluation, and treatment planning. A backward elimination starts with the candidate variable approach and they are sequentially applied to test the model. So this helps to determine which variables should be considered in the final model or full model development for the limited sample size and limited dataset. In appropriate selection of the variables, the main problem of poor performance of the clinical trials is the predictive modeling; they should be specific to the particular disease.

REFERENCES

1. Aerts HJWL, Velazquez ER, Leijenaar RTH, Parmar C, Grossmann P, Carvalho S, Bussink J, et al. Decoding tumour phenotype by noninvasive imaging using a quantitative radiomics approach. Nat Commun 5, no. 1 (2014): 1–9.
2. Norden AD, Dankwa-Mullan I, Urman A, Suarez F, Rhee K. Realising the promise of cognitive computing in cancer care: Ushering in a new era. JCO Clin Cancer Inf 2 (2020): 1–6.
3. Bi WL, Hosny A, Schabath MB, Giger ML, Birkbak NJ, Mehrtash A, Allison T, et al. Artificial intelligence in cancer imaging: Clinical challenges and applications. CA Cancer J Clin 69, no. 2 (2019): 127–157.
4. Cheng HD, Cai X, Chen X, Hu L, Lo X. Computer-aided detection and classification of microcalcifications in mammograms: A survey. Pattern Recognit (2003).
5. Castellino RA. Computer aided detection (CAD): An overview. Cancer Imaging 2 (2005): 16–28.
6. LeCun M, Bengio Y, Hinton G. Deep learning. Nature 521, no. 7553 (2015): 436–444.
7. Liang M, Tang W, Xu DM, et al. Lowdose CT screening for lung cancer: Computer-aided detection of missed lung cancers. Radiology 1 (2016): 279–288.
8. Li XA, Tai A, Arthur DW, et al. Variability of target and normal structure delineation for breast cancer radiotherapy: An RTOG Multi-Institutional Multiobserver Study. Int J Radiat Oncol Biol Phys 1 (2009): 944–951.
9. Maldonado F, Duan F, Raghunath SM, et al. Noninvasive computed tomography-based risk stratification of lung adenocarcinomas in the National Lung Screening Trial. Am J Respir Crit Care Med 1 (2013): 56–78. 10.5815/ijigsp.2013.06.03.
10. Marsadraee R, Oswal D, Alizadeh Y, Caulo A, van Beek E Jr. The 7th lung cancer TNM classification and staging system: Review of the changes and implications. World J Radiol 2 (2012): 12–17.
11. Manser S, Lethaby A, Irving LB, et al. Screening for lung cancer [serial online]. Cochrane Database Syst Rev 1 (2013): 10–12.
12. Massion PP, Walker RC. Indeterminate pulmonary nodules: Risk for having or for developing lung cancer? Cancer Prev Res 1 (2014): 1173–1178.
13. McKee Regis S, Borondy-Kitts AK, et al. NCCN guidelines as a model of extended criteria for lung cancer screening Natl Compr Canc Netw 4 (2018): 444–449.
14. Miyers MH, Ries LAG. Cancer patient survival rates: SEER program results for 10 years of follow-up. CA Cancer J Clin 1 (1989): 10–15.

15. Murugaesu N, Wilson GA, Birkbak NJ, et al. Tracking the genomic evolution of esophageal adenocarcinoma through neoadjuvant chemotherapy. Cancer Discov 5 (2015): 821–831.

16. Nielsen M, Kareore G, Loog M, et al. A novel and automatic mammographic texture resemblance marker is an independent risk factor for breast cancer. Cancer Epidemiol 4 (2011): 381–387.

17. Nishikawa RM. Computer-aided detection and diagnosis. In: Bick U, Diekmann F, eds. Digital Mammography. Berlin, Germany: Springer; 2010.

18. National Lung Screening Trial Research Team, Aberle DR, Adams AM, et al. Reduced lung-cancer mortality with low-dose computed tomographic screening. N Engl J Med 5 (2011): 395–409.

19. Rasch C, Barillot I, Remeijer P, Touw A, van Herk M, Lebesque JV. Definition of the prostate in CT and MRI: A multiobserver study. Int J Radiat Oncol Biol Phys 1 (1999): 57–66.

20. Rasprecht O, Weisser P, Bodelle B, Ackermann H, Vogl TJ. MRI of the prostate: Interobserver agreement compared with histopathologic outcome after radical prostatectomy. Eur J Radiol 3 (2012): 456–460. https://doi.org/10.1016/j.acra.2012.02.009.

21. Rumi-Porta R, Bolejack V, Giroux DJ, et al. International Association for the Study of Lung Cancer Staging and Prognostic Factors Committee, Advisory Board Members and Participating Institutions. The IASLC lung cancer staging project: The new database to inform the eighth edition of the TNM classification of lung cancer. J Thorac Oncol 1 (2014): 5–12.

22. San R, Limkin EJ. A radiomics approach to assess tumor-infiltrating CD8 cells and response to anti-PD-1 or anti-PD-L1 immunotherapy. An imaging biomarker, retrospective multicohort study. Lancet Oncol 9 (2018): 1180–1191. https://doi.org/10.1016/j.jtho.2018.11.023.

23. Scholtz JE, Lu MT, Hedgire S, et al. Incidental pulmonary nodules in emergent coronary CT angiography for suspected acute coronary syndrome: Impact of revised 2017 Fleischner Society guidelines. J Cardiovasc Comput Tomogr 12 (2018): 28–33.

24. Schwartz LH, Seymour L, Litiere S, et al. RECIST 1.1—Standardisation and disease-specific adaptations: Perspectives from the RECIST Working Group. Eur J Cancer 62 (2016): 138–145.

25. Schabath MB, Haura EB. Epidemiology of non-small cell lung neoplasms. In: Cameron RB, Gage DL, Olevsky O, eds. Modern Thoracic Oncology. Volume 2: Trachea, Lung, and Pleura. Singapore: World Scientific Publishing; 2018.

26. Schbath MB, Thompson ZJ, Gray JE. Temporal trends in demographics and overall survival of non-small-cell lung cancer patients at Moffitt Cancer Center from 1986 to 2008. Cancer Control 21 (2014): 51–56.

27. Suwen N, Acou M, Sima DM, et al. Semi-automated brain tumor segmentation on multi-parametric MRI using regularized non-negative matrix factorization [serial online]. BMC Med Imaging 17 (2017): 1–14.

28. Swhiner B, et al. Classification of mass and normal breast tissue: A convolution neural network classifier with spatial domain and texture images. IEEE Trans Med Imaging 15 (1996): 598–610.

29. Seltzer SE, Getty DJ, Tempany CM, et al. Staging prostate cancer with MR imaging: A combined radiologist-computer system. Radiology 202 (1997): 219–226.

30. Zhou H, Vallieres M, Bai HX, et al. MRI features predict survival and molecular markers in diffuse lower-grade gliomas. Neuro Oncol 19 (2017): 862–870.

Part IV

HMI and IoT in Cloud Setup

11 First-Mile Ridesharing Using Autonomous Shuttle Service and IoT Cloud Platform

Shyam Sundar Rampalli
TUMCREATE,
Singapore

Pranjal Vyas and Anuj Abraham
A*STAR,
Singapore

Justin Dauwels
TU Delft,
Netherlands

CONTENTS

DOI: 10.1201/9781003268796-15

TABLE 11.1
Recent Works on Ridesharing Problem

Ridesharing Problems	Methodology	Remarks
Dial-a-ride problem [1]	A set of designated vehicles is given, which can provide shared service for ride requests from the passengers	It is formulated as a single objective function
Carpooling problem (static ridesharing) [2]	Both an exact and heuristic method is used	It is based on two integer programming (IP) formulations
Dynamic ridesharing problem [3]	A generational genetic algorithm to solve such an IP formulation	The objective function contains four optimization goals at the same time (multi-objective function)
Dynamic ridesharing problem with rolling horizon [4]	Optimization-based approaches in practical environments	Developed simulation environment based on travel demand model data
Large-scale real-time ridesharing [5]	Developed a kinetic tree algorithm that is capable of better scheduling dynamic requests and adjusting routes	Model is based on large-scale taxi dataset

11.1 INTRODUCTION

Machine interaction and Internet-of-Things (IoT) have made the presence in almost every industry and transportation is no more exception. IoT used for transportation can help in tracking vehicles and passengers, connect infrastructures, and plan better rides for people. Ridesharing is one common transportation mode that aims in reducing congestion and hence is the need for today's urban traffic scenarios. Table 11.1 shows the recent work carried out on the ridesharing problem over several years.

In general, there are two types of ridesharing: static and dynamic. In static ridesharing, both the vehicles and passengers are known in advance, and once the ridesharing strategy is computed in advance, no further change will be made, whereas in dynamic ridesharing, each trip arrives online and a vehicle is assigned for an arrived trip without the knowledge of trips in the future. The most common characteristics of the ridesharing problem considered are the independent vehicles, automatic matching, and cost-sharing between the vehicle and the passengers. Other factors include carpooling, dynamic assignment in real time based on internet-enabled mobile devices. To achieve this, many mathematical optimization models are developed, to minimize either the total travel distance or time and costs of vehicles' trips or the passengers' trips and the total number of vehicles required. Other studies were also seen to maximize the number of served travel requests.

In this chapter, we focus on first-mile (FM) ridesharing problem. FM problem [6] in urban transport systems refers to the connection between the passengers' homes and public transport hubs such as metro stations or bus interchange. This FM gap compels passengers to choose a private mode of transport over public transport systems [7]. The problem lies in providing reliable public transport with an option that is profitable to the operator and reliable to the passenger at the same time. An efficient fleet that works on demand and served by shuttles of limited capacity rather than the traditional loop lines served by buses that have high capacity and higher operating cost might be economically viable and efficient. The problem lies where there is low or unreliable demand. In such cases, loop lines that are often served by buses are not envisioned as the best way to operate [8].

Autonomous vehicles (AVs) supposedly can make more intelligent routing decisions in a real-time traffic scenario along with efficiency in fleet management, which could be an even more efficient and reliable choice. This chapter defines an FM transportation problem that is similar to a capacitated vehicle routing problem (VRP) with multiple depots [9]. Demand from certain stops that are described in detail below will be fed to transport hubs or metro stations by AV-based requests made in a time window. Passengers who use this service are required to reach these designated stops to use the service.

Extensive research has been done in the field of operations research related to vehicle routing and related problems. Some of these approaches can also be used for first- and last-mile transport systems. In [6], the authors proposed one such application of mixed-integer linear programming (MILP) to solve the FM problem using AVs. A similar MILP model is implemented in [10] to solve the last-mile problem that included scheduling. Another relevant use case is the demand-responsive transport (DRT) problem [11, 12]. The research in [13] discusses a greedy local optimization heuristic algorithm to solve a typical FM DRT problem optimizing the vehicle miles traveled (VMT). A constrained DRT problem using the traditional MILP formulation used a two-stage approach where a heuristic is implemented first followed by MILP formulation for vehicle routing.

This chapter presents a two-level solution approach for FM transit service. The significant contributions are listed below:

- A two-level heuristic approach and a MILP model are proposed for clustering passengers based on demand and spatial information.
- The vehicle routing has been performed using Tabu search.
- A comparative study has been carried out by combining either of these two approaches.
- Simulation result has been presented for real data in a Singapore-based scenario.

The rest of the chapter is organized as follows: Section 11.2 describes the FM ridesharing problem and various assumptions considered. Section 11.3 presents the MILP approach for associating the passenger's request with the nearest vehicle. Section 11.4 proposes a two-level heuristic approach for clustering the pick-up locations based on demand and spatial information. Section 11.5 describes vehicle routing methodology for the traveling salesman problem, which uses the results of the MILP approach and two-level heuristic approach for completing the vehicle routing. Section 11.6 presents results with real data from an urban area in Singapore. Section 11.7 discusses the conclusion and future work.

11.2 PROBLEM DESCRIPTION

FM ridesharing problem is a variant of the VRP that involves dispatching vehicles for transporting the passengers from their respective locations to the nearest public transport hubs. The objective here is to minimize the overall transportation cost by reducing the total VMT by the entire fleet. The solution will provide the passengers to reach their destination in a shorter time compared to traditional loop lines with a fixed route. The passengers who wish to use this service need to book a ride in advance. All passengers' requests are aggregated within a time span, which is then served in the next time span. The parameter time span can be decided by the operator based on certain factors such as fleet-size availability, demand, and operational area.

Let $\mathbb{N}$ be the set of n pick-up locations, $\mathbb{M}$ be the set of m vehicles, and $\mathbb{D}$ be the set of drop-off locations. Let $\mathbb{G}$ be the union of $\mathbb{N}$, $\mathbb{M}$, and $\mathbb{D}$ with total elements be $p = n + m + d$. In FM problem, $\mathbb{D}$ has a single element (i.e. $d = 1$) that is the public transport hub. Each pick-up location has passenger request (demand at that pick-up location) that must be served by a single or multiple vehicles. Let the total requests be r. Each vehicle has a fixed capacity to accommodate q passengers. The total request r can be less or more than the total vehicle capacity mq. If the total requests exceed the full vehicle capacity, then the requests are served until the total fleet capacity is exhausted. The remaining requests shall be considered in the next iteration (i.e. time span). The objective here is to minimize travel cost and serve every passenger until the vehicle capacity exhausts. Let $\mathbb{S} = s_1, s_2, \cdots$ be the feasible solution space. Each solution $s_i \in \mathbb{S}$ needs to satisfy basic constraints:

- Every route of a vehicle ends at a public transport hub node.
- The load of the vehicle cannot exceed the total capacity at any time.

The next section presents a MILP model to associate pick-up locations with their nearest available vehicle.

11.3   MIXED INTEGER LINEAR PROGRAMMING-BASED CLUSTERING: APPROACH I

This section describes a technique for associating pick-up locations with the nearest available vehicles. This problem can be formulated as multi-knapsack problem using MILP. The vehicles can be assumed as analogous to different knapsacks and pick-up locations as the items to be placed in these knapsacks. The request at each pick-up location can be considered item weights and the distances of pick-up locations with each vehicle as values. On the contrary, the objective here will be to minimize the total cost in the knapsacks (i.e. sum of distances) rather than maximizing it, as pick-up locations have to be associated with the nearest available vehicle. Each node (i.e. pick-up location) is envisioned to be shared by more than one vehicle. Hence, the items are considered individual passengers rather than pick-up locations. The minimization of the objective function will result in a zero value as none of the items (i.e. passengers) will be allocated to the knapsacks (i.e. vehicles). Therefore, an extra constraint of limiting a minimum number of passengers for the vehicles is imposed. For this problem, it is the vehicle capacity itself.

Let E and Y be a matrix of size $m \times r$. Each row and column of the matrix represents vehicles and passengers, respectively. The elements e_{kl} of matrix E are the distances between passengers and vehicles and matrix Y represents the decision variables y_{kl} where $k \in \{1, \ldots, m\}$ and $l \in \{1, \ldots, r\}$. The decision variables are defined as follows:

$$y_{kl} = \begin{cases} 1, \text{ if } k\text{th vehicle picks up } l\text{th passenger in the route} \\ 0, \text{ otherwise.} \end{cases}$$

The objective function is defined in Eq. (11.1) as the sum of distances between passengers and vehicles such that the passenger gets allocated to the nearest available vehicle. The objective function and constraints are defined as follows:

$$\text{minimize} \sum_{k=1}^{m} \sum_{l=1}^{r} e_{kl} y_{kl} \tag{11.1}$$

subject to

$$\sum_{k=1}^{m} y_{kl} \begin{cases} = 1, \text{if } r \leq mq \\ \leq 1, \text{if } r > mq \end{cases} \quad \forall l \in \{1,\dots,r\} \tag{11.2}$$

$$\sum_{k,l=1}^{m,r} y_{kl} = \begin{cases} r, \text{if } r \leq mq \\ mq, \text{if } r > mq \end{cases} \tag{11.3}$$

$$\sum_{l=1}^{r} y_{kl} \leq m \quad \forall k \in \{1,\dots,m\} \tag{11.4}$$

Constraints can be formulated for two conditions: (a) when the total requests are less than the whole vehicle capacities, each passenger must get allocated to a vehicle while some of the vehicle seats may remain vacant and all the requests shall be fulfilled; (b) on the contrary, when the total requests are more than full vehicle capacities, the passengers are allocated until the vehicle capacity exhausts. Therefore, the remaining requests shall be considered for the next round and the total request fulfilled is equal to the total vehicle capacity. This constraint is illustrated in Eqs. (11.2) and (11.3). Eq. (11.4) refers to the constraint where the total number of passengers allocated to a vehicle will always be less than the vehicle capacity. The next section presents an alternative two-level heuristic approach for associating vehicles to the respective pick-up locations.

11.4   *K*-MEANS-BASED CLUSTERING: APPROACH II

This section presents the K-means algorithm-based technique for pick-up location allocation. K-means is one of the popular unsupervised clustering algorithms used for clustering spatial and non-spatial datasets. The goal of the K-means clustering algorithm is to group the data points into non-overlapping subgroups. The inter-cluster data points are kept as similar as possible while we have tried to keep the clusters as different as possible. The similarity between the data points is usually determined using distance-based measurements such as Euclidean-based distance or correlation-based distance. The decision of similarity measure is application-specific. Following are the steps for the K-means algorithm:

Step 1: Specify the number of clusters K.
Step 2: Initialize the centroids by randomly selecting K points from the dataset.

Step 3: Assign the data points to the closest centroid.

Step 4: Recalculate the centroids for each cluster by computing the average of all the data points.

Step 5: Keep iterating till there is no change in the centroids, i.e. assignment of the data points to the clusters is not changing.

The above K-means clustering algorithm can be used for associating pick-up locations to the nearest vehicles. The vehicle locations can be considered analogous to centroids and pick-up locations as the data points. The proposed technique uses a K-means clustering algorithm for assigning the pick-up locations to vehicles based on distance information. These pick-up locations are again clustered based on the number of requests as the passengers assigned to vehicles should not exceed the vehicle capacity.

The proposed technique uses **Algorithm 1** that computes the set of pick-up locations to be visited for each vehicle. Let L and V be the list of pick-up locations and vehicles from sets $\mathbb{N}$ and $\mathbb{M}$, respectively. Let W be the final array of lists of pick-up locations allotted to each vehicle. Let T be a list that contains the information about vehicle associated with the respective pick-up location. This list is obtained after applying K-means algorithm on L as data points and V as centroids.

Algorithm 1 Computation of pick-up location assignment

1: **function** ASSIGN_PICKUP_LOCATION(L,V)
2:     **while** $sizeof(L) \neq 0 \parallel sizeof(V) \neq 0$ **do**
3:         $T \leftarrow K\_means(L,V)$
4:         **for** $h = 1$ to $sizeof(L)$ **do**
5:             **if** $req(L(h)) > 0$ **then**
6:                 **if** $load(T(h)) > req(L(h))$ **then**
7:                     $W \leftarrow \{T(h), L(h)\}$
8:                     $L \leftarrow remove(L(h))$
9:                 **else:**
10:                     **if** $load(T(h)) > 0$ **then**
11:                         $req(L(h)) \leftarrow req(L(h)) - load(T(h))$
12:                         $W \leftarrow \{T(h), L(h)\}$
13:                     **end if**
14:                     $V \leftarrow remove(T(h))$
15:                 **end if**
16:             **else**
17:             **end if**
18:         **end for**
19:     **end while**
        return W
20: **end function**

Algorithm 1 returns a list W that gives the information of pick-up locations associated with their respective vehicles. The algorithm stops when either the vehicle capacity exhausts, or all the requests are fulfilled. List T consists of information about the vehicle assigned to a particular pick-up location. If the load of the assigned vehicle $(load(T(h)))$ for the pick-up location is more than the request $(req(L(h)))$ at that location, then the vehicle and pick-up location pair $(\{T(h), L(h)\})$ is added in W. Also, pick-up location $L(h)$ is removed from L. Otherwise, the remaining request that is not served is updated by reducing it by the load of the assigned vehicle and adding the vehicle and pick-up location pair $(\{T(h), L(h)\})$ in W. Since the total capacity of the vehicle is exhausted, that vehicle $(T(h))$ is removed from V. Figure 11.1 illustrates a single iteration example of this algorithm where a sample dataset is considered with blue and orange points representing pick-up locations and vehicles, respectively. Figure 11.1(b) shows the result after the application of the K-means algorithm where pick-up locations are associated with the nearest vehicle. Since the vehicle has limited capacity, some of the pick-up locations are filtered out to be considered for the next iteration, as shown in Figure 11.1(c).

The following section discusses the use of traveling salesmen problem (TSP) for individual vehicle routing by using the results from Approach I and II.

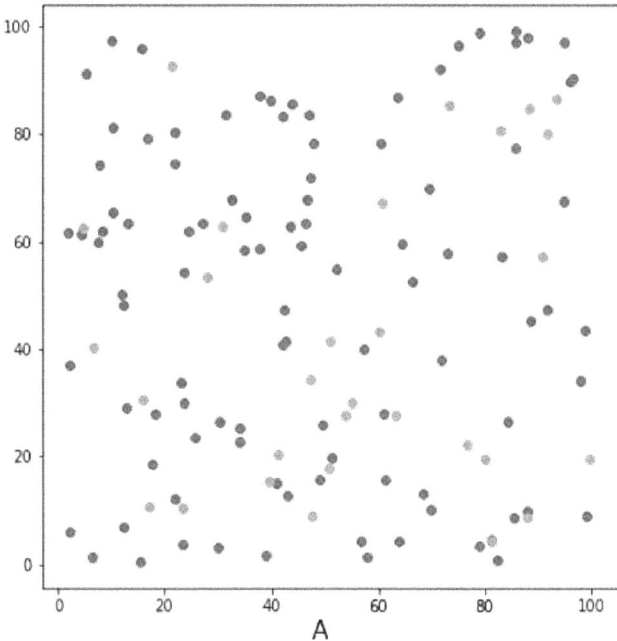

FIGURE 11.1 Sample iteration for Approach II. (a) Sample dataset with vehicles and pick-up locations. (*Continued*)

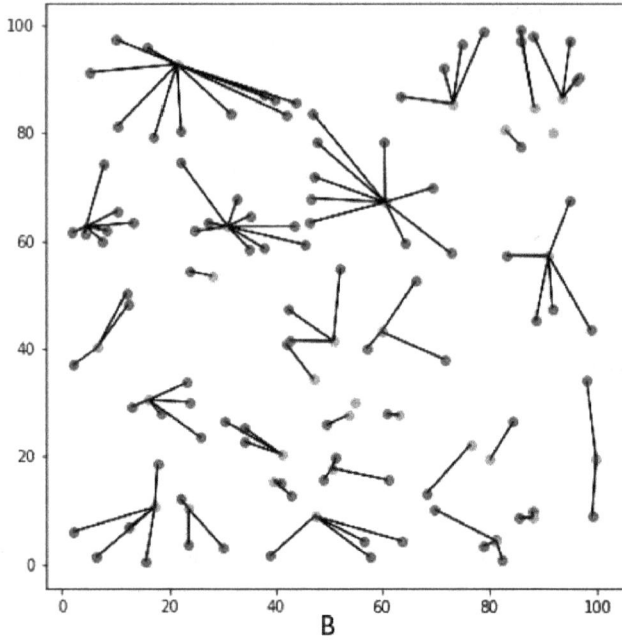

FIGURE 11.1   Sample iteration for Approach II. (b) Output of *K*-means algorithm.

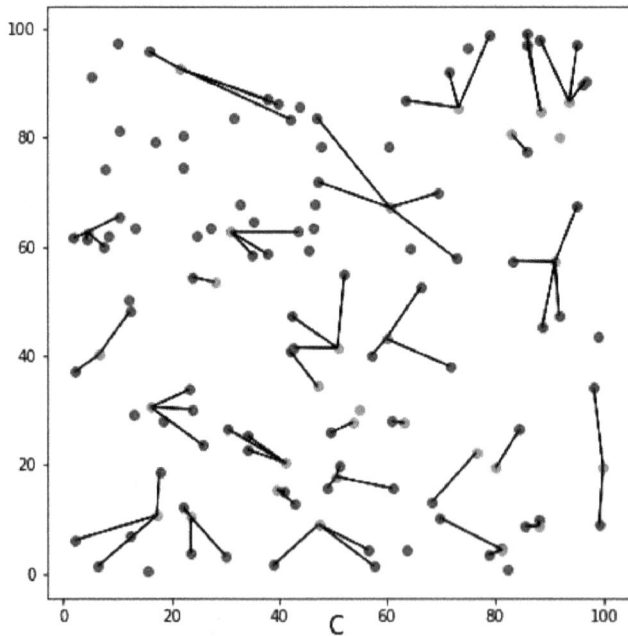

FIGURE 11.1   Sample iteration for Approach II. (c) Output after considering vehicle capacity.

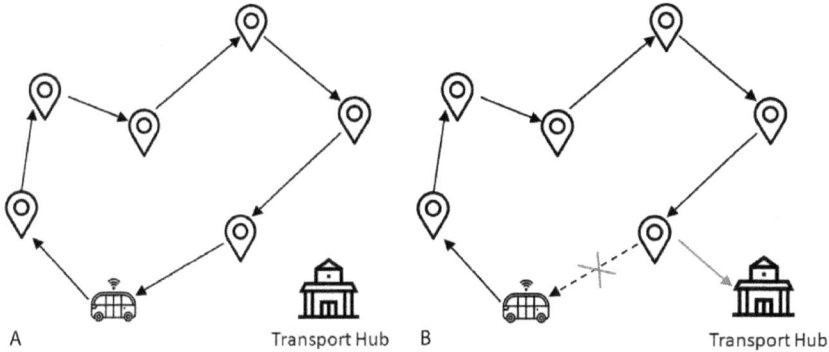

FIGURE 11.2 Vehicle routing example using TSP. (a) TSP with the same origin and destination. (b) TSP with different origin and destination.

11.5 VEHICLE ROUTING METHODOLOGY

In this section, we describe a technique for vehicle routing for FM transit. These techniques require a list of pick-up location associated with each vehicle that is obtained from Sections 11.3 and 11.4. Each vehicle has to visit a specified number of pick-up locations in a way that minimizes the distance traveled by the vehicle. A vehicle routing algorithm is implemented for each vehicle individually for its associated pick-up locations. Next subsection describes an algorithm that uses TSP for vehicle routing.

11.5.1 VEHICLE ROUTING USING TSP

This algorithm works on applying TSP for vehicle routing. Conventional TSP [14] is used when the origin and destination of the vehicle are the same. FM transit service starts with the vehicle as an origin and public transport hub as a destination. The service is analogous to the TSP with different origin and destination. Given a start location (i.e. vehicle's current location) and the pick-up locations the vehicles have to visit, and a route can easily be determined using TSP. This will result in a closed route that ends at the vehicle location, as shown in Figure 11.2(a). After obtaining an efficient closed vehicle route using TSP, the route connection between the last pick-up location and the vehicle location is removed. A new connection is created between the last pick-up location and the public transport hub, as illustrated in Figure 11.2(b). This results in an open-ended route with origin as vehicle location and destination as a public transport hub. The algorithm is implemented using the Tabu search meta-heuristic [15] approach. Tabu search is a meta-heuristic procedure that explores a set of solutions by repeatedly moving within a given problem space [16]. In the next section, a block diagram of the proposed workflow is presented.

11.5.2 BLOCK DIAGRAM DESCRIPTION

The block diagram of the proposed techniques is illustrated in Figure 11.3, which shows the workflow for solving the FM problem. The overall FM problem is divided

```
┌──────────┐      ┌─────────────────────────┐      ┌─────────────────────────┐
│  START   │─────▶│ Initialize n pick-up     │─────▶│ Pick-up location assignment: │
└──────────┘      │ locations and m vehicle  │      │ Approach I or Approach II │
                  │ locations                │      └─────────────────────────┘
                  └─────────────────────────┘
   ┌──────────────────────────────────────────────────────────────────────┘
   │ ┌────────────────────────────┐      ┌────────────────────────────┐      ┌──────────┐
   └▶│ Vehicle Routing using TSP  │─────▶│ Compare performance indices │─────▶│   END    │
     └────────────────────────────┘      └────────────────────────────┘      └──────────┘
```

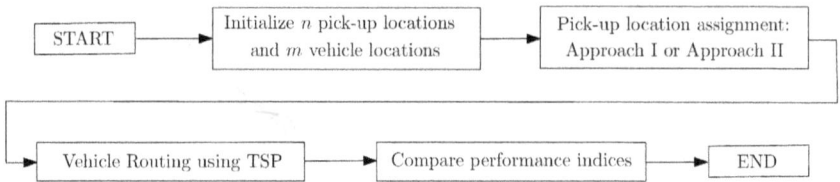

FIGURE 11.3 Block diagram for FM problem.

into two parts. In the first part of the problem, an optimization problem is solved for assigning pick-up locations to the nearest available vehicle. The second part of the problem is responsible for applying vehicle routing for each vehicle to visit its associated pick-up locations and find an optimal route from the vehicle location to the public transport hub. Section 11.6 discusses the simulation results and comparison between various combinations of the mentioned approaches.

11.6 DATA DESCRIPTION AND RESULTS

The proposed techniques for solving FM ridesharing problem are implemented in Python using the modules scikit-learn [17] and Google OR-Tools [18]. Simulations have been carried out on a PC running Intel core i7-7500U, 2.7 GHz processor with 24 GB RAM. An area of interest (AOI) has been considered in Singapore as a motivation to analyze the results with real data.

11.6.1 AREA OF INTEREST (AOI)

Ang Mo Kio (AMK) is a planning area and a residential town located in the North-East of Singapore. It has 2 mass rapid transit (MRT) train stations (Ang Mo Kio MRT and Yio Chu Kang MRT) and is further classified into 12 subzones. In this chapter, seven subzones that are comparatively near to Ang Mo Kio MRT are selected for the analysis as shown in Figure 11.4.

The AV shuttles are assumed to have parking locations in the AOI within the traditional parking locations where they can share space with other vehicles. AV shuttles are assumed to be parked at electronic surface parking areas in the AOI. According to the data obtained from Singapore's open data portal [19], there are 62 parking locations in the AOI. These shuttles will pick up passengers from pick-up locations, which are either existing bus stops or selected taxi pick-up/drop-off points and drop them at MRT. The next section discusses results obtained from implementing the proposed approaches.

11.6.2 SIMULATION RESULTS AND DISCUSSIONS

In this section, we compare the results of the proposed techniques. The proposed techniques combined with vehicle routing techniques are (a) Approach I with TSP

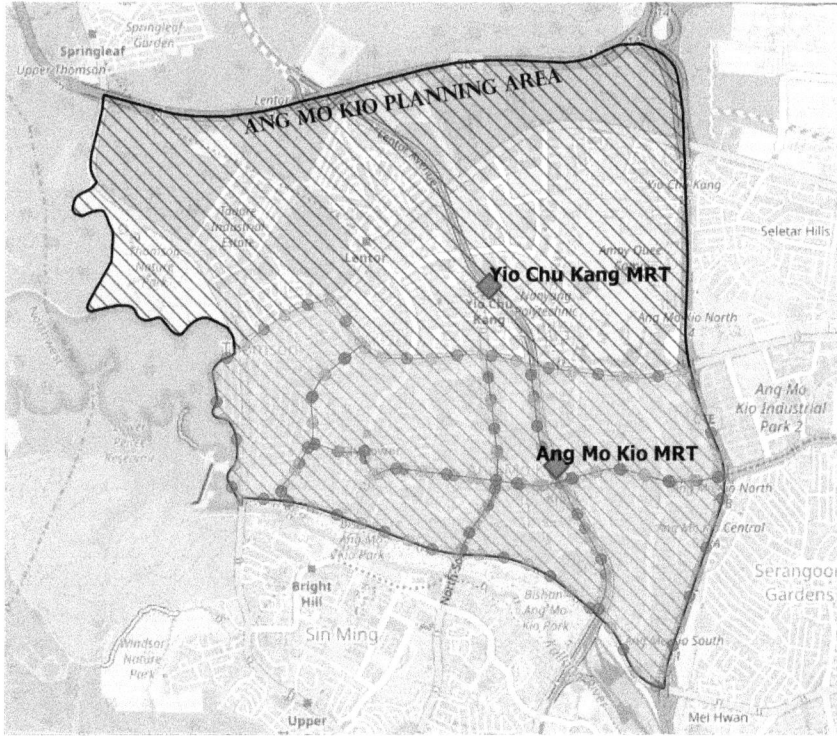

FIGURE 11.4   AMK planning area with AOI around Ang Mo Kio MRT.

and (b) Approach II with TSP. Four performance indices have been considered for the comparison. These performance indices are described as follows:

- **Total vehicle miles traveled:** This is the summation of the total distance traveled by all the vehicles. The lower value of total VMT signifies lesser transportation costs.
- **Percentage of requests fulfilled:** This parameter informs about the percentage of requests served. If the total request is less than the total vehicle capacity, then the percentage of requests fulfilled will be 100% and vice versa.
- **Computation time:** It indicates the computation time required for the technique to calculate all the routes for the vehicles efficiently.
- **Percentage of fleet utilized:** The fleet utilization is also an important parameter to evaluate the performance of proposed techniques. When the total request is less than the total vehicle capacity, it can be easily comprehended that lesser the fleet utilization, more efficient is the technique.

The simulation is carried out for three fleet sizes of vehicles $m = \{10,20,30\}$. The number of pick-up locations is varied as $n = \{30,60,90\}$. Maximum vehicle capacity

TABLE 11.2

Performance Indicators for $m = 10$ and Number of Pick-Up Locations ($n = 30, 60, 90$)

	$m = 10, n = 30$		$m = 10, n = 60$		$m = 10, n = 90$	
	Approach II + TSP	Approach I + TSP	Approach II + TSP	Approach I + TSP	Approach II + TSP	Approach I + TSP
Total VMT	41,223	28,000	52,370	35,862	98,135	34,023
% Demand fulfilled	100	100	91	79	82	54
Time taken (s)	7.13	7.11	9.07	10.45	17	10.6
% Fleet utilized	80	80	90	100	90	100

is considered $q = 12$ and maximum request at a pick-up location is considered 5. Vehicle and pick-up locations are generated randomly from the AOI. Tables 11.2–11.4 tabulate the simulation results for fleet size 10, 20, and 30, respectively. From the tables, it is evident that the total VMT is less for Approach I with TSP than Approach II with TSP. There is not much difference in the computation time of the proposed approaches. In both approaches, if the total request is less than the total vehicle capacity, then 100% demand is fulfilled and less than 100% when the total requests are more than the total vehicle capacity. The fleet size utilization is maximum when the total request is more than total vehicle capacity.

TABLE 11.3

Performance Indicators for $m = 20$ and Number of Pick-Up Locations ($n = 30, 60, 90$)

	$m = 20, n = 30$		$m = 20, n = 60$		$m = 20, n = 90$	
	Approach II + TSP	Approach I + TSP	Approach II + TSP	Approach I + TSP	Approach II + TSP	Approach I + TSP
Total VMT	52,270	30,102	76,457	54,834	98,990	73,872
% Demand fulfilled	100	100	100	100	100	100
Time taken (s)	7.70	6.87	13.48	14.42	17	20.5
% Fleet utilized	65	48	86	77	94	98

TABLE 11.4
Performance Indicators for $m = 30$ and Number of Pick-Up Locations
$(n = 30, 60, 90)$

	$m = 30, n = 30$		$m = 30, n = 60$		$m = 30, n = 90$	
	Approach II + TSP	Approach I + TSP	Approach II + TSP	Approach I + TSP	Approach II + TSP	Approach I + TSP
Total VMT	59,792	31,230	88,828	54,900	113,528	75,345
% Demand fulfilled	100	100	100	100	100	100
Time taken (s)	6.99	7.39	14.94	14.41	21.16	21.88
% Fleet utilized	54	34	74	54	87	73

11.7   CONCLUSION AND FUTURE WORK

VRPs, especially ridesharing problems, have turned into a popular topic due to the advancements in mobility and the transportation sector. This chapter outlines distance-based optimization techniques to solve the FM transit problem. The FM transit problem is divided into two parts. First, the pick-up locations are associated with the nearest available vehicle using Approaches I and II as mentioned, and then the individual vehicle routing is performed using TSP. The FM problem is solved using a combination of pick-up location association and followed by vehicle routing as proposed in this chapter. The results suggest that the total distance is minimum for Approach I with TSP, which provides a feasible solution. The proposed techniques ensure that maximum requests are served until the total vehicle capacity gets exhausted. Therefore, this approach encourages the usage of public transport for FM transit using AV shuttles using IoT cloud.

AVs and vehicle routing have far-reaching applications with a lot of scope for improvement. Some potential future research directions include (a) an attempt to include more constraints such as battery management of AVs, time windows for pick-up and drop-off, and fleet optimization of AVs for the FM use case; (b) an extension of search algorithm to emergency dial-a-ride (DAR) systems, such as medical, fire, and rescue services; (c) extend prediction techniques to large-scale problems, in order to tackle the routing problems with traffic congestion and investigate the influence of passenger comfort under congestion; (d) investigation of the latest vehicular technologies, such as electric, hybrid in the context of demand-responsive transportation with heterogeneous demand, fleet, dynamic traffic conditions; (e) to enhance the computational speedup of prediction algorithms for the VRPs; (f) sensitivity analysis on length of operating horizon and other key parameters such as penalty cost, buffer time, and fixed cost.

ACKNOWLEDGMENT

The authors gratefully acknowledge the contributions of Ms. Priyanka R. Mehta of Energy Research Institute, Nanyang Technological University, Singapore, for her timely support. This material is based on research/work supported by National Research Foundation under Virtual Singapore Program Award No. NRF2017VSG-AT3DCM001-018. Any opinions, findings, and conclusions or recommendations expressed in this material are those of the author(s) and do not necessarily reflect the views of National Research Foundation.

REFERENCES

1. F. Cordeau and G. Laporte, "The dial-a-ride problem: Models and algorithms," *Annals of operations research*, vol. 153, no. 1, pp. 29–46, 2007.
2. V. M. R. Baldacci and A. Mingozzi, "An exact method for the car pooling problem based on lagrangian column generation," *Operations Research*, vol. 52, no. 3, pp. 422–439, 2004.
3. W. Herbawi and M. Weber, "The ridematching problem with time windows in dynamic ridesharing: A model and a genetic algorithm," in *2012 IEEE Congress on Evolutionary Computation. 1em plus 0.5em minus 0.4em IEEE*, 2012, pp. 1–8.
4. M. S. N. Agatz, A. Erera, and X. Wang, "Dynamic ride-sharing: A simulation study in metro Atlanta," *Transportation Research Part B: Methodological*, vol. 45, no. 9, pp. 1450–1464, 2011.
5. R. J. Y. Huang, F. Bastani, and X. S. Wang, "Large scale real-time ridesharing with service guarantee on road networks," in *In Proceedings of the VLDB Endowment. 1em plus 0.5em minus 0.4em Citeseerx*, 2014, p. 2017–2028.
6. S. Chen, H. Wang, and Q. Meng, "Solving the first-mile ridesharing problem using autonomous vehicles," *Computer-Aided Civil and Infrastructure Engineering*, vol. 35, no. 1, pp. 45–60, 2020.
7. T. Perera, A. Prakash, C. N. Gamage, and T. Srikanthan, "Hybrid genetic algorithm for an on-demand first mile transit system using electric vehicles," in *International Conference on Computational Science. 1em plus 0.5em minus 0.4em Springer*, 2018, pp. 98–113.
8. S. Chandra and L. Quadrifoglio, "A model for estimating the optimal cycle length of demand responsive feeder transit services," *Transportation Research Part B: Methodological*, vol. 51, pp. 1–16, 2013.
9. J. R. Montoya-Torres, J. L. Franco, S. N. Isaza, H. F. Jiménez, and N. Herazo-Padilla, "A literature review on the vehicle routing problem with multiple depots," *Computers & Industrial Engineering*, vol. 79, pp. 115–129, 2015. [Online]. Available: http://www.sciencedirect.com/science/article/pii/S036083521400360X
10. H. Wang, "Routing and scheduling for a last-mile transportation system," *Transportation Science*, vol. 53, no. 1, pp. 131–147, 2019.
11. R. Chevrier, P. Canalda, P. Chatonnay, and D. Josselin, "Comparison of three algorithms for solving the convergent demand responsive transportation problem," in *2006 proceedings of IEEE Intelligent Transportation Systems Conference*, Sep. 2006, pp. 1096–1101.
12. J. Brake, J. D. Nelson, and S. Wright, "Demand responsive transport: Towards the emergence of a new market segment," *Journal of Transport Geography*, vol. 12, no. 4, pp. 323–337, 2004.

13. T. Perera, A. Prakash, and T. Srikanthan, "A scalable heuristic algorithm for demand responsive transportation for first mile transit," in *2017 IEEE 21st International Conference on Intelligent Engineering Systems (INES)*. 1em plus 0.5em minus 0.4em IEEE, 2017, pp. 157–162.
14. I. Brezina Jr and Z. Čičková, "Solving the travelling salesman problem using the ant colony optimization," *Management Information Systems*, vol. 6, no. 4, pp. 10–14, 2011.
15. S. Basu, "Tabu search implementation on traveling salesman problem and its variations: A literature survey," 2012.
16. F. Glover and E. Taillard, "A user's guide to Tabu search," *Annals of Operations Research*, vol. 41, no. 1, pp. 1–28, 1993.
17. F. Pedregosa, G. Varoquaux, A. Gramfort, V. Michel, B. Thirion, O. Grisel, M. Blondel, P. Prettenhofer, R. Weiss, V. Dubourg *et al.*, "Scikit-learn: Machine learning in python," *Journal of Machine Learning Research*, vol. 12, pp. 2825–2830, 2011.
18. L. Perron and V. Furnon, "Or-tools," Google. [Online]. Available: https://developers.google.com/optimization/
19. "Singapore government's open data portal," https://data.gov.sg/, accessed: 2020-02-20.

12 Human Interaction with Vehicles for Improved Safety
Internet of Vehicles

Raghavendra Pal
Department of Electronics Engineering,
Sardar Vallabhbhai National Institute of Technology,
Surat, India

CONTENTS

12.1 INTRODUCTION

Imagine that you are driving on a highway in a fog situation. You must rush to your destination due to some work. The foggy weather makes you drive slowly. You wish that somehow you know the location of every vehicle within a 5-km range so that you can avoid accidents while driving fast. All of a sudden, the real-time location of all the vehicles around you appears on your car screen. Now you can drive at your desired speed and reach the destination in time. This can be possible due to the communication between the human and machines in the smarter world. Internet of Vehicles (IoV) is the technology that enables such interaction and when included with 5G, this technology can do wonders. IoV is the advanced version of vehicular ad hoc networks (VANET).

IoV is the technology that enables Internet of Things (IoT) in the cars and other vehicles. It helps in traffic flow and avoids accidents with the help of data integration.

FIGURE 12.1   5G-V2X scenario [3].

The main goal of IoV and vehicle-to-everything (V2X) technologies is to connect vehicles, people, and things. Various sensors such as acceleration sensors, speed sensors, distance sensors, etc., improve the driving aids that enables adaptive cruise control [1]. Figure 12.1 shows the 5G-V2X scenario.

Guiding the traffic flow is an excellent example of the combination of the human's abilities and machines' intelligence. Inclusion of sensors and transceivers makes this a global network. The IoV supports interaction among humans, things, and vehicles. This is done with the help of the technologies developed by the researchers especially on security, quality of service (QoS), channel access, layers, etc. These are collectively called Information and Communication Technology (ICT). When the IoV is integrated with 5G, the amount of data that can be transmitted over the internet becomes huge [2].

Cellular with V2X (C-V2X) provides higher reliability and longer communication range. Its chipset solution is in synchronization with fifth generation and advanced driver assistance systems (ADAS); it provides a direct communication mode. Communication modes 3 and 4 of LTE-V Release 14 and 15 to support fifth-generation V2X support direct vehicle-to-vehicle (V2V) communications. Mode 4 is considered an alternative to dedicated short-range communication (DSRC) and IEEE 802.11 protocol for C-V2X.

12.1.1 IoV ARCHITECTURE

Application layer, network layer, and perception layer are three layers of IoV architecture. Application layer supports storage of data, its analysis, decision-making related to the surroundings, weather, etc. Network layer handles all the data transmission either inside or outside the vehicle. The perception layer contains the sensors that collect data to be transmitted by network layer. It includes global positioning system (GPS) receiver, cameras, radio frequency identification (RFID), etc. Due to the complexity of C-V2X system, it requires cognitive support computing. Hence, some of the researchers have divided it into four layers application, coordination computing control (CCC) layer, network access and transport (NAT), and network environment sensing and control (NESC) layer.

NESC layer corresponds to the perception layer since here vehicles sense the surroundings. NAT layer performs data analysis, transmission, and processing. It helps in remote monitoring as well. It is associated with the network layer.

Technical support for computing is provided by the CCC layer. Application layer has more or less similar functions to the architecture discussed previously.

The rest of the chapter describes the various technologies used in the area of vehicular communications. Section 12.2 describes the DSRC. 5G-V2X is described in Section 12.3. Section 12.4 discusses about the current research challenges in this area. Finally, Section 12.5 summarizes the chapter.

12.2 DEDICATED SHORT-RANGE COMMUNICATION (DSRC)

The research in the area of vehicular communication is being conducted for more than a decade now. It provides safety as well as entertainment services to the drivers, i.e. accident information, local map information, traffic jam, best route, etc. Medium access control (MAC) layer and physical layer (PHY) parameters for vehicular communication are defined by IEEE in the standard named IEEE802.11p [3]. IEEE802.11e [4] and IEEE802.11a [5] are the two standards that are the basis of IEEE802.11p. These are the standards for wireless LANs. IEEE802.11a is based upon the carrier sense multiple access with collision avoidance (CSMA/CA) and it defines the distributed coordination function (DCF). IEEE802.11e is based upon the prioritizing the data to maintain QoS. United States Federal Communication Commission (FCC) has allotted a 75 MHz of spectrum in the 5.8-GHz band. It contains seven channels of 10 MHz. The rest 5 MHz is used as a guard band. This spectrum is named DSRC [6]. It is shown in Figure 12.2.

The safety of the vehicles is provided by a device called On-Board Unit (OBU). OBU transmits and receives the data of other vehicles. The data may be related to a safety application or any infotainment service. If any vehicle receives a safety-related message, its driver can act accordingly and the OBU can inform other vehicles as well. The OBU can also send the information regarding the map of a local area so that the driver can see all the routes and select its desired path. Infotainment

10 MHz Non-Safety	172	5.855-5.865 GHz
10 MHz Non-Safety	174	5.865-5.875 GHz
10 MHz Non-Safety	176	5.875-5.885 GHz
10 MHz Critical-Safety	178	5.885-5.895 GHz
10 MHz Traffic Efficiency	180	5.895-5.905 GHz
10 MHz Traffic Efficiency	182	5.905-5.915 GHz
10 MHz Traffic Efficiency	184	5.915-5.925 GHz

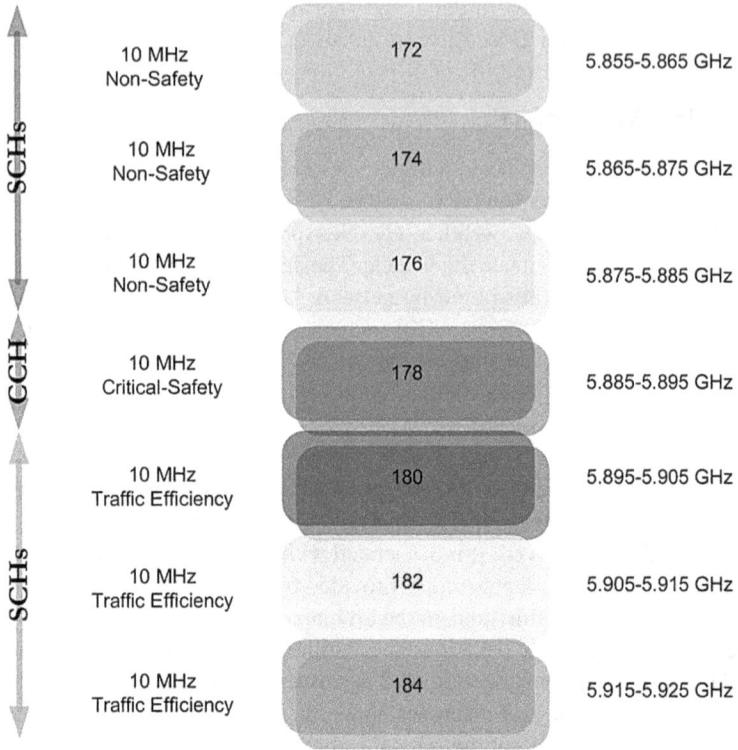

FIGURE 12.2   Dedicated short-range communication band [7].

messages are a combination of these types of messages. Other examples are real-time traffic information. These messages are given lesser priority than the safety messages. Hence, there are two broad categories of the message used in the vehicular communication, safety, and infotainment messages. However, only these two categories are not sufficient to enable the OBU to take the decision related to the priority of messages. There are several subcategories of these message categories, i.e. if an accident has already taken place, the message carrying this information will be prioritized in comparison to the message that contains the information that tells us about a possible accident? The answer is obviously yes. Both of these are safety messages, but one is given higher priority than the other one. This priority is based upon the type of the messages and it is called the static priority.

To avoid confusion in deciding the priority among the messages of the same type, a few additional parameters are also considered: speed of vehicle, packet size, and the remaining lifetime of the packet along with the distance between the transmitter and receiver. The priority that is based upon these parameters is called dynamic priority. Both the static and dynamic priorities are used to calculate the overall priority of the message.

Whenever a vehicle receives a packet, it stores it in a queue so that it can be transmitted whenever the channel is idle. If another packet is generated at the vehicle or the vehicle receives another packet, this packet is also enqueued in the queue. If the

TABLE 12.1

Some MAC and PHY Layer Parameters in Vehicular Communications

Parameters	Values
AC	VO, VI, BE, BK
CW_{min}	15
T_{slot}	16 μs
Short interframe space (SIFS)	32 μs
AIFS	Depends on AIFSN and T_{slot}
CW_{max}	1023
Transfer rate	3–27 Mbps
Bandwidth	10 MHz
AIFSN	Depends upon AC

queue size is large, the data structure that is used to store packet becomes important. It is because the efficient data structure enables the enqueue and dequeue operation fast. Some parameters which affect the delivery time are provided in Table 12.1.

Researchers have been working in this area. IEEE has defined broad priorities in its standard IEEE802.11p [3] as shown in Figure 12.3. Different priorities are defined using the different values of contention window (CW) and arbitrary interframe space (AIFS). The lesser the value of CW and AIFS, the higher is the priority of the particular category. However, these classifications are very broad and need to be elaborated further. One can see in the figure that there are two sections of CW and AIFS. One is control channel (CCH) interval, and another is service channel (SCH) interval. There are two main classifications of the messages. One is safety and another is non-safety. Hence, the time in vehicular communication is divided into two sections. One is called control channel interval (CCHI) and another is called the service channel interval (SCHI). These times are of 50 ms each. Similarly, the seven channels in the DSRC band are divided into two categories. One is CCH and another is called SCH. On the one hand, CCH is only one, i.e. there is only one channel dedicated for control purpose. The same channel is also used for the transmission of the safety messages. On the other hand, there are six SCHs. These are used in the SCHI for the transmission of messages other than the safety messages. CCHI is also used for the transmission of the beacons. These are the control packets that allow all the vehicles to receive their neighborhood information. At the beginning of the CCHI, every vehicle transmits its beacon to the neighboring vehicles. Since the network in the vehicular communication is ad hoc, beacon transmission is required. Researchers in [7] have developed a new algorithm that ensures the successful transmission of beacons.

Authors in [8] have further classified the messages in vehicular communication as shown in Figure 12.4. They have divided the priorities into four classes. The message generation type is also defined, i.e. whether the message is event-driven or periodic. Authors in [9–15] have proposed algorithms that transmit either the safety message or non-safety message or both effectively.

All the packets are stored in a data structure in the vehicle. Whenever a packet is received by the vehicle or a packet is generated by the vehicle, it first stores the

		0	1	2	3
	Access Category number	Background traffic	Best Effort	Video	Voice
SCH Interval	AIFS(µs)	112	48	32	32
	AIFSN	7	3	2	2
	CWmax	1023	1023	15	7
	CWmin	15	15	7	3
CCH Interval	AIFS(µs)	144	96	48	32
	AIFSN	9	6	3	2
	CWmax	1023	15	7	7
	CWmin	15	7	3	3

FIGURE 12.3   Different access categories.

packet in that data structure. After that when it finds the channel idle, the packet of the highest priority is transmitted first. When a packet is received by the vehicle, the packet needs to go all the way to its place in the queue according to its priority. This may take some time when the queue is implemented in the hardware. Furthermore, the dynamic priority of a vehicle changes with time, so a packet that currently has low dynamic priority may attain high dynamic priority after some time. This also causes the frequent rescheduling of queue. So, there is a need for a data structure that allows fast searching and enqueue and dequeue operations. Authors in [9] have proposed the data structure based upon the static and dynamic priorities of a message, which reduces the time for enqueue and dequeue operations.

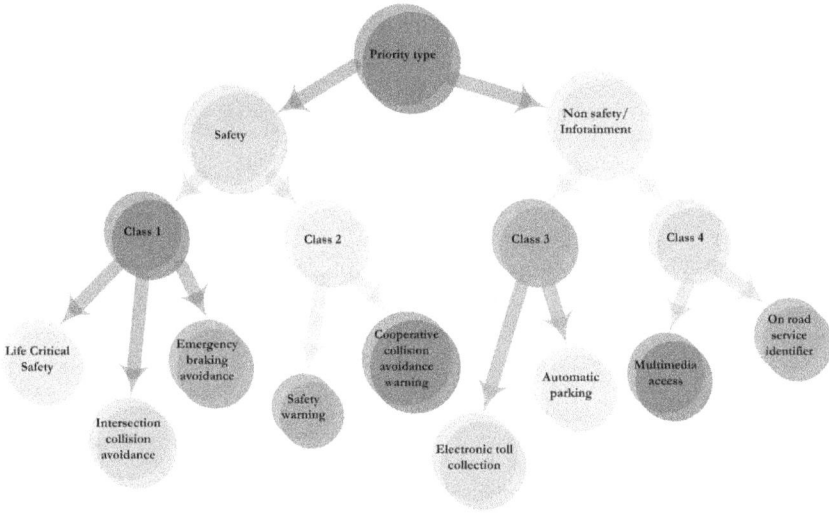

FIGURE 12.4 Message classifications according to [8].

12.2.1 MESSAGE CLASSIFICATIONS AND PRIORITY ASSIGNMENT IN VEHICULAR COMMUNICATIONS

The major classifications of messages in vehicular communications have been provided in Figure 12.4. However, the classifications can further be divided into subclasses. For example, life-critical safety message can be of many types, i.e. accident occurred, extreme weather conditions, when a vehicle is out of control or if the driver has lost the balance, fuel tank leakage warning, etc. Similarly, on-road service identifiers can be of many types. The possible classifications of safety messages are shown in Figure 12.5 and that of non-safety messages are shown in Figure 12.6. However, there can be many other messages. The static priority assigned to each message is solely based upon the type of the message. Priority of each message type may or may not be different. However, in most cases, it is different. In case the static priority is not different for two more message types, the dynamic priority is used to find out which message is to be transmitted first. Dynamic priority is also used to identify the priority of two or more same types of messages.

Figures 12.5 and 12.6 define the static priority only. To calculate dynamic priority, some parameters are considered. These parameters are speed of the vehicle, distance between the transmitting and receiving vehicle, message size, deadline of the message, etc. These parameters are combined to form an equation to calculate the dynamic priority as in [16]. This equation is intelligently defined. For example, let us assume a simple equation to calculate the dynamic priority of a message

$$\text{Dynamic Priority} = \frac{\text{Distance between sender and receiver} + \text{speed of the vehicle}}{\text{message size} + \text{data rate}}$$

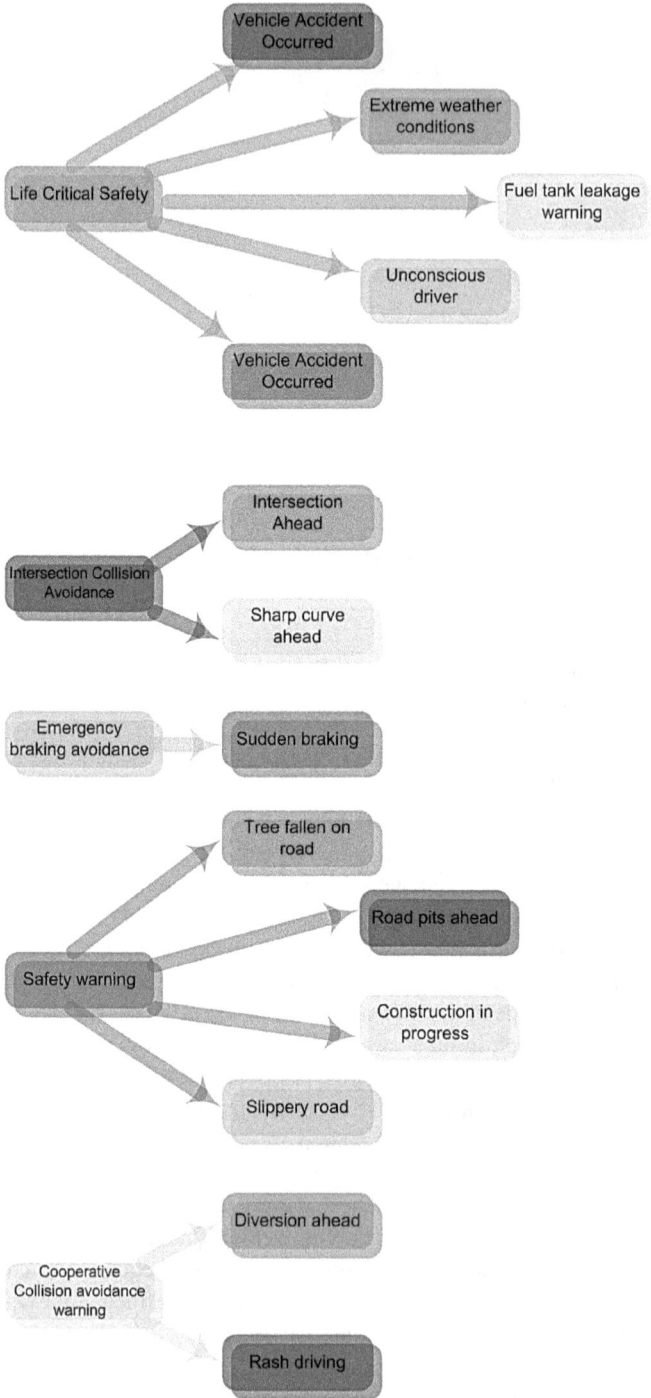

FIGURE 12.5 Further classification of safety message types in vehicular communications.

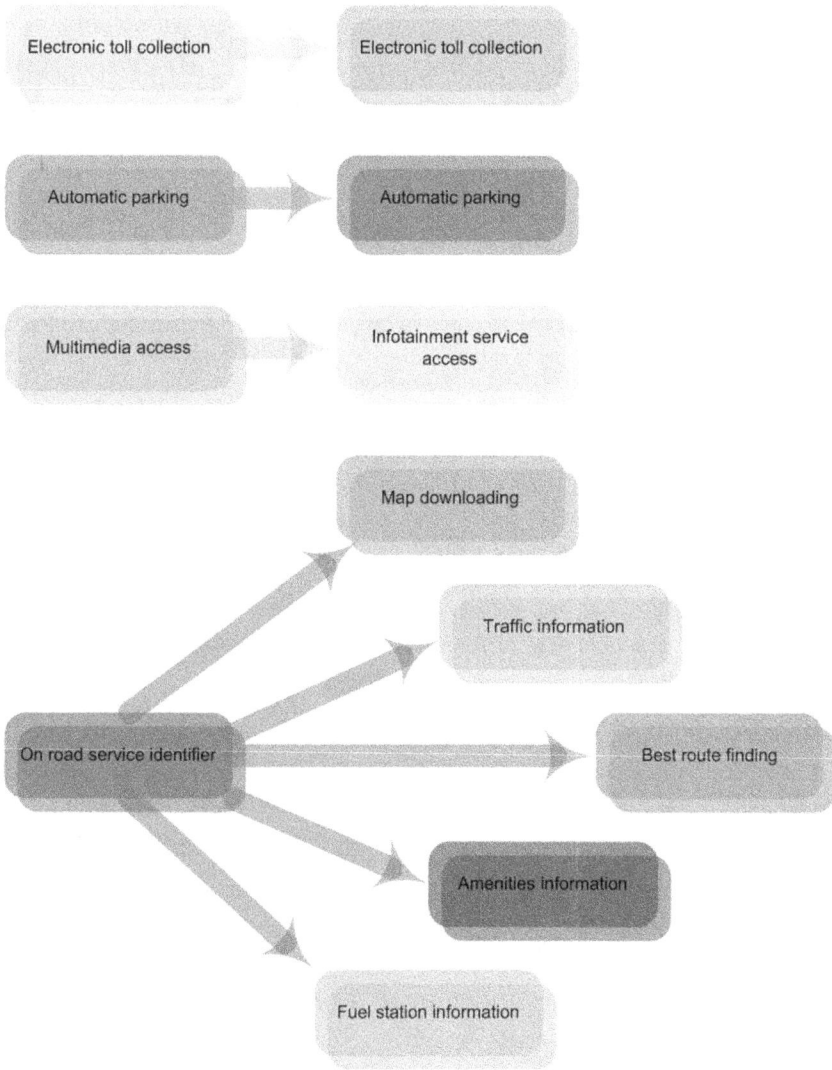

FIGURE 12.6 Further classification of non-safety message types in vehicular communications.

Look at the numerator part of the equation, as distance between sender and receiver increases, chances of them getting out of each other's range are high. Hence, as the value of distance is high, the dynamic priority should be increased. Hence, distance and dynamic priority are directly proportional to each other. Similarly, if the message size is less, it will take comparatively less time to transmit the message; hence, this message can be transmitted with high priority. For this reason, dynamic priority and message size are inversely proportional to each other. The explanation for the other two variables is similar.

12.3   5G-V2X

The 5G has the mmWave technology that operates in 30–300 GHz spectrum. Under ideal circumstances, the IEEE802.11p and DSRC standards are outperformed by the mmWave systems. Core network (CN) and radio access network (RAN) functions are the components of network slice in 5G. Operator is free to compose different network slices parallelly. Communication of one slice should not negatively interfere with another slice. It affects the functionalities of the control plane (CP) such as authentication, mobility, and session management. The functionalities of user plane (UP) are made auto configurable.

To enable the network slicing technology, the CP and UP are decoupled. Due to this decoupling of both functionalities, these can be placed in different convenient locations. On the one hand, to reduce latency, the UP functions are placed close to the user. On the other hand, the CP functions are placed near the central point. It makes the controlling of the network easy and decreases its complexity.

Network function virtualization (NFV) chains in a slice can be configured by the software-defined network (SDN) controller. It can be scaled, relocated according to the network demands. UP and CP functionalities can be configured automatically to handle data traffic. This slicing makes the wide variety of applications satisfied for the demands over the 5G infrastructure. [17]. V2X provides increased perception of the environment to enable autonomous driving; hence, it provides the cooperation between the vehicle and the 5G control subsystem [18].

In [19], authors have done an extensive study in this area. According to it, an OpenFlow protocol is proposed as an alternative to 5G-V2X [20]. It enables direct and secure communication with SDN controller. The network slices logically separate CP and UP according to the 3GPP specifications. With this, the various V2X communication modes are taken. Authors in [21–23] have presented comprehensive literature surveys of recent advances in vehicular communications.

12.3.1   Various 5G-V2X Modes

V2V 5G communication is supported by the C-V2X radio interface. It has three different modes as shown in Figure 12.7.

1. **C-V2X:** This is uplink or downlink communication within a base station. It can take road side units (RSUs) into account as well.
2. **Cellular supported by V2V:** In this mode, base station provides control information to vehicles so that they can carry out V2V communication efficiently.
3. **V2V unassisted by cellular:** In this, vehicles do not get any direct assistance by access point. All the resources are handled by cellular network.

12.3.2   Challenges and Future Directions

In the present scenario, the cellular technology provides uplink data rate of up to 75 Mbps and downlink data rate of up to 300 Mbps. On the one hand, the connectivity

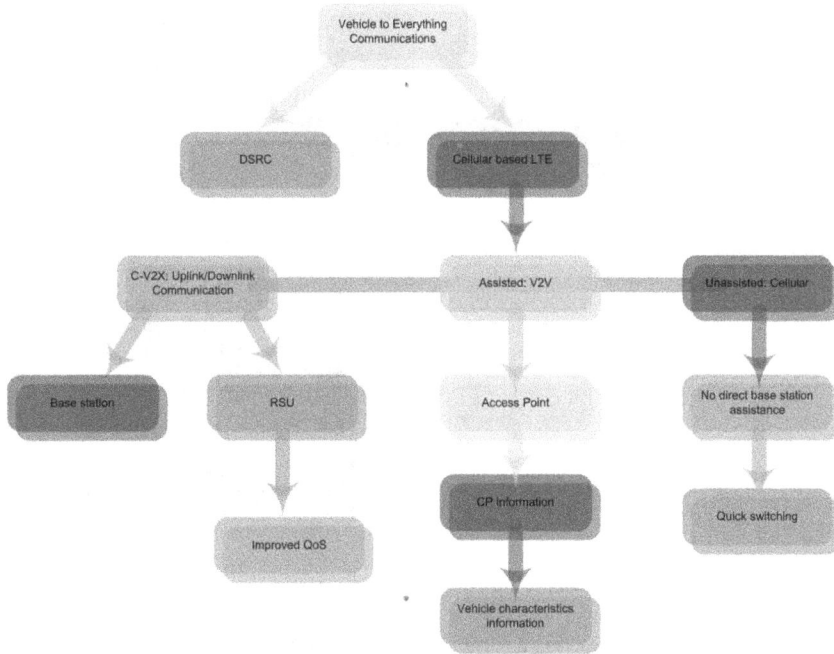

FIGURE 12.7   Different V2X modes.

range is up to 2 km and the delay is 1.5–3.5 seconds. This is mainly suitable for vehicle to infrastructure (V2I) communication and slightly for V2V communication. On the other hand, the connectivity range provided by the DSRC is 300–1000 m. Delay can be as low as 500 μs. The data rate provides up to 27 Mbps. It is suitable for V2V application that requires low bandwidth. Some of the major issues in IoV and their possible solution are shown in Figure 12.8.

12.4   TECHNOLOGICAL CHALLENGES AND FUTURE DIRECTIONS

Some of the technological challenges include the integration of the work of multiple groups working in this area and a lack of common architecture. Another major problem is the installation of the infrastructure. Since 5G uses higher frequency bands, hence the signals will not reach as far as in the 4G or 3G. The effect of various obstacles is much more. Hence, there will be the requirement of too much additional infrastructure. Figure 12.9 shows some common problems of the DSRC and LTE-V2C technologies.

Furthermore, there is another challenge that is the limitation of bandwidth. Cognitive radios (CRs) or the SDNs is the solution to this problem. This technology uses the communication channels of nearby bands to transfer data in case of the non-availability of the allotted channels. However, this technology has its own challenges. The main users of the cognitive channels are known as primary users (PUs) and their transmission should not be interfered with in any case. Hence, the

Issues	Solutions
Big Data: There is a lot of data of many sensors. In some applications, the data rate can reach upto 1 GBps	Big data analytics along with mobile could computing will help in handling such issue. [22]
Security and Privacy: This is a network consisting of many technologies. It may cause network intrusion.	Proper data security and privacy system is required. [23]
Reliability: Sensors can provide incorrect data due to transmission errors.	Appropriate communication protocols are required for improvement of reliability
Standards: There is a lack of a single international standard.	Academia and industries should collaborate to design a common standard.

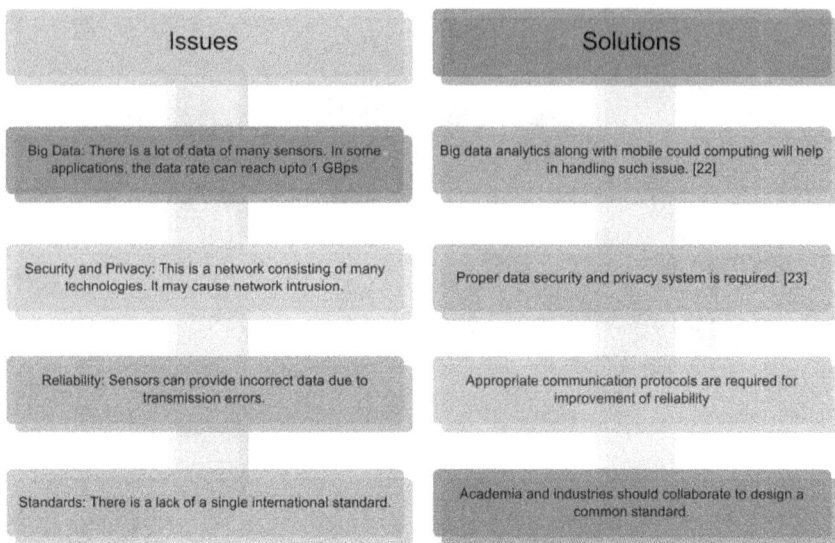

FIGURE 12.8   Major issues in IoV and their possible solutions.

researchers are working in this area to provide a solution to reduce the possibilities of interruption of the PUs' transmission. Currently, there is an increase in the use of machine learning algorithms. These algorithms might be the possible solutions to this problem since the channels can be optimally selected by analyzing the previous data of channel occupancy. Authors in [24] have described a deep learning algorithm

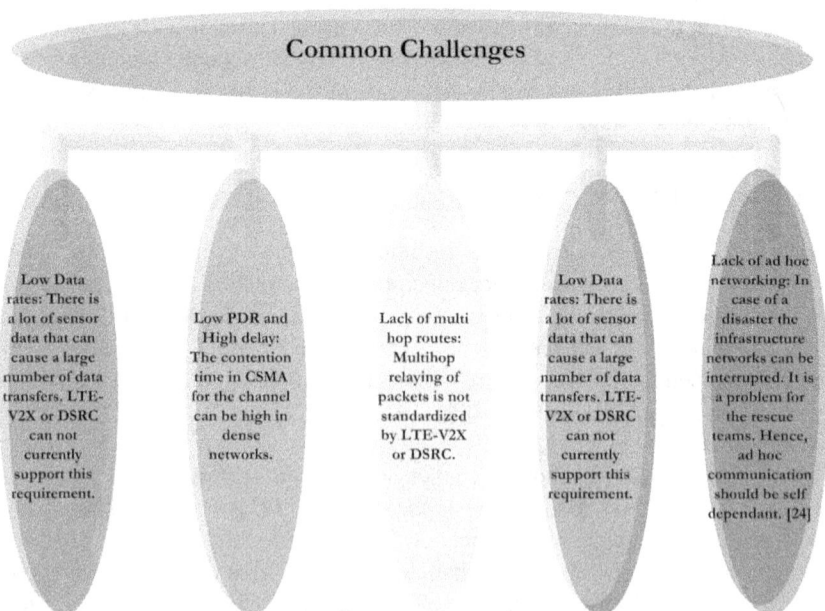

Common Challenges

Low Data rates: There is a lot of sensor data that can cause a large number of data transfers. LTE-V2X or DSRC can not currently support this requirement.

Low PDR and High delay: The contention time in CSMA for the channel can be high in dense networks.

Lack of multi hop routes: Multihop relaying of packets is not standardized by LTE-V2X or DSRC.

Low Data rates: There is a lot of sensor data that can cause a large number of data transfers. LTE-V2X or DSRC can not currently support this requirement.

Lack of ad hoc networking: In case of a disaster the infrastructure networks can be interrupted. It is a problem for the rescue teams. Hence, ad hoc communication should be self dependant. [24]

FIGURE 12.9   Some common challenges to the DSRC and LTE-V2X technologies.

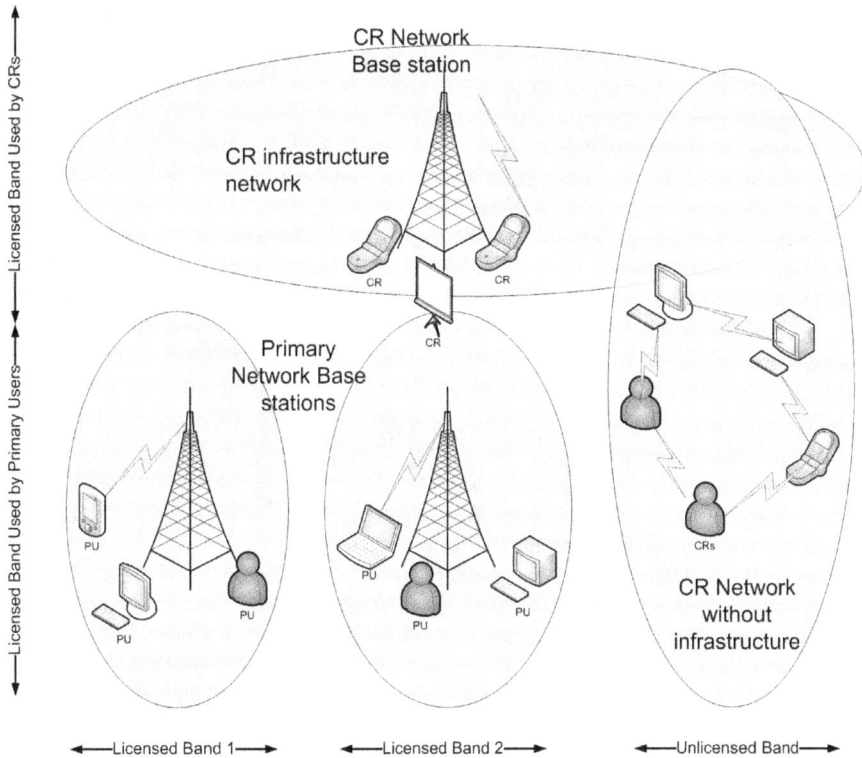

FIGURE 12.10 CR network architecture.

that uses the reinforcement learning to tell the agent to select the optimal channel among a group of available cognitive channels. Here, optimal channel means the channel where the possibility of PU arrival is less, and the capacity and signal to noise ratio (SNR) are high. Authors in [25] have given an extensive study of CR networks. Figure 12.10 shows the CR network architecture.

12.5 SUMMARY

This chapter discussed the two main technologies in the area of connected vehicles. First is the VANET and another is the IoV. The classification of the messages in the vehicular communication and their importance are discussed in detail. The different priorities of the messages are given to different messages. The way in which the priorities are given is shown in this chapter. The concept of static and dynamic priorities is also explained. The rescheduling of queues frequently is necessary for the transmission to go on smoothly. Hence, an efficient data structure is required to cope with the frequent scheduling of messages so that the delay does not get too high. The 5G-V2X is also discussed in this chapter. Its development groups and the issues and challenges in this area are also covered with various possible solutions of those.

REFERENCES

1. M. Chen, Y. Tian, G. Fortino, J. Zhang, and I. Humar, "Cognitive Internet of vehicles", Comput. Commun., vol. 120, pp. 58–70, May 2018.
2. F. J. Martin-Vega, M. C. Aguayo-Torres, G. Gomez, J. T. Entrambasaguas, and T. Q. Duong, "Key technologies, modeling approaches, and challenges for millimeter-wave vehicular communications", IEEE Commun. Mag., vol. 56, no. 10, pp. 28–35, Oct. 2018.
3. IEEE Standard for information technology– Local and metropolitan area networks– Specific requirements– Part 11: Wireless LAN Medium Access Control (MAC) and Physical Layer (PHY) Specifications Amendment 6: Wireless Access in Vehicular Environments, 2010.
4. A. Grilo and M. Nunes, "Performance evaluation of IEEE 802.11e", *The 13th IEEE International Symposium on Personal, Indoor and Mobile Radio Communications*, Pavilhao Altantico, Lisboa, Portugal, vol. 1, pp. 511–517, 2002.
5. H. Lee, J. Lee, S. Kim, and K. Cho, "Implementation of IEEE 802.11a Wireless LAN", *2008 Third International Conference on Convergence and Hybrid Information Technology*, Busan, pp. 291–296, 2008.
6. B. J. Kenny, "Dedicated short-range communications (DSRC) standards in the United States", Proc. IEEE, vol. 99, no. 7, 2011.
7. R. Pal, N. Gupta, A. Prakash, and R. Tripathi, "Adaptive mobility and range based clustering dependent MAC protocol for vehicular ad-hoc networks", Wireless Pers. Commun., vol. 98, pp. 1155–1170, 2018. DOI: https://doi.org/10.1007/s11277-017-4913-9.
8. N. Gupta, A. Prakash, and R. Tripathi, "Medium access control protocols for safety applications in vehicular ad-hoc network: A classification and comprehensive survey", Veh. Commun., pp. 223–237, 2015. https://doi.org/10.1016/j.vehcom.2015.10.001.
9. R. Pal, A. Prakash, R. Tripathi, and K. Naik, "Scheduling algorithm based on preemptive priority and hybrid data structure for cognitive radio technology with vehicular ad hoc network", IET Commun., vol. 13, no. 20, pp. 3443–3451, 2019.
10. R. Pal, A. Prakash, R. Tripathi, and K. Naik, "Regional super cluster based optimum channel selection for CR-VANET", IEEE Trans. Cogn. Commun. Netw., vol. 6, no. 2, pp. 607–617, June 2020. DOI: 10.1109/TCCN.2019.2960683.
11. R. Pal, A. Prakash, and R. Tripathi, "Triggered CCHI multichannel MAC protocol for vehicular ad hoc networks", Veh. Commun., vol. 12, pp. 14–22, 2018. DOI: https://doi.org/10.1016/j.vehcom.2018.01.007.
12. R. Pal, A. Prakash, R. Tripathi, and D. Singh, "Analytical model for clustered Vehicular Ad hoc network analysis", ICT Express, vol.4, no. 3, pp. 160–164, 2018. DOI: https://doi.org/10.1016/j.icte.2018.01.001.
13. P. V. Prakash, S. Tripathi, R. Pal, and A. Prakash, "A slotted multichannel MAC protocol for fair resource allocation in VANET", Int. J. Mobile Comput. Multimedia Commun., IGI Global, vol. 9, no. 3, pp. 45–59, 2018. DOI: 10.4018/IJMCMC.2018070103.
14. P. Singh, R. Pal, and N. Gupta, Clustering based single-hop and multi-hop message dissemination evaluation under varying data rate in vehicular ad-hoc network. In: Choudhary R., Mandal J., Auluck N., Nagarajaram H. (eds) Advanced Computing and Communication Technologies. Advances in Intelligent Systems and Computing, vol 452. Springer, Singapore, 2016.
15. U. Prakash, R. Pal, and N. Gupta, "Performance evaluation of IEEE 802.11p by varying data rate and node density in vehicular ad hoc network", *2015 IEEE Students Conference on Engineering and Systems (SCES)*, Allahabad, 2015, pp. 1–5. DOI: 10.1109/SCES.2015.7506457
16. B. B. Dubey, N. Chauha, N. Chand, and L. K. Awasthi, "Priority based efficient data scheduling technique for VANETs", Wireless Networks, vol. 22, no. 5, pp. 1641–1657, 2016. https://doi.org/10.1007/s11276-015-1051-8.

17. C. Campolo, A. Molinaro, A. Iera, and F. Menichella, "5G network slicing for vehicle-to-everything services", IEEE Wireless Commun., vol. 24, no. 6, pp. 38–45, Dec. 2017.
18. H. Cao, S. Gangakhedkar, A. R. Ali, M. Gharba, and J. Eichinger, "A 5G V2X testbed for cooperative automated driving", Proc. IEEE Veh. Netw. Conf. (VNC), Columbus, OH, USA, Dec. 2016, pp. 1–4.
19. C. R. Storck and F. Duarte-Figueiredo, "A survey of 5G technology evolution, standards, and infrastructure associated with vehicle-to-everything communications by internet of vehicles", IEEE Access, vol. 8, pp. 117593–117614, 2020.
20. C. R. Storck and F. Duarte-Figueiredo, "A 5G V2X ecosystem providing Internet of vehicles", Sensors, vol. 19, no. 3, p. 550, 2019. https://doi.org/10.3390/s19030550.
21. W. Xu, H. Zhou, N. Cheng, F. Lyu, and W. Shi, "Internet of vehicles in big data era", J. Autom. Sin., vol. 5, no. 1, pp. 19–35, Jan. 2018.
22. M. Muhammad and G. A. Safdar, "Survey on existing authentication issues for cellular-assisted V2X communication", Veh. Commun., vol. 12, pp. 50–65, Apr. 2018.
23. L. Zhao, X. Li, B. Gu, Z. Zhou, S. Mumtaz, V. Frascolla, H. Gacanin, M. I. Ashraf, J. Rodriguez, M. Yang, and S. Al-Rubaye, "'Vehicular communications: Standardization and open issues", IEEE Commun. Stand. Mag., vol. 2, no. 4, pp. 74–80, Dec. 2018.
24. R. Pal, N. Gupta, A. Prakash, R. Tripathi, J. J. P. C. Rodrigues, "Deep reinforcement learning based optimal channel selection for cognitive radio vehicular ad-hoc network", IET Commun., vol. 14, no. 19, pp. 3464–3471, Dec. 2020.
25. K. Kumar, A. Prakash, and R. Tripathi, "Spectrum handoff in cognitive radio networks: A classification and comprehensive survey", J. Network Comput. Appl., vol. 61, pp. 161–188, 2016.

13 Prioritized Route Patterns for Autonomous Vehicles Using an IoT Cloud in Urban Scenarios

Anuj Abraham, Pranjal Vyas, and Chetan B. Math
A*STAR,
Singapore

Justin Dauwels
TU Delft,
Netherlands

CONTENTS

13.1 INTRODUCTION

In the last four decades, the significance of the vehicle routing problem (VRP) has increased due to higher traffic volumes and the advancement of autonomous vehicle (AV) technology. Many researchers in the area of static and dynamic vehicle routing algorithms are paying immense attention to VRP and its variants. In the

DOI: 10.1201/9781003268796-17

static routing problem, all information is known at the time of planning the routes, whereas, in the dynamic routing problem, the detour of vehicles happens due to the uncertainties and complexities in the real-time system that occur simultaneously to the routes being carried out. For dynamic assignment of routes, extensive research was performed in previous studies and shows that it is extremely difficult to devise one single algorithm that can cater to diverse scenarios.

A typical solution to dynamic vehicle routing is a combinatorial optimization that deals with traveling salesman problem (TSP), minimum spanning tree (MST), knapsack, bin packing problem, etc. [1, 2]. In this methodology, the search space solution is finite and bounded with discrete constraints. Khouadjia et al. proposed the shortest path formulation based on heuristic algorithms, but it can lead to a suboptimal solution and does not guarantee global optimized values for some specific scenarios [3]. Moreover, efficient vehicle routing algorithms for large-scale fleet management systems were proposed by Nagavarapu et al., where a simulator platform was developed for validation and testing in virtual environments. In general, it is seen that the main objectives considered for VRP are to minimize the total travel time and cost, the total required fleet size, overall total traveling distance, and waiting time imposed on vehicles [3].

In the past few years, efforts in intelligent transport systems show that the Internet of things (IoT) platform can play a significant role in traffic management by connecting the sensor devices over the internet to exchange information wirelessly. An IoT application provides an added insight into both the operation and performance of all aspects of the sensor devices and has been of interest to researchers [4]. Recently, intelligent transport solutions combine mobility components with technologies. These include machine interaction that enables transportation applications to get continuous access to connectivity platforms. Real-time information systems use IoT to connect with driver services, mobile, and online devices. For innovative urban transportation technologies, an increase of automation and road capacity, improved safety and services, and optimized fuel consumption are the key factors considered [4]. Advanced wireless sensor technologies and software for vehicle-to-everything (V2X) environments are used to support intelligent transportation services. Various data from global positing systems (GPS), sensors, etc. from the probe vehicle can be collected using vehicle-to-infrastructure (V2I) communication. A V2I communication in semi/fully automated vehicles expands the horizon for solving VRP in the urban scenario [3]. On-board units (OBUs) and roadside units (RSUs) that broadcast their state information through dedicated short-range communication (DSRC) use wireless access for vehicular environments (WAVE) as a wireless access technology [5]. Also, V2I communication can be performed either by using DSRC or cellular V2X (C-V2X) technologies. The recent research on vehicle mobility coordination system based on a V2I communication was developed and describes an architecture that is capable of operating either in AVs or manual driven vehicles [6].

In view of this, several studies were also carried out to improve urban mobility for an intelligent transportation system by monitoring real-time traffic updates on traffic congestion and unusual traffic incidents via roadside units [7]. Investigation on the development of other system architectures for moving vehicles or roadside infrastructure which can dynamically connect to nearby cloudlets was performed. These are particularly relevant to smart cars and intelligent transportation for updates on accidents, safety, and other applications [8].

This chapter focuses on the development of a simulator platform for prioritizing AVs route patterns based on actual traffic flow data via IoT cloud platforms using wireless communication technologies such as DSRC or Cellular (4G or 5G). Traffic flow data provided by Land Transport Authority (LTA), Singapore consists of vehicle counts on each lane at normal and peak hours approaching the intersection, lane speed limits, turning proposition, traffic signal phase timings, and pedestrian light activation rate at intersections. An effective way of estimating the travel times is formulated based on the gradient boosting machine (GBM) method [9]. The significance of having phase information via wireless communication technologies, in addition to estimated travel times, is addressed in this chapter. This information is provided as input for the dynamic route assignment such that an AV passes the intersections smoothly without stopping. This allows AV to reach its destination in the smallest travel time. Here, we assume that information exchange is 100%, i.e., no packet drops.

The network layout for AMK area of interest (AOI) in Singapore was built in VISSIM for simulation. First, the calibration and validation process are carried out for fine-tuning of specific parameters such as minimum gap time, the minimum headway, and the maximum/minimum speed of the vehicle, etc. in VISSIM. Once the calibration process is performed, the traffic flow data and signal heads are validated based on the data provided. The traffic flow data for static routing in VISSIM includes: (1) turn movements for each junction and (2) input flow in vehicles per hour. The traffic flow data needed for static routing can be obtained from the installed detectors loops and traffic monitoring systems.

Dynamic assignment parameters include: (1) origin-destination (OD) matrix, (2) location of zones and parking lots, (3) traffic composition, (4) vehicle speed at free-flow (speed limit of the road), and (5) travel time. Obtaining the OD matrix needed for dynamic assignments is often challenging. The input of the OD matrix not only involves collected data obtained from traffic counts but also a process of determining the values according to a selected cost function. Similarly, traffic signal control includes: (1) cycle length; (2) green, amber, and red times for each traffic signal group at intersections; and (3) pedestrian signal timing at intersections.

The rest of the chapter is organized as follows: the specific descriptions of the proposed approach to IoT system architecture are introduced in Section 13.2. Function modules in the simulator design platform are given in Section 13.3. This section includes information on the traffic simulator with the real-world traffic scenarios, followed by a description on the application simulator. Section 13.4 deals with the VRP followed by the implementation of prioritized route patterns for AVs in Section 13.5. This section details the steps involved in dynamic assignment route, LTA data, OD matrix data, and learning predictor. Lastly, the chapter is concluded in Section 13.6.

13.2 IoT SYSTEM ARCHITECTURE

The system architecture consists of an IoT cloud for data storage and data processing. The basic architecture is shown in Figure 13.1. The data from traffic light systems, mainly the phase timings, are sent to the cloud for storage. An AV sends a route planning query to the IoT cloud using wireless communication (cellular or DSRC). The query is handled by the learning predictor application running on the IoT cloud. The query

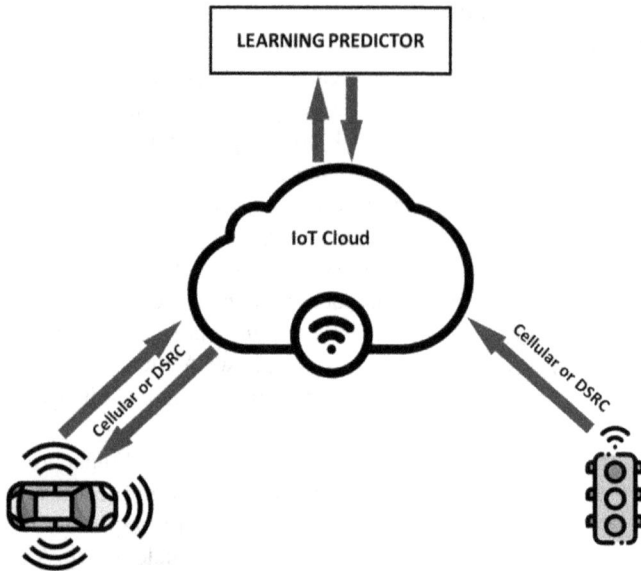

FIGURE 13.1   IoT system architecture.

contains information such as destination, present location, current speed, and acceleration. The learning predictor uses the query, map, and real-time and historical data collected by traffic light systems to find an optimal route for the AV. This route is sent to the AV vehicle as a suggestion. The route data has information such as path and travel times.

13.3   SIMULATOR DESIGN

13.3.1   Traffic Simulator

A traffic simulator is required to create an accurate urban traffic model that matches with the real-world traffic scenarios. VISSIM is chosen because of the following reasons: (1) it precisely replicates the trajectories of any vehicle in the simulation; (2) it simulates the right and left hand side movements; (3) it fulfills the advanced transit and traffic signal requirements; (4) it produces the vehicular and infrastructure (e.g., signals) information at high-resolution interval as little as 100 ms; (5) it provides real-time data exchange with external programs; and (6) it manipulates infrastructure components in a traffic network (e.g., cars dynamics and routes) as a response to stimuli obtained from external programs. Though the traffic simulator known as Simulation of Urban MObility (SUMO) satisfies most of these conditions, it does not possess the capability of simulating left-hand driving. Since the simulations are done majorly for Singapore road networks, VISSIM is chosen over SUMO.

13.3.2   Application Simulator

There are a number of applications that can be designed to solve traffic problems. These may be broadly classified as information exchange, safety applications, or vehicle rerouting. In this work, we chose MATLAB for application simulation. The

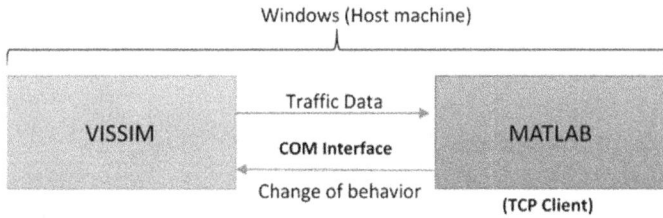

FIGURE 13.2   Function modules in simulator design platform.

application simulator also contains the logic of the V2I application that is being tested. It should seamlessly interact with traffic simulators. Multiple choices are available that satisfy this criterion. However, for the integrated simulator, MATLAB was chosen because of its mature and convenient communication interface with VISSIM.

The basic block diagram of functional modules used in the simulator design is shown in Figure 13.2. The combination of the software has been carried out in a manner that allows an online exchange of data among them [5]. The component object module (COM) interface helps in communication between MATLAB and VISSIM [5, 10]. Traffic data can be fetched in MATLAB and control signals (the optimal route for vehicles) or the infrastructure (change in phase timings of traffic signals) can be given to VISSIM using the COM interface.

The vehicles, traffic scenarios, and traffic signals are simulated using VISSIM. The data from VISSIM is collected by MATLAB. The collected data is used by the learning predictor (random forest [RF] algorithm) running on MATLAB to find the optimal route. Similar to an IoT architecture, the data from the traffic light system is collected at MATLAB and the RF algorithms run in MATLAB to find the optimal route. The optimal route is later suggested to the vehicles in VISSIM where it generates the OD matrix for selection. Note that we assume ideal wireless communication and IoT cloud with very high reliability and low delays.

13.4   VEHICLE ROUTING PROBLEM

The VRP is one of the most popular challenges in modern-day transportation as it generalizes TSP and solves real-life scenarios. VRP defines a set of vehicles at a depot and delivers to a set of customers where all the customers are served only once.

In VRP, the main objective is to determine an optimal route by improving the travel time or some other cost. There are many constraints influencing the problem at different conditions such as vehicle load, the time interval between each customer that has to be delivered, backhauls, the time frame within each customer needs to be serviced, and a route is designed for door-to-door mobility and users' uncertainty [11]. These are classified into capacitated VRP (CVRP), VRP with time windows (VRPTW), VRP with backhauls (VRPB), VRP with pick-up and delivery (VRPPD), dial-a-ride problem (DARP), and dynamic VRP (DVRP) [12]. Figure 13.3 shows the different variants of VRP according to a survey study.

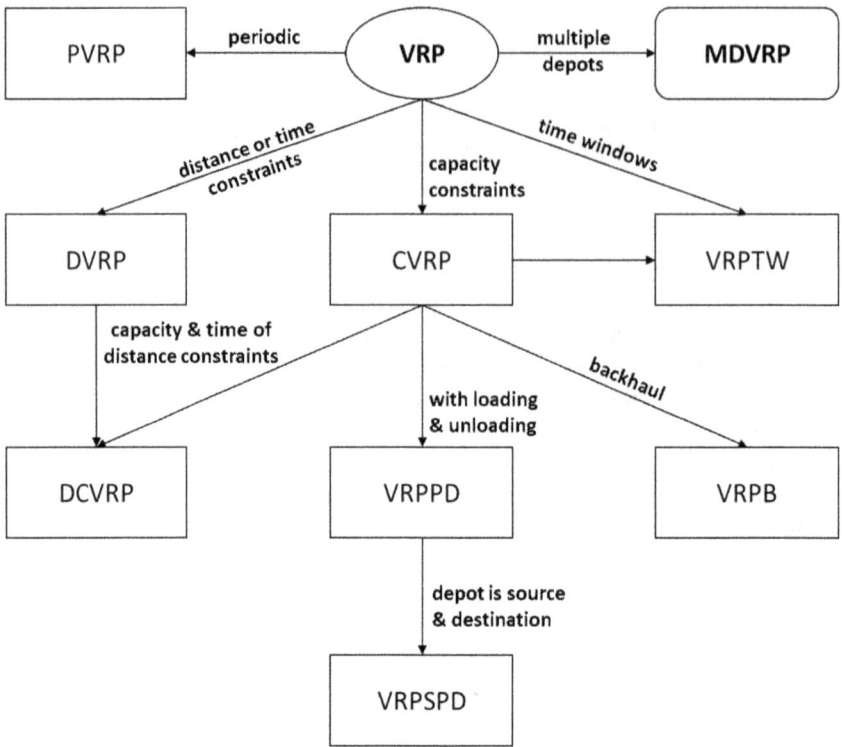

FIGURE 13.3 Different types of VRP [12].

In this chapter, DVRP yields an optimal path between a set of OD matrix with variance in signal phase timing between junctions and vehicle volumes. Figure 13.4 represents the concept of DVRP.

Figure 13.4 explains the execution of single-vehicle entry. Initially, the route is defined as A, B, C, D, and E after two new dynamic requests as time constraint a

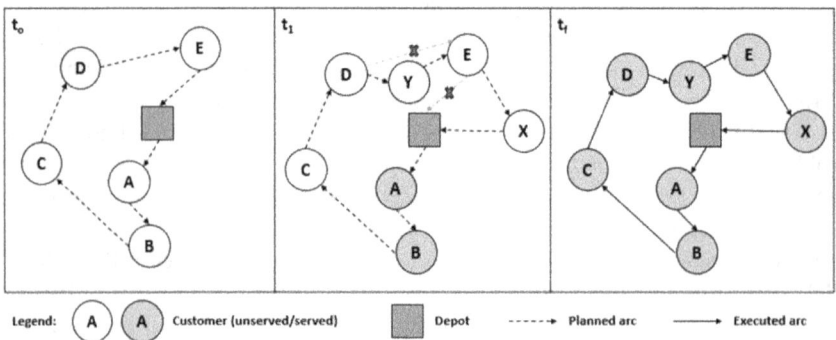

FIGURE 13.4 Dynamic VRP [13].

new route has been designed such as A, B, C, D, Y, E, and X. This example reveals the optimal path for dynamic vehicle routing and real-time communication is necessary between the vehicles [13]. The proposed work develops a simulator platform for prioritizing AVs route pattern based on real data.

With reference to static elements, dynamic routing problem faces some challenges such as congestion due to emergency vehicle affecting the cost factor of travel time and uncertainty of traffic conditions such as peak and normal hours within a week.

13.5   IMPLEMENTATION OF PRIORITIZED ROUTE PATTERNS

This work uses VISSIM 11 for the dynamic route assignment and generates travel cost for different zones combinations. The main inputs to dynamic route assignment depend on the generation of OD matrix, vehicle composition, and its cost function.

13.5.1   STEPS INVOLVED IN DYNAMIC ASSIGNMENT ROUTE

In VISSIM, dynamic assignment/route builds a model according to route choice behavior, avoiding the static routes and using the OD matrix as flow input. OD matrix contains information such as the number of trip flows where each cell represents the travel demand between each origin (row) and destination (column) zone. This is represented as a matrix column according to the number of zones and parking lot in each zone. In this chapter, we consider the parking lot as zone connector. Parking lot defines the point where the vehicle actually appears or leaves the road network and there are three types:

1. Zone connector;
2. Abstract parking;
3. Real parking lot.

Each zone can have only one parking lot that is placed in between two nodes. Nodes placed in the junctions or path where it could diverge and edge are route sequences or links. Some parameters considered for dynamic route assignment in VISSIM are Kirchhoff assignment (defines the polarization of path distribution), logit function (defines parking lot choice model), correction of overlapping paths (affects the vehicle distribution on paths that are partially identical), travel time on paths and edges, and volume on edges. The flow chart representing the steps involved in dynamic vehicle route is described in Figure 13.5.

13.5.2   LAND TRANSPORT AUTHORITY DATA

In this work, the AMK road network is considered where five signal junctions are used for the study of a high prioritized route pattern. The traffic flow data [14] contains information such as:

1. Traffic light and pedestrian signal timing;
2. Pedestrian light activation rate at intersections;

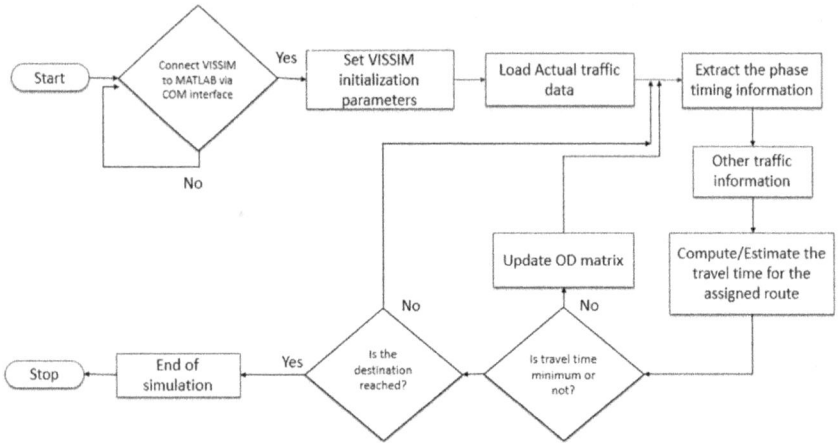

FIGURE 13.5   Steps involved in dynamic vehicle route.

3. Turning proportions at intersections;
4. Vehicle counts for each lane at normal and peak hours.

Figure 13.6(a) shows lane marking and signal phase timings for a particular inter-section in AMK road section. The signal phases are represented according to the ring-and-barrier diagram [14]. Figure 13.6(b) shows the vehicle counts data for traffic control signal TCS 9302 intersection in AMK road section.

13.5.3   OD Matrix Data

In this chapter, OD matrix considers four zones in AMK region that are simulated in traffic simulator. The OD matrix that represents the number of trips between the zones is tabulated in Table 13.1. The total matrix travel time considered is 5 minutes.

For example, a scenario is considered where AV is prioritized based on the traffic data, mainly at peak traffic and normal traffic periods in the road network. Similarly, another scenario is analyzed where both the traffic flow data and signal flow control for smooth movement of AV is illustrated in Figure 13.7. For both the scenarios, the objective is to minimize the travel time and determine an optimal route for AV in virtual environments.

13.5.4   Learning Predictor

RF is a well-known supervised collective learning method and is used to do travel time prediction [15]. A real-time traffic data set for an urban road network of AMK road networks in Singapore is considered to perform this simulation. Traffic data can be updated periodically in real-time to estimate the shortest travel time/path. RF learning predictor is one of the best travel time prediction techniques that use tree-based classifiers [16]. In training phase, RF creates multiple classification and regression trees (CART)-like trees [17, 18], each trained on the original training data and

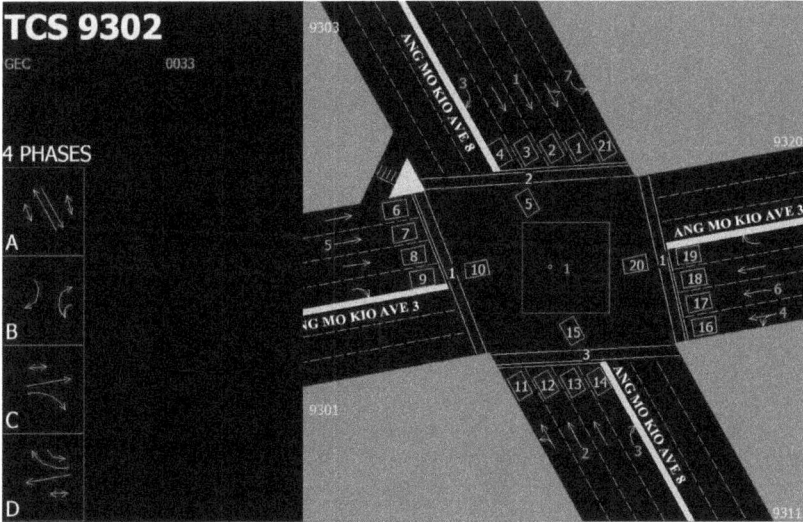

FIGURE 13.6 (a) Lane markings signal phase timing at intersections in AMK road section and (b) vehicle count data.

TABLE 13.1

OD Matrix

	Zone 1	Zone 2	Zone 3	Zone 4
Zone 1	0	20	40	59
Zone 2	77	0	56	44
Zone 3	56	56	0	23
Zone 4	78	89	32	0
Total	211	165	128	126

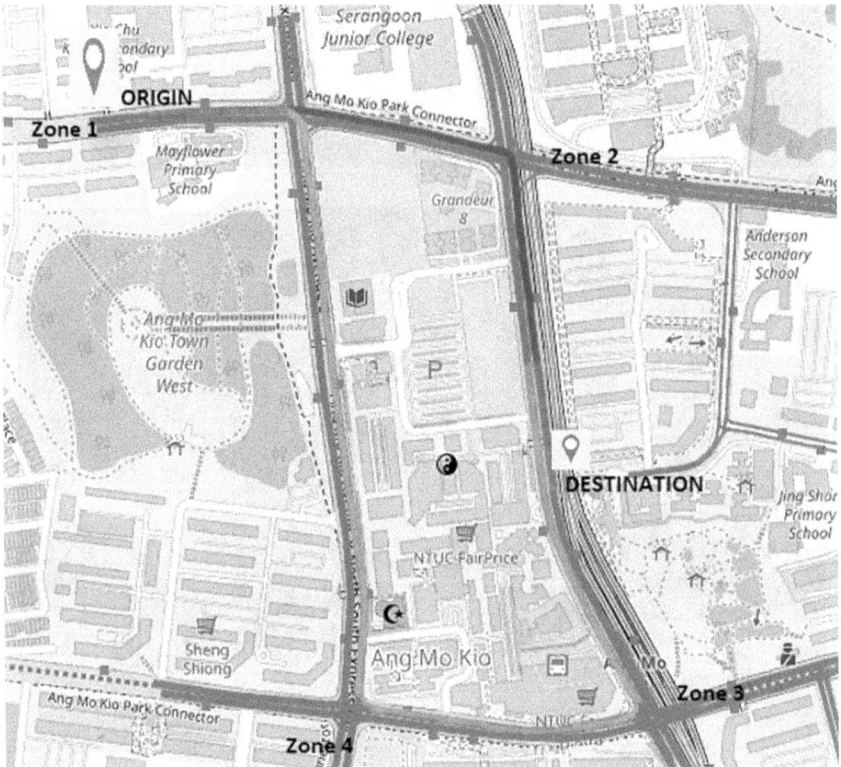

FIGURE 13.7   Dynamic vehicle routing scenario in AMK section based on traffic flow data and traffic signal control in VISSIM.

searches only across the randomly selected subset of the input variables to determine a split. Keywords related to RF are as follows:

1. *Bagging*: In bagging or bootstrap aggregation, each base model is trained with the training data.
2. *Decision Tree*: A decision tree is a hierarchical data structure with nodes and branches. The entry point of a decision tree, which is at the top of the

hierarchy, is called the root node. The root node will have child nodes, which in turn may have their successors.

By limiting the number of trees, the computational complexity of the algorithm is controlled and the correlation between trees is decreased. As a result, RF can easily handle high-dimensional data by using large number of trees. The main advantages of using RF in traffic prediction are: (1) it runs efficiently on large datasets such as traffic data; (2) gives predictions with very low prediction errors; (3) it is relatively robust to noise; and (4) it has less computational complexity. Defining mathematically, the variance of a RF, (σ^2_{RF}) can be given as shown in Equation (13.1),

$$\sigma^2_{RF} = \rho_i \sigma^2_i + \frac{1-\rho_i}{m} \sigma^2_i \qquad (13.1)$$

where m is the total number of trees, ρ_i is the correlation between the individual trees, and σ^2_i is the variance of the individual trees.

Recent studies on forecasting methods [20] prove that the travel time and traffic condition predictions that are based on data-driven approaches are appropriate to determine the shortest travel time. To achieve improved accuracy, the RF is trained with a fixed value of 500 trees [16]. But irrespective of the problem, the number of trees is to be varied such that we get better accuracy with an objective of RF variance minimized to zero. To find the optimum number of trees for training, one can perform sensitivity analysis. In this chapter, the number of trees in RF was determined by ten-fold cross-validation, which is between 50 and 1000. In our study, the number of trees is taken as 200 (best threshold value), as the smaller the trees, the lesser the computational cost. Also, it was seen that for a higher value of trees chosen, it gives the same minimum test error of 5% with additional computation cost. Thus, a trade-off is created between the accuracy and complexity.

The two important tuning parameters of RF are the number of trees m and the random independent variables σ^2_i considered at each split. In our work, the prediction accuracy is represented in terms of mean absolute percentage error (MAPE). At each split, the tree only considers some randomly selected variables and average the predicted values to get the new forecast. All models were trained with three independent variables to generate a forecasting result by using historical data, i.e., the three most recent travel times observed ($t - 3$, $t - 2$, and $t - 1$) in the network are used to predict travel time for the next interval (t). Thus, MAPE which is used as the evaluation criteria is given in Equation (13.2),

$$\text{MAPE} = \frac{1}{n} \sum_{t=1}^{n} \frac{|A_t - F_t|}{A_t} \qquad (13.2)$$

where n is the total number of samples, A_t is the actual value of the travel time, and F_t is the forecasted value of travel time.

It is observed that MAPE is reducing with increase in number of trees m. The RF supervised ensemble learning method is built using the "TreeBagger" class that creates a bag of decision trees, as used in MATLAB [19], is involved in this chapter.

Also, "TreeBagger" selects a random subset of predictors to use at each decision split as in RF. For regression problems, the class "TreeBagger" supports mean and quantile regression forest.

From Figure 13.7, it is seen that the four zones, parking lots at each zone connector, and OD information are predefined in VISSIM. The initial static route is represented in blue color based on the shortest path algorithm. The routing cost obtained from the shortest path is 1.2 km. The sequence of the served zones is: **origin → zone 2 → destination**.

Then, the prioritized AV route pattern is determined dynamically based on traffic volume and signal heads. It is observed from the vehicle flow data that congestion occurs at the intersection near to zone 1 and zone 2 in a certain time instant. Therefore, the estimated travel time while incorporating the traffic data and phase timing information at the intersection is 350 s, whereas the optimal dynamic route sequence determined with a minimal travel time of 276 s is: **origin → zone 4 → zone 3 → destination**.

The dynamic path assignment is performed through zone 1, zone 4, and zone 3, respectively, (indicated in green color) to minimize the travel time rather than the shortest path. Hence, it is observed that the total travel times are minimal when the phase information is considered using wireless communication in IoT platform and the optimal route is determined.

13.6   CONCLUSIONS

A simulation platform has been developed to test, validate, and illustrate prioritized route patterns of AVs in virtual environments using an IoT platform. The network layout for AMK region in Singapore was built in VISSIM. The travel time of AV is estimated from actual traffic data and optimized through traffic light controller in an IoT cloud using wireless communication technologies such as DSRC or cellular (4G or 5G). The primary objective of smallest travel time of AV incorporating other traffic condition in reaching the specified destination is performed and analyzed. Hence, it is observed that the total travel time is minimal when the phase information is considered via IoT cloud using wireless communication and optimal route is determined.

Future work involves investigating the other techniques to include more constraints such as real-time delays, road conditions, weather/climatic data, routing policies, time windows for pickup and drop-off, battery management, and fleet optimization of AVs into the simulation platform to illustrate more realistic vehicle routing scenarios. Also, an attempt will be performed to understand the influence of passenger comfort under traffic congestions.

ACKNOWLEDGMENTS

The authors gratefully acknowledge the contributions of Ms. Rajashree A. Sundaram, Ms. Priyanka R. Mehta, and Mr. Usman Muhammad of Energy Research Institute @ NTU, Nanyang Technological University, Singapore for their timely support.

This material is based on research/work supported by National Research Foundation under Virtual Singapore Program Award No. NRF2017VSG-AT3DCM001-018. Any opinions, findings, and conclusions or recommendations expressed in this material are those of the author(s) and do not necessarily reflect the views of National Research Foundation.

REFERENCES

1. Prins, C. 2004. A simple and effective evolutionary algorithm for the vehicle routing problem. *Computer & Operations Research*, 13(12):1985–2002.
2. Nagavarapu, S. C., Tripathy, T., and Dauwels, J. 2017. Development of a simulation platform to implement vehicle routing algorithms for large scale fleet management systems. *Proceedings of the 20th IEEE International Conference on Intelligent Transportation Systems (ITSC)*, Yokohama, Japan, 124–129.
3. Khouadjia, M. R., Sarasola, B., Alba, E., Jourdan, L., and Talbi, E. G. 2012. A comparative study between dynamic adapted PSO and VNS for the vehicle routing problem with dynamic requests. *Applied Soft Computing*, 12(4):1426–1439.
4. Alshehri, A. and Sandhu, R. 2016. Access control models for cloud-enabled internet of things: a proposed architecture and research agenda. *Proceedings of the IEEE International Conference on Collaboration and Internet Computing (CIC)*, Pittsburgh, PA, USA, 530–538.
5. Milanés, V., Godoy, J., Pérez, J., Vinagre, B., González, C., Onieva, E., and Alonso, J. 2010. V2I-based architecture for information exchange among vehicles. *IFAC Proceedings Volumes*, 43(16):85–90.
6. Sarrab, M., Pulparambil, S., and Awadalla, M. 2020. Development of an IoT based real-time traffic monitoring system for city governance. *Global Transitions*, 2:230–245.
7. Gupta, M., Benson, J., Patwa, F., and Sandhu, R. 2020. Secure V2V and V2I communication in intelligent transportation using cloudlets. *IEEE Transactions on Services Computing*, 13(14):1–12.
8. Tangirala, N. T., Abraham, A., Choudhury, A., Vyas, P., Zhang, R., and Dauwels, J. 2018. Analysis of packet drops and channel crowding in vehicle platooning using V2X communication. *Proceedings of the IEEE Symposium Series on Computational Intelligence (SSCI-2018)*, Bangalore, India, November 18–21, 281–286.
9. Mason, L., Baxter, J., Bartlett, P. L., and Frean, M. 1999. Boosting algorithms as gradient descent. *Proceedings of the 12th International Conference on Neural Information Processing Systems*, 512–518.
10. Choudhury, A., Maszczyk, T., Asif, M. T., Mitrovic, N., Math, C. B., Li, H., and Dauwels, J. 2016. An integrated V2X simulator with applications in vehicle platooning. *Proceedings of the IEEE International Conference on Intelligent Transportation Systems (ITSC)*, 1017–1022.
11. Ghannadpour, S. F., Noori, S., Tavakkoli-Moghaddam, R., and Ghoseiri, K. 2014. A multi-objective dynamic vehicle routing problem with fuzzy time windows: model, solution and application. *Applied Soft Computing*, 14(1): 504–527.
12. Jairo, R. M.-T., Franco, J. L., Isaza, S. N., Jiménez, H. F., and Herazo-Padilla, N. 2015. A literature review on the vehicle routing problem with multiple depots. *Computer & Industrial Engineering*, 79:115–129.
13. Pillac, V., Gendreau, M., Guéret, C., and Medaglia, A. L. 2013. A review of dynamic vehicle routing problems. *European Journal of Operational Research*, 225(1):1–11.
14. Traffic Signal Timing Manual. 2017. [*LTA data*] https://ops.fhwa.dot.gov/publications/fhwahop08024/chapter4.htm.

15. Svetnik, V., Liaw, A., Tong, C., Christopher Culberson, J., Robert Sheridan, P., and Bradley Feuston, P. 2003. Random forest: a classification and regression tool for compound classification and QSAR modeling. *Journal of Chemical Information and Computer Sciences*, 43(6):1947–1958.
16. Narayanan, A. K., Pranesh, C., Nagavarapu, S. C., Kumar, B. A. and Dauwels, J. 2019. Data-driven models for short-term travel time predictio. *Proceedings of the IEEE International Conference on Intelligent Transportation Systems (ITSC)*, Auckland, New Zealand, 1941–1946.
17. Breiman, L. 2001. Random forests. *Machine Learning*, 45:5–32.
18. Breiman, L., Friedman, J., Stone, C. J., and Olshen, R. A. 1984. *Classification and Regression Trees*. New York: CRC press.
19. MATLAB. *Treebagger*. 2019. https://www.mathworks.com/help/stats/treebagger.html.
20. Oh, S., Byon, Y.-J., Jang, K., and Yeo, H. 2015. Short-term travel-time prediction on highway: a review of the data-driven approach. *Transport Reviews*, 35(1):4–32.

14 Teaching IoT Smart Sensors Programming for a Smarter World

A Tool to Enable the Creation of a Remote Lab

Hugo Martins
University of Trás-os-Montes e Alto Douro (UTAD),
Vila Real, Portugal

Nishu Gupta
Department of Electronic Systems, Faculty of
Information Technology and Electrical Engineering,
Norwegian University of Science and Technology,
Gjøvik, Norway

Manuel José Cabral dos Santos Reis
UTAD/IEETA,
Vila Real, Portugal

CONTENTS

14.1 INTRODUCTION

In the last decades, society has assisted to an astonishing development of technological devices, ever increasing in its "smartness" capabilities. Included in these devices

DOI: 10.1201/9781003268796-18

is the class of those enabling the acquisition of different environmental parameters, such as temperature, number of vehicles on road, among others. Besides contributing to a smarter world, the goal is to automate the work and connect the devices using the internet. The acquired data are then analyzed by local processing and eventually stored locally. Then, data are sent from local storage to cloud storage. Finally, using the gathered information, an appropriate action is taken. The devices used in this process are typically known as IoT (Internet of Things) devices, meaning that they can form an internet where all kinds of devices ("things") of this type are connected to [1]. In 2020, it is estimated that the number of connected IoT devices was above 50 billion [2].

Special care must be taken in the teaching and development of the necessary skills to efficiently program these devices. The learning process, traditionally, is achieved with the help of several physical laboratorial facilities, being particularly evident during the teaching of university courses. Here, a tool to help in this process of developing the "smartness" (i.e., the software) component is presented. This tool can be used to remotely access the available resources, thus creating an "online lab." Figure 14.1 presents a general overview of this "online lab."

This "online lab" has particular usefulness in the current pandemic context. The developed methodology enables students to have access to real hardware devices, thus helping them developing their skills and knowledge at distance. We would like

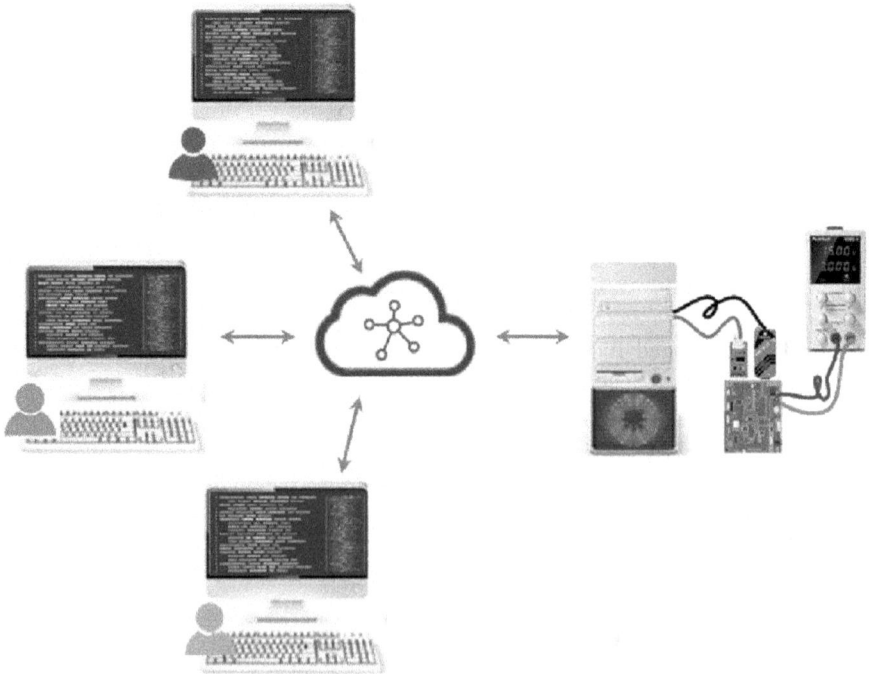

FIGURE 14.1   Global view of the "online lab."

to start by emphasizing the following advantages of using this approach, at least from the perspectives of teachers and students:

- *Reducing costs and increasing the number of different types of architectures*: the needed programming kits and hardware components have a considerably high price. With this approach, there is no need for having one programming kit and hardware components for each student. The extra money saving can be invested in the purchase of different programming kits and hardware components for different types of architectures (e.g., PIC, ARM, MSP). Additionally, this can also lead to a greater flexibility. For example, one student may be using a kit for programming a PIC and another for programming an ARM, at the same time, and the same student may experiment with different architectures.
- *Increasing the availability of the devices*: there is no need for any extra time period to check for any abnormalities related to the previous usage (e.g., broken wires, damaged components). In case this is needed, this check must be performed by someone specially trained. Aside from maintenance periods, a device is always ready.
- *One device shared by many students*: it is true that a device may be physically requested by different students, but in order for one student to work with it, first, it must be delivered by the previous student using it, thus causing loss of time. In the current pandemic situation, this loss of time is aggravated (equipment quarantine, disinfection, etc.). With the approach proposed here, the device will be available as soon as the slot of time available for the previous student ends.
- *Widening of the utilization period*: because it is no longer necessary to have physical access to the laboratory, the access may be scheduled on a 24-h/24-h base (excluding the maintenance periods) instead of much shorter periods, typically from 9 am to 8 pm.
- *Increasing focus and productivity*: by providing parts of code already working, the student would only worry about developing the part that offers her/him the greatest difficulties and the one she/he wants to learn, thus helping the student get to focus on what is really needed and increasing her/his productivity.

14.2   RELATED WORK

Physical and virtual labs are used to teach students and provide them with the chance to experiment, in a "hands-on experience" approach, the fundamental concepts in almost all the areas of knowledge. The number of available examples of such an approach is huge. For example, a remote lab for experiments with a team of mobile robots, using the LEGO Mindstorms technology and the Matlab programming language, was presented in [3], and a low-cost solution for the teaching of mechatronics control concepts was presented in [4]. The benefits of such an approach have been published and then implemented in other areas where this approach was not typically previously used [5]. These benefits have also been identified by the US Defense

Advanced Research Projects Agency (DARPA), leading to the presentation of the Open Control Platform (OCP), an open middleware solution for embedded systems [6]. Also in the control area, in [7], the authors use techniques from the area of supervisory control for discrete event systems trying to produce management systems capable of being self-adaptive at "applications level (functionalities) as well as deployment level (software tasks, execution platform)." They use the Heptagon/BZR and ReaX programming languages to build up a design environment. The contributions of their research include the creation of generic behavioral models, both at the application and deployment levels. A distance learning laboratory course on embedded systems, provided by the German University of Hagen, was presented in [8]. The authors used a virtual machine, including all necessary software for this lab to simplify support and provide the same programming environment for all participants.

Peter, Momtaz, and Givargis [9] addressed the programming of cyber-physical systems (CPS), where the authors have presented a programming framework that allows students to implement and test CPS control programs using a high abstraction level in a web page. Students can incrementally design a CPS and experience different challenges (e.g., channel delays, model uncertainties, real-time behavior), but without using low-level programming or tools. They tested the framework in an embedded system design class, and the results not only show the ability of a JavaScript-based programming and execution environment to design, program, and run CPS on different levels of abstraction but also show an increase in the approval of the students and a significantly improved understanding of modeling and programming.

In [10], the authors proposed the creation of a "smart" physics lab/cabinet. This lab is fully automated and can be operated in three modes: "standard," "automatic," and "automatic energy saving." The devices in the lab are controlled using a smartphone, PC, and remote control. These devices include a variety of sensors, indicators, electronic components based on Arduino UNO, MEGA, and Wi-Fi module ESP8266-12E. An ATMEL microcontroller controls three modules: "informational," "executive," and "demonstrational." The "demonstrational" module can be used for mechanics, molecular physics, electrodynamics, and optics demonstrations. Several sensors can be used to monitor the environment of the lab and outside it (including temperature, humidity, pressure, illumination, level of carbon dioxide, and other gases in the air). These sensors and actuators can also be used to remotely control peripheral devices (including TV, projector, lamps, sockets, and curtains).

In [11], a user-oriented management system for ubiquitous computing environments, named iPlumber, is presented. This proposal explores the meta-design approach, being different from the typical low-benefit "zero-configuration" systems or high cognitive cost "end-user programming" tools. Basic-level software sharing, foraging, low-cost software configuration, advanced-level cooperative software co-design, and error handling are among the management activities supported. They have tested the usability of the system through a user study, with a total of 33 subjects to test the management activities from an open exhibition environment and a controlled university environment.

A framework, supporting developers in modeling smart things as web resources and exposing them through RESTful Application Programming Interfaces (API) and developing applications on top of them, was proposed in [12]. This framework consists of a "web resource information model, a middleware, and tools for developing and publishing smart things digital representations on the web." The authors tested and evaluated this framework in the "SmartSantander European Project" smart city scenario. An open-source framework for civil structural health monitoring (SHM) using networks of wireless smart sensors is presented in [13]. The framework is based on a modular, reusable, and extensible service-oriented architecture. To demonstrate the efficacy of the proposed framework, the authors presented the development of "Decentralized Data Aggregation," a decentralized application for measurement, aggregation, and compression of sensor data using the proposed services. In [14], the authors present a model, middleware architecture, and engineering of applications in a smart home environment. They defined ambient ecologies for modeling ubiquitous computing applications, specified design patterns and programming principles, and developed infrastructures to provide a paradigm of application engineering and tools to support ambient ecology designers, developers, and end users.

In [15], the author recognizes that it is very difficult to design intelligent devices, and he proposed a tool for designing smart devices. This tool allows realizing the soft of an application, which would be implemented in a microcontroller in order to obtain smart devices. The tool uses the IEC 1131-3 norm to describe the functions and automatically generates a C program.

Some of the problems that are addressed in this work were also partially discussed in [16, 17, 18, 19, 20]. In [16], the authors presented a contribution to the development, in real time, of wireless sensor networks, capable of creating dense air pollution maps for compact urban areas. They have presented a prototype of a reconfigurable device (chip), containing a programmable 10-bit analog-to-digital converter, a programmable clock generator, partial blocks of an artificial neural network, and the tool for programming and testing this device. The problem of automatic generation of user interfaces for remote laboratories was addressed, for example, in [21]. The authors used the smart device specifications, and they also proposed a tool to automatically generate the user interface of the chosen experiment(s).

The emergence of IoT in the engineering curriculum, toward building human resources that contribute to the current issues related to smart cities, is addressed in [22], and as a result, the authors proposed a framework for inclusion of IoT as a course in the undergraduate curriculum.

Burhan et al. [2] proposed six elements needed to deliver the functionality of IoT: identification, sensing, communication, computation, services, and semantics. The majority of the published literature research programming tools that follow apply to the third level, computation, and above. For example, in [23], the authors proposed an agent-based computing paradigm to support IoT systems analysis, design, and implementation. The agent-oriented approach, based on the agent-based cooperating smart object (ACOSO) methodology and the related ACOSO middleware, enables to develop smart and dynamic IoT systems of diverse scales. They applied this approach to the development of a Smart University Campus.

Very recently, smartphones were used as a transfer mechanism to simplify the deployment and rekeying of LoRaWAN devices [24]. By using this approach, the authors have removed the need for a laptop, a wired connection, and programming software, allowing devices to be provisioned out in the field without the need for calibration or specialized tools.

14.3 THE DEVELOPED TOOL

Like many of the models used in the related works presented above, the tool presented here was developed using a client-server model. Figure 14.1 presents a global view of the online laboratory. As can be seen, the IoT device to be programmed, located on the rightmost part of this figure, is connected to a server computer, through a programmer (e.g., PICkit 3/4, ST-Link V2). The developers (students or professionals in the field) use the client application installed on their personal computers to connect to this server, through the cloud/internet, and thus are able to upload to this server the code they are developing and testing it remotely. As can also be seen, the same device can be simultaneously shared by a group of users. Figure 14.2 shows a general overview of the developed tool. As can be seen from this figure, the client(s) and the server are connected using the MQTT broker provided by HiveMQ [25]. This broker is designed for cloud native deployments, and to make use of the cloud resources. It reduces the network bandwidth required for sharing of data, and allows the connection of any devices and backend systems, using push technology designed for IoT applications. This mechanism will help in sending and receiving data from and to the connected devices in a very fast way. It has an open API, which allows the integration of IoT data into enterprise systems and pre-built extensions (such as Kafka, SQL, and NoSQL databases). It also allows the deployment of the MQTT broker on private, hybrid, and public clouds (such as AWS and Microsoft Azure). It is based on the MQTT open IoT standard (which facilitates the integration with Eclipse Paho, custom-built MQTT libraries, and libraries direct from HiveMQ).

The client and the server were both developed in Java, and we have used JSON to share data between them. In a typical working session, the server and the client start their execution by connecting themselves to the MQTT broker. Next, when the client connects to the server, the server checks for the serial ports and the tools that are connected to it. Then, the server sends these data, using an MQTT publish, to the client that is connected to this server. From this moment on, the server will be waiting for the commands that the client will issue. It should be noted that in fact we can have more than one client connected to the same server. This is particularly useful in

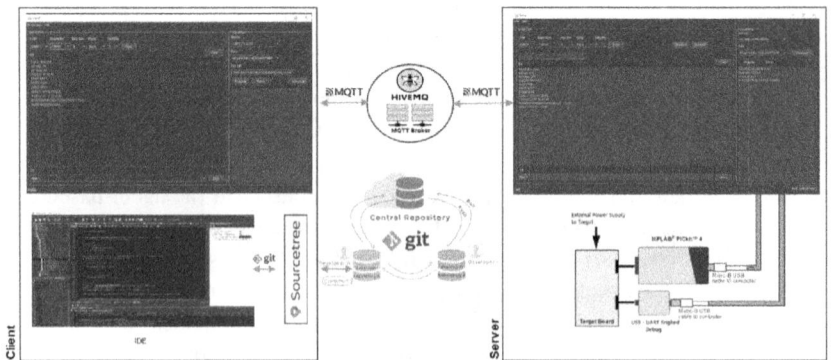

FIGURE 14.2 General overview of the approach and tool used.

real development situations. For example, when a student is having problems, he/she can ask for help, and his/her teacher/supervisor can connect to the same server and see in real time what is happening and solve the problem.

14.3.1 THE CLIENT

The client application may locally save the log file (in the computer where the client is running), with the results of the working session simulations and programming.

In order to save time and made more flexible the development, on the client side, there is an option to select the ".hex" file that the client wants to load to the device for programming it. Typically, it is suggested to use the path to this ".hex" file that is used by the compiler during the compiling time, so the user will not need to change or copy it to another place (folder) every time he/she introduces any (small) changes in the project.

The sending of the ".hex" file is initiated by the transmission of an MD5 hash, in order to ensure the integrity of this file. The server will calculate its own MD5 hash from the received file, and compare it with the received one. If they are equal, everything is fine. If they are different, the file needs to be resent.

After receiving a ".hex" file, it must be sent to the device, through the programmer. In the case of the Microchip, this is done by using the "P" command. Please refer to the "server" subsection (below) to see how this work, and see concrete implementation examples.

The "reset" command is very useful when developing and programming hardware. As suggested by its name, this command enables the restarting of the device, in case it enters a dead-lock state, and it is available in the client application.

The client may also change in real time the serial (COM) port he/she is listening, during the execution/debugging. This is useful if the device has several output ports. For example, one serial port may be attached to the GPS data and another to a modem that is sending these data, and the user wants to check these data.

It should be noticed that one client can be connected to only one server, and thus only one programmer (only one of the devices or hardware bases presented below in section "Results").

In summary, the user develops the program he/she wants to be saved to the IoT sensor device (or hardware platform as mentioned below), by using his/her local personal computer. Then, he/she will use the client application to send the compiled version of this program to the server. The server will use the programmer to save this new version of the software to the IoT sensor device, thus reprogramming it. The user can then use the client application to see if the device is functioning properly, debug the code, among other options.

14.3.2 THE SERVER

The server has control over the serial (COM) ports, in order to enable the debug and log.

The server, and the client also, can change the universal asynchronous receiver-transmitter (UART) parameters (like baud rate, number of bits per char, parity, stop bits, etc.), as well as open and close the serial ports.

It is the server that communicates with the programmer (e.g., PICkit 3/4, ST-Link V2); using other words, the programmer is directly attached to the USB ports of the server, for programming the devices being developed/tested.

There is a list (hardcoded list) for the available architectures that can be used (e.g., Microchip-PIC8/16/24/32, ST-ARM-STM8/32). This list is hardcoded because there is no standard scheme for communications with the tools that each architecture provides to control the programmer. For example, Microchip provides "ipecmd.exe," and by using the command "-T" we receive a list of all the Microchip programmers that are connected to the computer ports (e.g., PICkit 4, ICD). However, the data that are provided by this program are different from the data that will be provided by another architecture (e.g., ST-ARM). As such, the parsing of these data must be hardcoded. Note that the use of an XML configuration file does not introduce any flexibility to this particular issue. This fact can be seen as a limitation in the flexibility of our model, but this applies only to the server-side application; this is something completely transparent to the client-side application (the client "sees" only the available architectures).

As can be seen from the example in the previous paragraph, the server application communicates and controls the programmer through the program provided in the architecture. For example, Microchip provides the "ipecmd.exe" and the server application issues commands to this program, which actually sends them to the programmer. Listings 1, 2, and 3 show the concrete implementation of the command to obtain a list of all the Microchip programmers that are connected to the computer ports. Listing 1 shows the excerpt of the server main program, which in turn will execute the code in listing 2 that will return an error code (0 if everything is OK) and a string ("report"). Next, the "report" string is parsed to get the full list, by executing the code in listing 3. Figure 14.3 shows a close-up of a screenshot of the server application, with an example of the tools that are actually available (in this particular case, only one). As stated above, at the end of the server initialization, this information must be sent to the client application. This

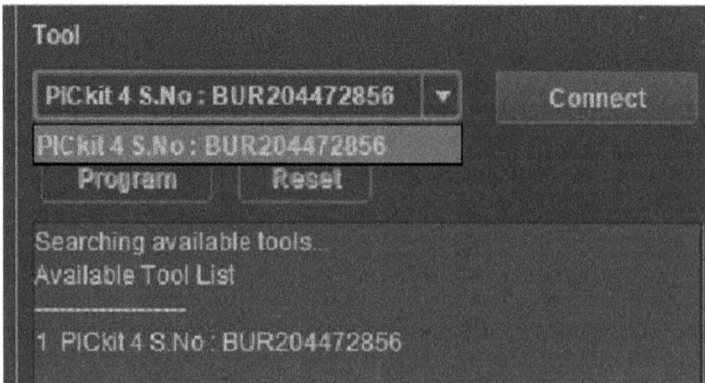

FIGURE 14.3   Close-up of a screenshot of the server application, showing the tools that are currently available (i.e., the result of executing the Microchip "ipecmd.exe -T" command).

is done by sending a JSON object to the client; listing 4 shows an example of an actual JSON object.

Listing 1. An excerpt from the server main program to get a list of all the Microchip programmers that are connected to the computer ports ("ipecmd. exe -T").

Listing 2. Implementation example of the function to execute the commands by the operating system.

Listing 3. Implementation of the parsing function to obtain the list of the available tools connected to the server.

Listing 4. An example of a JSON object sent by the server to the client application at the end of the initialization process.

14.4   USING THE TOOL

We have been using this tool with Git and Sourcetree while teaching these concepts to our students. Once again, please refer to Figure 14.2.

Git (https://git-scm.com/) is today's reference of a free- and open-source distributed version control system, designed to handle everything from small to very large projects with speed and efficiency. Git is easy to learn and has a tiny footprint. It was originally developed by Linus Torvalds (creator of the Linux kernel) and it is compatible with a wide variety of operating systems and integrated development environments (IDE). It has a distributed architecture, a repository containing the complete history of all changes (local and remote), and keeps a copy of all code development work. In its main characteristics, we can include safety, flexibility, version control, and high performance.

Sourcetree (https://sourcetreeapp.com/) simplifies the interaction with Git repositories, so the developer can focus on coding. Sourcetree's simple Git graphical user interface enables visualization and management of the repositories. It is very easy to be used by beginners, as it is desired in the early learning stages. Its Git client simplifies distributed version control and also includes features for review change-sets, stash, cherry-picking between branches, among others. It also enables getting information on any branch or commits with a single click. It works with Git and Mercurial.

From a teaching and learning perspective, the usage of a tool like Git will enable the following possibilities:

- Place original devices programming code, drivers, etc., in the repository, as a starting point for students to start their projects.
- Develop team projects with different groups of students, where each student is responsible for developing a particular feature, thus creating different branches.
- Keep track of the different development stages in the repository, for a later analysis by the teacher of the progress of each student in their respective branch.
- Learn about using a powerful tool and take advantage of its resources. For example, in case of encountering a problem during the development with

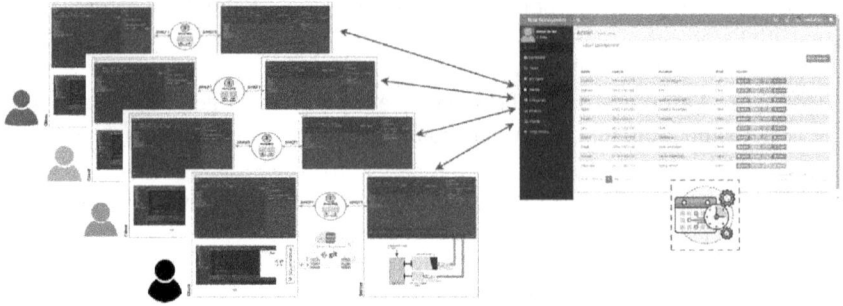

FIGURE 14.4 Example of a web-based service that the student accesses to book a period of time to use the lab.

the current version, the student will be able to revert the code to the previous stable version; by doing so "the fear of making changes to the code will be lost."

However, from the teaching and learning perspectives, our approach will also need a way to control the access to a specific device or "virtual lab." In other words, it will need a mechanism the students may use to choose the "time slices" of the day/week where he/she has access and control over the device. This access can be manually granted and controlled directly by the student's teacher/supervisor (i.e., the teacher keeps track of the student who has been granted access to the device on a daily basis). It will also be the responsibility of the teacher to see the assignment and distribution of the passwords that enable the remote access to the lab. Alternatively, it can be used as a web-based service (a typical web page), like the one depicted in Figure 14.4, where a student, first, chooses the "time slice" from the time schedule/timetable of available periods. Next, the student chooses one hardware base from those available. The password management will be also made by this service, where the assignment and destruction of the passwords will be assured by this system.

14.5 RESULTS

In order to test our tool, we have developed and made available to the students two hardware bases. The first one, which we have called "GPS tracker," has the following main components: MCU PIC24FJ512GA606, Accelerometer MMA8452Q (I2C), Battery Monitor MAX17048 (I2C), GNSS Module SAM-M8 (UART), GSM/GPRS Module SIM800C (UART), Debug Console (UART), Flash Memory W25Q16JVZPIQ (SPI), and RGB led. The second one, which we have called "irrigation system," has the following main components: MCU PIC32MX370F512H, 2 Relay, 4 Inputs, RGB Led, Flash Memory S25FL216K0PMFI040 (SPI), and GSM/GPRS Module SIM800C (UART).

With these hardware platforms, we propose to our students the development of two different projects. With the "irrigation system" IoT device, and as suggested

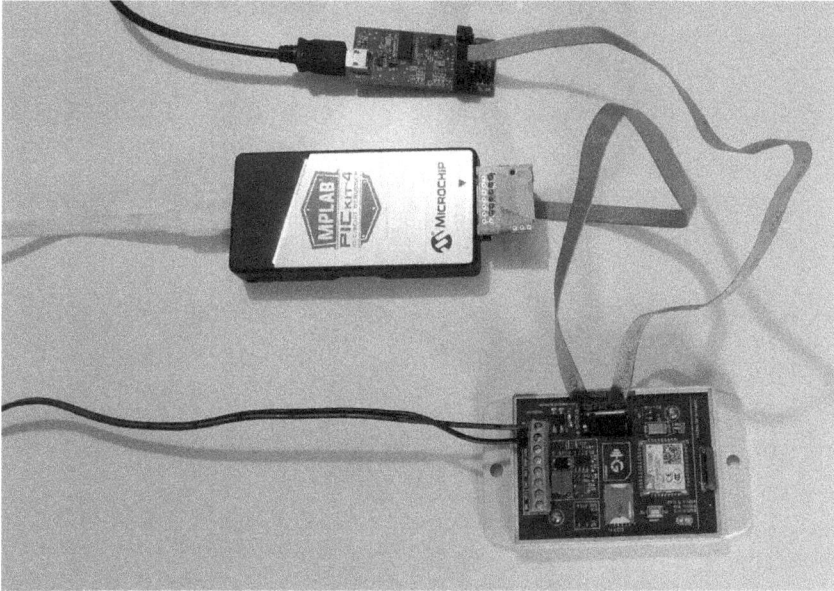

FIGURE 14.5 Photograph of the available "irrigation system" IoT device (please refer to the main text for details).

by its name, we propose to our students to develop an irrigation system, with two waterlines. In each waterline, there is a valve, connected to a relay. The system has a water reservoir, with a water level detector, and two soil moisture sensors (one for each line), all connected to the device's inputs. It is proposed that the students develop the programming of this IoT device to control the irrigation according to the water content in the soil, read from the humidity sensors. The humidity levels must be sent by SMS or MQTT/GPRS to a server. When there is no water in the reservoir, an alert message must also be sent. The settings of the IoT device (humidity levels, irrigation time, etc.) can be changed/reprogrammed (and recorded in the flash memory) through the use of SMS or MQTT/GPRS. The irrigation process can be manually controlled (started and stopped), issuing SMS or MQTT/GPRS start and stop irrigation commands. Figure 14.5 presents a photo of this "irrigation system" IoT device.

With the "GPS tracker" IoT device, we propose the development of a tracker. This device can be attached, for example, to a dog or a cat and the idea is that our students provide a location service to the owner of dogs or cats. It should be possible to choose the starting date and time at which the device starts gathering information. It should also be possible to define an operating daily window, e.g., from 9 am until 8 pm. The interval between acquisitions should also be configurable with a resolution of 1 minute. Thus, depending on the particular situation, the device can be switched on immediately after being "attached," or a date and time can be set. A very important feature should be the ability to change the operation of the device during its normal operation through a configuration message (SMS or MQTT/GPRS). The parameters

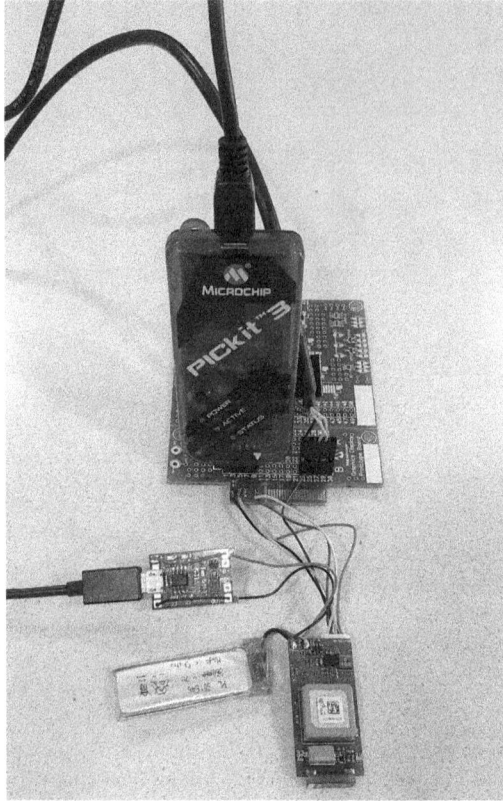

FIGURE 14.6 Photograph of the available "GPS tracker" IoT device (please refer to the main text for details).

that can be changed are the daily window for location acquisitions; time between each acquisition (in case it is found that the acquisition time is not appropriate to the actual situation); allow advance in time—for example, with a 10-minute advance the device will sample at 11:10 instead of 11:00. The location coordinates should be sent by SMS or MQTT/GPRS to a server. The owner of dogs or cats can access this server to see where his/her pet is currently located in real time. Figure 14.6 presents a photo of this "GPS tracker" IoT device.

It should be noticed that, with both IoT devices ("GPS tracker" and "irrigation system"), the students choose the hardware components that he/she want to program, with the other components being automatically disabled or programmed with the default programs.

We have tested this tool with two PhD students, from the PhD degree in Electrical and Computer Engineering of the University of Trás-os-Montes e Alto Douro, Portugal, and two MSc students, from the MSc degree in Electrical and Computer Engineering of the same university. Their opinion is unanimous: this tool facilitates the learning. In informal chats, they used expressions like the following ones to express what they feel about the tool:

- "the tool is very easy to use";
- "(it) put me in touch with new tools like Git and Sourcetree";
- "there was a specific situation where I was not being able to revert an error in the programming and it was very easy to recover the previous version, where there were no errors (I think I lost the fear of making mistakes and of not being able to easily recover from those errors)";
- "I was able to better manage my time, because I was able to use the device at night";
- "It was great to be able to share my device with the teacher and instantly solve the problem (a badly initialized variable)";
- "Not having to wait for a colleague of mine to hand over the device and then having to lift it physically saved me a lot of time and made it easier to organize my schedule";
- "I even asked my colleague for help in real time";
- "When I took this type of device home, I was always afraid of damaging something: either the broken wires or the soldering of some component ... in addition to the time it took to assemble everything on my workbench...";
- "I feel that I can really focus more on the development of the firmware itself—it seems that a level of abstraction has been added...";
- "I am more focused on programming and less on problems with wires, power supplies, etc.";
- "This is amazing! I wish it could be used on other courses."

One of the PhD students is already working in this field in an international enterprise and he let us know that they are using this approach on a daily basis to develop their projects. He stated that this tool is particularly useful in situations where remote and real-time debugging are required, i.e., in situations where debugging has to be performed right where the device is operating (e.g., in failure or malfunctioning situations, which were not predicted during the development phase, and now they have to be solved in-loco).

Following some of the suggestions provided by the students and our own experience, in the near future, we want to implement other functionalities, in particular those related to the server and client applications.

14.6 CONCLUSIONS

We have presented a tool to help in the development of the "smartness" component of sensor devices. This tool can be used in the teaching and learning of the necessary skills. The set of presented tools helps in the establishment of an "online lab," which has particular usefulness in the current pandemic context.

From the teaching/learning perspective, the following advantages can be highlighted, as explained above: one device can be shared by many students, increasing the devices' availability, widening of the devices' utilization period, reducing costs, and increasing the number of different types of architectures available.

One of the students is working at an international enterprise and he confirmed that they are using the approach presented here to develop new devices.

In the future, we intend to integrate all the functionalities provided by the different programs in our model within a single application, for example, integrate into our application the functionalities offered by Sourcetree and Git.

REFERENCES

1. Oppitz, M., Tomsu, P.: Internet of things. In: Inventing the Cloud Century. pp. 435–469. Springer (2018).
2. Burhan, M., Rehman, R.A., Khan, B., Kim, B.S.: IoT elements, layered architectures and security issues: A comprehensive survey. Sensors 18(9) (Sep 2018). https://doi.org/10.3390/s18092796
3. Casini, M., Garulli, A., Giannitrapani, A., Vicino, A.: A remote lab for experiments with a team of mobile robots. Sensors 14(9), 16486–16507 (Sep 2014). https://doi.org/10.3390/s140916486
4. Stark, B., Li, Z., Smith, B., Chen, Y.: Take-home mechatronics control labs: A low-cost personal solution and educational assessment. In: Proceedings of the ASME Design Engineering Technical Conferences, vol. 4, pp. 1–9 (2013). https://doi.org/10.1115/DETC2013-12735
5. Sujatha, C., Jayalaxmi, G.N., Suvarna, G.K.: An innovative approach carried out in data structures and algorithms lab. In: Proceedings of the 2012 IEEE International Conference on Engineering Education: Innovative Practices and Future Trends (AICERA). pp. 1–4 (2013). doi: 10.1109/AICERA.2012.6306734.
6. Paunicka, J., Mendel, B., Corman, D.: The OCP – An open middleware solution for embedded systems. In: Proceedings of the 2001 American Control Conference. (Cat. No.01CH37148), 2001, pp. 3445–3450 vol.5, doi: 10.1109/ACC.2001.946163.
7. Sylla, A.N., Louvel, M., Rutten, E., Delaval, G.: Modular and Hierarchical Discrete Control for Applications and Middleware Deployment in IoT and Smart Buildings. In: Proceedings of the 2018 IEEE Conference on Control Technology and Applications (CCTA), 2018, pp. 1472–1479, doi: 10.1109/CCTA.2018.8511406.
8. Lenhardt, J., Kleimann, A.: Distance learning laboratory course on embedded systems. In: Proceedings of the 2018 12th European Workshop on Microelectronics Education (EWME), 2018, pp. 89–93, doi: 10.1109/EWME.2018.8629469.
9. Peter, S., Momtaz, F., Givargis, T.: From the browser to the remote physical lab: Programming Cyber-physical Systems. In: Proceedings of the 2015 IEEE Frontiers in Education Conference (FIE), 2015, pp. 1–7, doi: 10.1109/FIE.2015.7344228.
10. Tychuk, R.B., Petrovych, S.D.: Creation of "smart" cabinet of physics in a technical college. Information Technologies and Learning Tools 66(4), 78–92 (2018). https://doi.org/10.33407/itlt.v66i4.2133
11. Guo, B., Zhang, D., Imai, M.: Enabling user-oriented management for ubiquitous computing: The meta-design approach. Computer Networks 54(16, SI), 2840–2855 (Nov 15 2010). https://doi.org/10.1016/j.comnet.2010.07.016.
12. Paganelli, F., Turchi, S., Giuli, D.: A web of things framework for RESTful applications and its experimentation in a smart city. IEEE Systems Journal 10(4), 1412–1423 (Dec 2016). ttps://doi.org/10.1109/JSYST.2014.2354835

13. Rice, J.A., Mechitov, K.A., Sim, S.H., Spencer, Jr., B.F., Agha, G.A.: Enabling framework for structural health monitoring using smart sensors. Structural Control & Health Monitoring 18(5), 574–587 (Aug 2011). https://doi.org/10.1002/stc.386

14. Goumopoulos, C., Kameas, A.: Ambient ecologies in smart homes. Computer Journal 52(8), 922–937 (Nov 2009). https://doi.org/10.1093/comjnl/bxn042

15. Bayart, M.: LARII: Development tool for smart sensors and actuators. In: In: Proceedings of IFAC, Volume 34, Issue 29, Pages 70–75, 2001. doi.org/10.1016/S1474-6670(17)32795-7.

16. Banach, M., Talaska, T., Dalecki, J., Dlugosz, R.: New technologies for smart cities – High-resolution air pollution maps based on intelligent sensors. In: Concurrency and Computation-Practice & Experience 32(13, SI), (Jul 10, 2020). https://doi.org/10.1002/cpe.5179

17. Bin Zikria, Y., Yu, H., Afzal, M.K., Rehmani, M.H., Hahm, O.: Internet of Things (IoT): Operating system, applications and protocols design, and validation techniques. Future Generation Computer Systems – The International Journal of Escience 88, 699–706 (Nov 2018).

18. El-Abd, M.: Balancing low-level vs. high-level programming knowledge in an undergraduate microprocessors course. In: Proceedings of the 2018 IEEE Global Engineering Education Conference (EDUCON), 2018, pp. 268–275, doi: 10.1109/EDUCON.2018.8363239.

19. Mange, D.: Teaching firmware as a bridge between hardware and software. IEEE Transactions on Education 36(1), 152–157 (Feb 1993). https://doi.org/10.1109/13.204836

20. Myers, B.A.: Human-centered methods for improving API usability. In: Proceedings of the 2017 IEEE/ACM 1st International Workshop on API Usage and Evolution (WAPI), 2017, pp. 2–2, doi: 10.1109/WAPI.2017.2.

21. Halimi, W., Salzmann, C., Jamkojian, H., Gillet, D.: Enabling the automatic generation of user interfaces for remote laboratories. In: Auer, M.E. and Zutin, D.G. (eds.) Online Engineering & Internet of Things. Lecture Notes in Networks and Systems, Vol. 22. pp. 778–793. Springer, Dordrecht (2018). https://doi.org/10.1007/978-3-319-64352-673.

22. Raikar, M.M., Desai, P., Vijayalakshmi, M., Narayankar, P.: Upsurge of IoT (Internet of Things) in engineering education: A case study. In: Proceedings of th 2018 International Conference on Advances in Computing, Communications and Informatics (ICACCI), 2018, pp. 191–197, doi: 10.1109/ICACCI.2018.8554546.

23. Fortino, G., Russo, W., Savaglio, C., Shen, W., Zhou, M.: Agent-oriented cooperative smart objects: From IoT system design to implementation. IEEE Transactions on Systems Man Cybernetics-Systems 48(11), 1939–1956 (Nov 2018). https://doi.org/10.1109/TSMC.2017.2780618

24. McPherson, R., Irvine, J.: Secure decentralised deployment of LoRaWAN sensors. IEEE Sensors Journal 21(1), 725–732 (Jan 1, 2021). https://doi.org/10.1109/JSEN.2020.3013117

25. HiveMQ MQTT broker. https://www.hivemq.com/, accessed: 2021-01-29.

APPENDIX

Listing 1. An excerpt from the server main program to get a list of all the Microchip programmers that are connected to the computer ports ("ipecmd.exe -T").

```
...
String report;
List tools;
...
```

```
String cmd = "ipecmd.exe -T";
int errorcode = execCmdSync(cmd);
if(!errorcode) {
        //process the data in the 'report' string var
        //'tools' var will have the list of available tools
        tools = ListConnectedTools();
}
...
```

Listing 2. Implementation example of the function to execute the commands by the operating system.

```
private static int execCmdSync(String cmd)
        throws java.io.IOException, InterruptedException
{
        int ret = 0;
        System.out.println ("exec cmd: " + cmd);
        report = "";
        Runtime rt = Runtime.getRuntime();
        Process proc = rt.exec(cmd);
        BufferedReader stdInput = new BufferedReader(
                new InputStreamReader(proc.getInputStream()));
        BufferedReader stdError = new BufferedReader(
                new InputStreamReader(proc.getErrorStream()));
        StringBuffer stdOut = new StringBuffer();
        StringBuffer errOut = new StringBuffer();

        //Read the output from the command:
        String s = null;
        while ((s = stdInput.readLine()) != null) {
                if(!s.isEmpty()){
                        s=s+"\r\n";
                        stdOut.append(s);
                }
        }
        //Read any errors from the attempted command:
        while ((s = stdError.readLine()) != null) {
                errOut.append(s);
        }
        ret = proc.exitValue();
        report = stdOut.toString();
        System.out.println("Exit value = " + ret);
        return ret;
}
```

Listing 3. Implementation of the parsing function to obtain the list of the available tools connected to the server.

```
public List ListConnectedTools() throws IOException,
        InterruptedException
{
```

```
        List tools = new List();
        Scanner scanner = new Scanner(report);
        while (scanner.hasNextLine()) {
                String line = scanner.nextLine();
                if(line.contains("S.No:")){
                    line = line.substring (line.indexOf(" ")+2);
                    System.out.println(line);
                    tools.add(line);
                }
        }
        scanner.close();
        return tools;
}
```

Listing 4. An example of a JSON object sent by the server to the client application at the end of the initialization process.

```
{
        "func": "status",
        "status": {
                "serial": {
                "baud": 10,
                "port": 0,
                "dbits": 3,
                "parity": 0,
                "sbits": 0,
                "ports": [
                        "COM3",
                        "COM1",
                        "COM7",
                        "COM11",
                        "COM13"
                        ]
                },
                "prog": {
                "tools": [
                        "PICkit 4 S.No: BUR204472856"
                        ]
                }
        }
}
```

Part V

Sustainable Approaches
Towards Artificial Intelligence

15 Sustainable and Smart Regions
Examples and Case Studies

Anil Kumar Gupta
Centre for Development of Advanced
Computing (C-DAC),
Bengaluru, India

Pranjal Chinchwade, Harsh Gupta,
Priyanka Kakade, Aditya Gupta,
and Rachna Somkunwar
Department of Computer,
Dr. D.Y. Patil Institute of Technology,
Pune, India

CONTENTS

DOI: 10.1201/9781003268796-20

15.1   INTRODUCTION

A smart city is a digital, intelligent, and wired city with an implication of Information Communication Technology (ICT) to enhance the standards of life for citizens [1]. They emphasize the environment, energy, traffic management and transportation, water, wastewater, education, housing and homeland security, information systems, interconnection with the government, local economy, human resources, and emergency preparedness. A city focuses upon an optimal and sustainable use of all resources while maintaining an appropriate balance among social, environmental, and economic costs. We can compare these cities to living things in which substances, agents, and processes interact to keep them functional. Similarly, a municipality should be nourished, cleaned, cultivated, and nurtured to thrive and grow [2]. The care and attention it receives will give it confidence and protect it from many threats. This concept represents an irrepressible platform for IT-service innovation. It provides us with a vision where providers use other technologies to occupy citizens to construct more productive urban organizations and systems that can enhance our lives [3].

Promoting a city is advantageous regarding the federal benefits as it creates competitiveness, enables the business sector, improves living standards, directs proper utilization of assets, and the like. Intelligent cities line up with unsustainable development goals [4]. They expect to ensure harmlessness and be beneficial in whatever aspect possible. These limitations ought to be directed to potential future ideas. Multiple cities have converted to innovative and intelligent cities. As it enhances, a town becomes capable of identifying its needs, finding ways to satisfy them, being creative, and creating a new intellectual life [5]. There are varied instances of smart and intelligent cities. Thus, we will give some recommendations on how these cities will execute more effectively in the future.

15.1.1 Factors of a Smart City

Cities worldwide are becoming more intelligent. The aim of building a city smart is to provide an effective and quality life [6]. These intelligent cities utilize recent technologies like digital technology, good connectivity, and telecommunication to keep the network connected, developing and solving the city problems. They are implementing various technologies for a greener and safer urban environment, with cleaner air and water, better mobility, efficient deployment of services, which ensures meeting the requirements and needs of the present and future generations [7].

Some core factors for developing a smart city (refer Figure 15.1).

15.1.1.1 Utilities

The utilities are the basics of every city, but also, it is an area with the most considerable environmental impacts [8]. Consequently, the current innovative city initiatives have focused on this area.

- **Water**—Without enough clean water, humans cannot sustain life. It means any gain in efficiency or quality is precious. Since water is a scarce resource, reducing waste is also often a focus area.
- **Energy**—Energy is a vital need for almost every function; these cities lead consumers of power on a large scale. Like water, violence is a scarce resource that we try to reduce with initiatives like dynamic electricity pricing and intelligent solutions for lighting that turn lights on and off as required.
- **Waste**—Elimination of waste is essential that makes a city healthy and pure. Especially when cities have tourists, garbage removal is the key. Cities want to minimize the resources used and look at solutions for optimizing

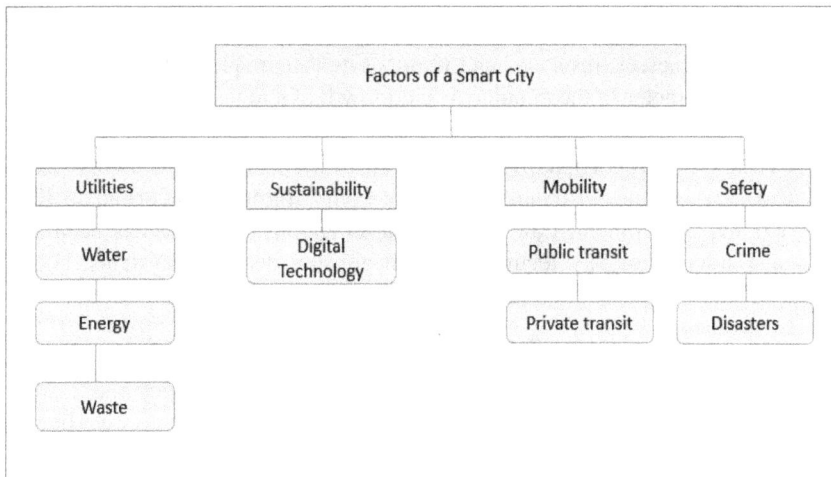

FIGURE 15.1 Characteristics of a smart city.

routes for garbage trucks and automatic notifications when garbage bins are full.

15.1.1.2 Sustainability

Cities are becoming more sustainable. These cities make use of various ICT technologies for their growth and development in social, economic, and environmental areas.

- **Digital technology**—Internet of Things (IoT) is an emerging technology that uses various built-in sensors and software connected to share data. This sharing of data is done on a wireless platform reducing energy usage to manage the traffic and improve all the operations and services.

15.1.2 MOBILITY

Even though more jobs, food, and entertainment are available in a city, they are scarcely close to residential areas [9]. Mobility is the most thriving region of city tech-innovation. Also, there are still two primary areas that form the basis:

- **Public transit**—Despite the innovations, we have considered that public transit is still a fundamental form of mobility in many major cities. Since breakdowns in the transit systems are often a prime cause of delays, maintenance is another application that improves mobility.
- **Private transit**—The personal vehicle has still not died as it is unique in providing the most incredible flexibility and mobility. Consequently, innovative city applications have targeted traffic regulation to avoid congestion. It can be through congestion pricing, taxing, or traffic regulation.

15.1.2.1 Safety

Keeping the businesses, residents, and visitors safe is a unique goal for any city and an essential parameter in the popularity and growth of a town [10].

- **Crime**—Disproportionally, crime increases when cities grow more extensive, and people will make no pause complaining about crime to the regional government. Law enforcement is another part that has been an early adopter of city technology with gunshot detection systems, body cams, and neighborhood policing solutions. But safety is more than combating crime.
- **Disasters**—They happen every day on a small scale. The fire department responds to fires, and ambulances take sick people to clinics and hospitals. Anything that can aid them in arriving earlier will help save lives. While these are smaller everyday or at least frequent occurrences, cities need to be ready for more significant shocks like volcanoes, flooding, droughts, earthquakes, and hurricanes depending upon their geographical location.

15.2 SMART CITIES: SUPPORTING GROWTH AND IMPROVING SERVICES

A smart city is sustainable, innovative, and a city that uses information and communication technologies and other ways to improve efficiency and quality of life. Various operations, services, competitiveness ensure that they meet the requirements of present and future generations along with social, economic, and environmental features [11].

Table 15.1 shows various case studies on how the cities are evolving toward growth and improving their services using new technologies.

15.2.1 ASIAN REGION

15.2.1.1 Singapore

As the city-state in the world, the situation of Singapore is unique. Singapore is now focusing on being the world's first smart nation in the Smart Nation Program developed in 2014, which concentrates on ICT, networking, and data to support practical living, provide various opportunities soon, and support healthier communities.

TABLE 15.1

Case Studies—Evolving Smart Cities

<table>
<tr><th colspan="4">Evolving Smart Cities</th></tr>
<tr><th>Sr. No</th><th>Country/City</th><th>Smart Projects</th><th>Launch Date</th></tr>
<tr><td>1.</td><td>Asian Region: Singapore</td><td>Smart Nation Program:
Innovative services: A web portal to keep track of traffic information (Section 15.2.1.1).
Visionary Environment: Smart waste bins sensor lids that collect information notified to a garbage-collection team (Section 15.2.1).</td><td>2014–2015</td></tr>
<tr><td>2.</td><td>China</td><td>Urbanization management</td><td>2019</td></tr>
<tr><td></td><td>Shenzhen Guangming</td><td>Low-carbon Eco-city</td><td>2018</td></tr>
<tr><td></td><td>Tianjin</td><td>Eco-city: It is an approach to planning new urban lands.</td><td>2011–2013</td></tr>
<tr><td>3.</td><td>European Region: Milan</td><td>E-WASTE project: To strengthen and optimize the entire process connected to recycling electrical waste and equipment (Section 15.3.1).</td><td>2014</td></tr>
<tr><td></td><td>Amsterdam Smart City</td><td>Smart meters and smart grids: Reduction of CO_2 (Section 15.4.2.1).</td><td>2019–2020</td></tr>
<tr><td></td><td></td><td>Smart Building: The BlueGen recovers heat for domestic hot water, brings total efficiency properly.</td><td>2019–2025 (Redevelopment)</td></tr>
<tr><td>4.</td><td>Mediterranean Region</td><td>Smart Energy: To gain reduction of greenhouse gas emissions in the city (Section 15.4).</td><td>2019–2030 (Redevelopment)</td></tr>
</table>

FIGURE 15.2   Your speed sign, Singapore.

15.2.1.2   Case Studies of Smart Services

a. **One motoring**

It is a comprehensive portal serving all drivers and vehicle owners in Singapore. People can track traffic information gathered from surveillance cameras installed on roads and taxis with Global Positioning System (GPS) on this web portal. Through Traffic Smart, drivers can see snapshots of roadways taken at every 5-minute interval. Due to security reasons, real-time moving video or close-up shots are not provided online. It also includes information on current electrical road pricing rates, sections where road works are in progress, traffic images of significant expressways, traffic news, travel time calculator, road maps, and street direction. It is operable on mobile devices. One motoring provides traffic information and offers information and guidance for citizens regarding buying, selling, and maintaining their vehicles.

b. **Speed sign**

Your speed sign is intelligent (Figure 15.2), a live electronic device that displays vehicles' real-time pace and displays alerts drivers who violate the speed limit [12]. It encourages drivers to stay under the limit and thus maintain safety on the roads.

15.2.1.3   A Case Study on Smart Environment

As a small-scale city, Singapore has limited land to gather and store rainwater. Shortage of water is a continuous challenge with growing demand. Singapore had to be inspired to carry out and develop capabilities in the area. The Ministry of Environment Water Resources has statutory boards, the environment agency and public utility board, water agency. It controls water and air pollution, handles waste management of each area, promotes energy efficiency and general hygiene, etc. Singapore power issues a mobile application that allows citizens to see their bills and payment status and gain a superior understanding of utility usage and submit meter readings. It leads consumers to audit their usage to handle their water consumption.

The intelligent waste dustbins are provided as a part of the innovative waste management program (refer to Figure 15.3). The sensing monitors connected to bin lids gathered information on contents and location and notified. It helps the management team optimize their route planning and keep the public spaces clean. The pollution level monitored by National Education Association (NEA) is present on its official website. For example, citizens can easily access information of 24-hour pound-force per square inch (PSI) value, the integrated air quality reporting index online [13].

FIGURE 15.3   Smart Waste Bins concept with sensing monitors.

15.2.1.4   China

The upcoming future of Asian region reports is a vital resource to explore unique and emerging policy opportunities to realize urban sustainability for the region. The report indicates policies and actions from a sustainable development view. The report recognizes the policy pathways for decision-makers and stakeholders to imagine the natural environments in Asian and Pacific cities and offers policy solutions across various cities to gain global development agendas. The solutions address major development obstacles—natural resources, climate change, threats, and inconsistency.

15.2.1.5   Case Study: China

Cities of China are vast and rebuilt daily by their residents, business, immigrants, civil society groups, planners, politicians, commuters, investors, and visitors. It brings their own identity, aspirations and demands a good quality of urbanization. It estimates that China's local population will grow from 528 million in 2005 to 927 million in 2025. Cities with having a population exceeding 1 million are probably to increase from 153 to 226 in that same time. In 2012, the Chinese National Bureau of Statistics also announced that China's rate of urbanization had surpassed 50%.

In 2012, they launched a program China smart city. The operation regards intelligent urban management, with a new development of urbanization techniques and some business needs. The urban management, safe operation, and convenient service of a brilliant city and municipal infrastructure are combined to realize managing cities and serving smartly to explore a new urbanization progression mode. To gain this success, it has identified 192 cities in 2013 and 2014, including 77 provincial capital cities/prefecture-level cities, 75 county-level cities, 34 new districts, and 8 cities. There are 2600 programs that have already been approved. Government funding and bank loans represent 50% of funding. The political equilibrium needs private investment, including international ones. As per the National Intelligent City Pilots Index

System, the general design of intelligent governance and service includes infrastructure renewal and security systems, urban administration, and agricultural planning, achieving proper energy regulation and industrial development. The program, China smart city, retrieved by the government, is a massive and systematic project. For its development, it will need some resources of any type and a process of enormous participation.

15.3 EXAMPLES OF LOW-CARBON ECO-CITIES OF CHINA

15.3.1 CASE STUDY: SHENZHEN GUANGMING

Shenzhen is a special economic zone when China first opened its doors to the nation and introduced its economy, deploys a new experiment for economic transformation and less carbon social, in the rank of its similar role in the late 1980s. The experimental process of ecological development and low carbon of Shenzhen has consisted of both the metropolitan-wide scale and the new city (district) of Guangming New District. It increased land efficiency by the readapt of the recent urban area at the metropolitan level has been regarded as a development priority. During the rehabilitation and reuse of urban land, the promotion of sustainable energy, green establishments, and transport should be a priority of the progress [14]. (Refer to Figure 15.4)

The Shenzhen low-carbon eco-city development indicator system has been related to various plans production at a hierarchical level to mentor economic, social, and environmental expansion. The system indicators have also been further powered to award and assess ecological community in Shenzhen, Guangming, to motivate the engagement of local communities in low-carbon eco-city expansion by following a sound and sustainable lifestyle. During the innovation of this development, the international experiences and national indicator systems were assumed as core

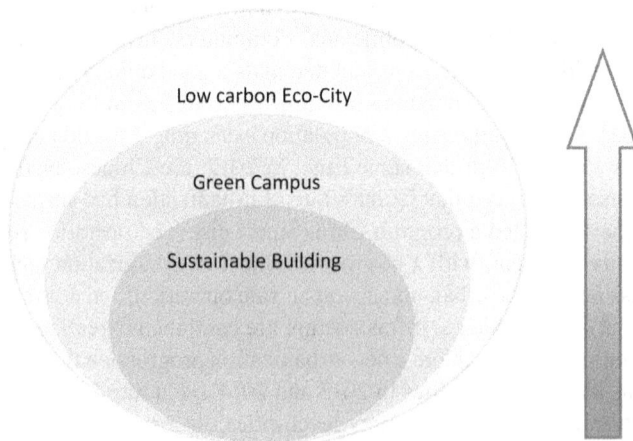

Low carbon Eco-City

Green Campus

Sustainable Building

FIGURE 15.4 Integrating low-carbon eco-city, green campus, sustainable buildings in China.

material for consideration while maintaining local social, economic, and climate conditions. The delivery of a low-carbon policy encourages the Shenzhen authority to adopt an innovative approach in urban planning. The Guangming Ecological District has then been the case for experimental practices. This new area used to be a state-owned farm placed in the northwest of Shenzhen. Its location makes it the gateway linking Guangzhou-Dongguan Shenzhen-Hong Kong. The defined regions for expansion within the new district cover 240, among which 57 are the newly built-up regions [15]. This region uses include industry, local housing, administration and office plots, road, and public utilities.

15.3.2 SINO-SINGAPORE TIANJIN PROJECT

Following the previous cooperation between China-Singapore on "Suzhou Industrial Park" in China, signed an agreement for the Sino-Singapore eco-city project. The deal raised it to an inter-administration partnership. The prime minister's entailed support has enabled Sino-Singapore Tianjin Eco-City (SSTEC) to gain momentum from political commitment while growing from Singapore's extensive experience and knowledge in combined urban planning and water resource management. The plan for Sino-Singapore was completed and accepted in 2010, and the related control programs have been finalized. Phase I of the project's development work has begun and is expected to be completed between 2010 and 2015. It aims to gain this vision by taking an approach to planning new urban lands in an environmentally sustainable manner. According to the program, SSTEC promotes combining land use, transport, and balancing housing supply.

Solar Thermal Energy Centre (STEC) promotes the "use of renewable energy and recycles of resources through unique technologies and eco-friendly policies across various sections", including water, transport, land and energy, among others. The Chinese authority is responsible for the overall development and coordination of the project. Security and Stability Advisory Committee (SSAC) was set up, especially for the project is responsible for government functions. The Chinese side takes responsibility for constructing necessary infrastructure, transportation networks, and land acquisition. The SSTEC organizational structure is a problematic partnership that includes multiple sector participants from both countries, ruled by the SSTEC Agreement Framework, correspondence among all groups, and the Socio-Economic Caste Census (SECC) Regulation.

15.4 EUROPEAN CITY: MILAN

The concept of the "smart city" has been entirely fashionable in the field of upcoming years. Its primary aim seems to be on the ICT infrastructure. However, much research has also been carried out on human capital/education, relational and social capital, and environmental interest as essential urban growth drivers. The European Union (EU) has devised a strategy for gaining urban growth in a "smart" sense for metropolitan areas. The EU and other institutions think tanks believe in an ICT-driven and wired form of development. The Intelligent Forum generates research on the effects of the ICT revolution, which is available worldwide. The Oslo Manual focuses on innovation in ICT instead and gives a toolkit to identify the indicators, thus shaping a good analysis framework for urban design researchers.

15.4.1 CASE STUDY OF RECENT PROJECTS AND EXPERIMENTS

The Milan city signed 14 projects to kick off a financing line of 94 million Euros from the Ministry of Education, Lombardy Region. The race toward a more excellent sustained, interactive, accessible, and inclusive city doesn't stop. Projects range in various areas:

> An E-WASTE project aims to strengthen and optimize the entire process connected to recycling Electrical waste and equipment to recover valuable material through a pilot process based on medium-sized and small companies. Another focus is to strengthen collaborative actions among different sectors in the recycling sector. In Italian regions, a critical mass is given to each area according to their ability to compare and share experiences. The project follows the Milan City Council's institutional objectives to increase collecting and waste- recycling, its optimization and regulation while minimizing inappropriate work related to waste, especially considering the category of waste treated. It has some potential effects on the economies, provides new job opportunities, and supports local-based companies in crisis. The Playful project focuses on harnessing information technologies to provide advanced tools for integrating and developing children's communication skills in pre-school age. The focus is to give a child and his family some customized tools to learn more and, if necessary, to follow specific therapies and allow families to interact with the context (school) and among them.
>
> The simulator project (Modular Integrated System for the risk management) aims to develop a Decision Support System based on an ICT mechanism designed for the territory's protection and safety measures. It is a system based on new technologies and techniques for prevention, forecasting, monitoring, and real-time management of risks due to anthropogenic (due to technological accidents, chemicals/industrial, and roadway accidents) or natural causes.
>
> The recent project, School-Sustainable Campuses as Urban Open-Lab Areas, has the primary goal of testing an advanced school system that can integrate in an intelligent and coordinated way. Various features based on the intelligent grid issue depend upon renewables-based generation systems from a thermal/electrical point. The following are some goals: To research, fix and test innovative photovoltaic structures to combine thermal systems and electric storage for buildings within the Campus, on households, and public places; to study and develop innovative charging strategies for electric vehicles; to monitor and manage throughout smart gadgets, located in subordinate stations, the network resources by synchronizing the generation of energy by innovative PV systems, and in general the distributed batch and the power needed by the loads; to determine the positive impacts on the NZEB network "Nearly Zero Energy Buildings."

15.4.2 NORTH EUROPEAN CITY: AMSTERDAM

Various North European cities are engaged in sustainability, an economy with low environmental impact, high living standards, and urban spaces' livability. The current European scaling of intelligent cities like Amsterdam is evidence of cities forerunners and entered in the rankings done by the research centers. After extensive research on the ranking systems and tools to measure city smartness globally, the device has been defined. It is an advantage for many innovative city initiatives in the United States, Argentina, and Iceland. As stated by the scaling, from the ten most intelligent cities, the first three belong to north Europe, Copenhagen, Stockholm, and Amsterdam.

15.4.2.1 Smart City: Amsterdam

The initiative taken for the city has proved that we have to integrate smart meters and grids, bringing profound changes. The first intervention in July 2009, the Central Street of town, thanks to the local entrepreneur's cooperation, developed into a commercial street that contained a reduction of CO_2 releases by 55% in contrast with recorded data.

The project provided the following:

- Recycled material used to create bus stops;
- Solar panels generate LED-based lighting rooftops;
- Solar generators power the collection of waste, compactor-based bins used;

The smart city project's continued success is the definition of a master plan that is fundamental for the development and individual measures.

15.4.2.2 Smart Mobility: Clean Air Action Plan

Reducing the traffic and electrifying the speed of vehicles will make the air in the city cleaner and extend the life expectancy of the average resident of Amsterdam. Noise pollution and climate will also have positive effects on it. Electric vehicles cause less noise pollution at locations where traffic is slow-moving, which improves the quality and the city's inhabitants. Suppose the energy required for electric transport is sustainably generated; implementing this action plan will reduce CO_2 emissions by 9% in 2030. That represents a substantial contribution in achieving Amsterdam's climate ambitions in the road map for a climate-neutral Amsterdam. The city cannot survive without traffic. It is essential for supplying shops and businesses in the town, for example. But it does not remain the same. The municipality promotes traffic and transport that has minimum effects on space and health. And it's not alone in that. Residents, visitors, and businesses, which use their feet, bicycles, or public transportation for 70% of their trips, are usually a good choice.

Some vehicles drive more room for green, playing, recreation, and healthy traffic. That goal specifies less space for parking spaces of cars, but there will still be space for the traffic flows, such as deliveries to shops. Smarter designing of the traffic gives a simple way to transfer from the car to public transportation, bicycle or e-scooter, on the region's margin. For example, the visitors will be motivated to travel differently, offering them a clean and realistic alternative such as e-Kart Hubs and Mobility as a Service.

Cleaner traffic refers to the ambition of reducing pollution. The number of districts where only emission-free vehicles are permitted to enter will gradually expand outward from the center. These are generally passenger cars, trucks, vans, and scooters that use batteries or a fusion of hydrogen and fuel cells in practice. Important milestones are planned in 2020, 2022, and 2025, and emission-free mobility will yield tremendous benefits for air and CO_2 emissions. This Clean Air Action Plan paves for the final step: clean traffic. The schedule for a low-traffic Amsterdam and the intelligent mobility program forms a coherent strategy for fewer, more elegant, and more pristine vehicles in Amsterdam.

15.4.2.3 Smart Building

In Amsterdam, there is 38% of CO_2 emissions associated with building sectors. If there is no redevelopment of the noteworthy buildings, the authorities indicate that this amount of % would rise to 40% by 2027. It then demands some proper measures focused on the historical-town area that experiments a new mechanism, BlueGen, refined by some fuel cells of ceramic (German art factory). The system setup produces maximum electricity required for the building's needs on-site. For private warm water, this retrieved heat is utilized and brings the total efficiency above 80%. As this section's electrical efficiency is more significant than other mechanisms until now, the municipality is now planning to extend the utilization of this system to an increasing number of buildings.

Here, the first action taken was a mediation on Income Tax Office (ITO) Tower in the area of Amsterdam. The structure is sustainable with regard to the utilization of the new tracking technologies and various control systems. The office is 38,000 m², put into completing analysis of energy utilization and CO_2 emissions. Thus, through a management system (mini-grid), a grid of sensors holding up a building automation system, some functions are implemented: lighting control, cooling and heating regulation, and safety; these actions result in low fuel utilization.

15.5 THE MEDITERRANEAN REGION

This city has some examples from the Mediterranean area. Since 2014, those that come under the podium of the Intelligent City Index scale in Italy are analyzed here. The index measures all Italian provincial capitals' smartness level through various indicators for different characteristics to issue a piece of equipment with calculating, understanding, and comparing multiple local conditions with a proper methodology. The essential attribute taken to scale is the ratio of intelligence in every city with various characteristics that make a smart city. The implication rank is technologically high because it aims at digital services and information technology use. Italian cities provide a unique context: ancient heritages and architectural buildings. There are some more examples that, trading with various local contexts, compute smooth, noninvasive, and decisive actions that lead to high-efficiency targets. Italy has a vast legacy: the historical centers. The real dare hereafter is to realize the new technologies that can help get the best from these cities and mechanisms through the web and digital sensors.

15.6 TOP TEN SMART CITIES RANKING IN THE WORLD FOR 2021

Table 15.2 represents the top ten leading smart cities across the world [16]. These cities have been selected in the top rankings based on their overall development and integration with IoT and various digital technologies that have upgraded the city into a smarter one [17].

TABLE 15.2
Top Ten Smart Cities Ranking in the World for 2021

Country/City	Index Rank Based on Core Factors of Smart City			
	Sustainability	Environment	Health	Mobility
Singapore	1	2	4	3
Amsterdam	2	4	4	3
London	3	2	3	2
Tokyo	4	6	8	5
Hong Kong	5	3	3	4
Milan	6	8	5	2
Copenhagen	7	5	4	8
Dubai	8	7	6	3
Berlin	9	2	3	7
Vienna	10	8	6	4

15.6.1   RANK 1: SINGAPORE

It is considered the topmost ranked smart city and one of the world's safest places (as per the index of 2019). The Smart Nation Program developed in 2014, which focuses on ICT and networking, provides various opportunities soon and supports healthier communities. It aims to mobility by using some advanced technologies, providing people to live more active and advanced lifestyles.

15.6.2   RANK 2: AMSTERDAM

The next city in the ranking index is Amsterdam focusing upon the reduction in traffic and launching electric vehicles causing a clean and pollution-free environment in the city.

15.6.3   RANK 3: LONDON

Being a world-class city for finance, industries, and technologies is also a leading city that handles the data to manage its transport, social, and economical systems, which makes it a smarter one.

15.6.4   RANK 4: TOKYO

Tokyo is considered a center for global finance and economy. They have deployed many projects based on advanced technology, energy efficiency, and security solutions [18]. Many project proposals have been done in 2020, out of which a few have been approved and are under development.

15.6.5   RANK 5: HONG KONG

Hong Kong focuses on technology and innovation. It plays an important role in smart city development. Hong Kong publishes a "Smart City Blueprint" in 2020, which aims for a smart city [19], smart people, and the environment. It can also draw an idea in near future to build a more livable and sustainable city.

15.6.6 RANK 6: MILAN

Focusing upon new technologies such as IoT is improving the urban mobility, smart environment, inclusion and smart citizenship.

15.6.7 RANK 7: COPENHAGEN

Copenhagen for its "connection" is awarded the World's Smart City Award. It is widely known for its data collection and analysis for creating a greener city, providing a quality life for its citizen as well.

15.6.8 RANK 8: DUBAI

Dubai has launched many projects and initiatives in recent years by making use of various IoT technologies and sensors [20]. The recent projects launched, i.e. various apps for smart devices, paperless stamps, etc., make the city smart and look forward to advanced and smart living.

15.6.9 RANK 9: BERLIN

Berlin is one of the smart cities, focusing upon the expansion of the metropolitan regions and increase in resource efficiency, climate neutrality by 2040.

15.6.10 RANK 10: VIENNA

Vienna is also known as a smart city for its development in various sectors such as safety, mobility, digitalization, and environment.

Thus, smart cities provide us with various solutions through ICT and innovation for improving a living, reducing traffic and pollution, and making a city greener. Thus, by making the required funds available to the city, building a smart city will continue in the future as well.

ABBREVIATIONS

ICT	Information Communication Technology
GPS	Global Positioning System
IoT	Internet of Things
NEA	National Education Association
PSI	pound-force per square inch
EU	European Union
ITO	Income Tax Office
SSTEC	Sino-Singapore Tianjin Eco-City
SSAC	Security and Stability Advisory Committee
SECC	Socio-Economic Caste Census

REFERENCES

1. Eleonora Sanseverino, Raffaella Sanseverino, Valentina Vaccaro, Ina Macaione, and Enrico Anello, November 2017. Smart Case Studies by Eleonora Sanseverino, Raffaella Sanseverino, Valentina Vaccaro, Ina Macaione and Enrico Anello.
2. Zhou, Nan, Williams, Christopher. An International Review of Eco-Cities, Indicators, and Case Studies LBL China Energy Group, 2013-03-01.
3. https://www.pwc.com/ph/en/consulting/assets/smart-cities-in-southeast-asia-report.pdf
4. Nerantzia Tzortzi, Joanna Sophocleous. The Green Wall: Sustainable Tool in Mediterranean Cities – The Case Sstudy of Limassol, Cyprus, -2018.
5. Lee, Sang Keon; Kwon, Heeseo Rain; Cho, HeeAh; Kim, Jongbok; Lee, Donju, International Case Studies of Smart Cities: Singapore, Republic of Singapore, 2016, doi: http://dx.doi.org/10.18235/0000409.
6. http://indiaenvironmentportal.org.in/content/424320/case-studies-in-improving-urban-air-quality/ Service Ddelivery and Supporting Growth. http://indiaenvironmentportal.org.in/content/424320/case-studies-in-improving-urban-air-quality/.
7. http://niua.org/c-cube/blog/content/clean-air-action-plans-cities-and-their-corresponding-challenges.
8. Igor Calzanda, SMART CITY Barcelona The Catalan Quest to Improve Future Urban Living.-Published: 2018.
9. https://www.google.com/www.mdpi.com/Characteristics-of-Smart-Sustainable-City-Development.
10. Anders Lisdorf, Demystifying Smart City: A New Perspective on How Cities Leverage the Potential of New Information Technologies, December 2019.
11. Sanjeev Banzal, Smart city- – Regulatory Issues and Policies, 2021.
12. http://www.atstraffic.com.sg/Radar-Speed-Sign.
13. Michael Reiner Kamm, Michael Gau, Johannes Schneider, Jan vom Brocke. Smart Waste Management Processes: A Case Study about Smart Device Implementation. 07 April 2020. Michael Reiner Kamm University of Liechtenstein-Michael Gau, Johannes Schneider, Jan vom Brocke University of Liechtenstein. Published: 07April 2020.
14. https://urbachina.hypotheses.org/10082.
15. www.springer/low carbon eco-city/green-campus/green-buildings.
16. Vaccaro, V., Riva Sanseverino, R., Riva Sanseverino, E., Macaione, I., & Anello, E. (2017). Smart cities: Case studies. In Smart Cities Atlas - Western and Eastern Intelligent Communities (pp. 47–140). (SPRINGER TRACTS IN CIVIL ENGINEERING).
17. www.asme.org/topics-resources/top-10-growing-smart-cities.
18. www.metro.tokyo.lg.jp/governor/speeches/2018/0221.
19. www.smartcity.gov.hk/modules/custom/custom_global_js_css/assets/files/ HKSmartCityBlueprint.Hong   Kong   2030/Public-private-sector/insights/smart-city-development.
20. u.ae//about-the-uae/digital-uae/smart-dubai.

16 Effective Machine Communication Using Quantum Techniques Provides Improvement in Performance and Privacy through IoT Application

Devendar Rao and Ramkumar Jayaraman
Department of Computing Technologies,
SRM Institute of Science and Technology,
Chennai, India

Laura Pozueco
Department of Computer Science,
Escuela Politécnica de Ingeniería,
University of Oviedo,
Gijón, Spain

CONTENTS

DOI: 10.1201/9781003268796-21

16.1   INTRODUCTION

Every individual in 21st century depends on technology for all activities that he/she perform in every day's life like business transaction or communication among every household device like Television, Thermostat, and CCTV camera through online [1]. The world population is around 7.8 billion (2021 population survey), out of which 3.8 billion users are accessing smart gadgets like smart phone connected with digital household utilities (approx 7–8 devices per user) for every day functionality [2, 3]. Let's frame the architecture for smart world through Internet of Things (IoT) application that performs effective machine interactivity like communication among edge server, Wi-Fi router, and cloud machine with a real-time scenario. In Figure 16.1, the parent wants to nurture a child at home while he/she performs official duty in working environment through smart equipment available at both ends. An unknown

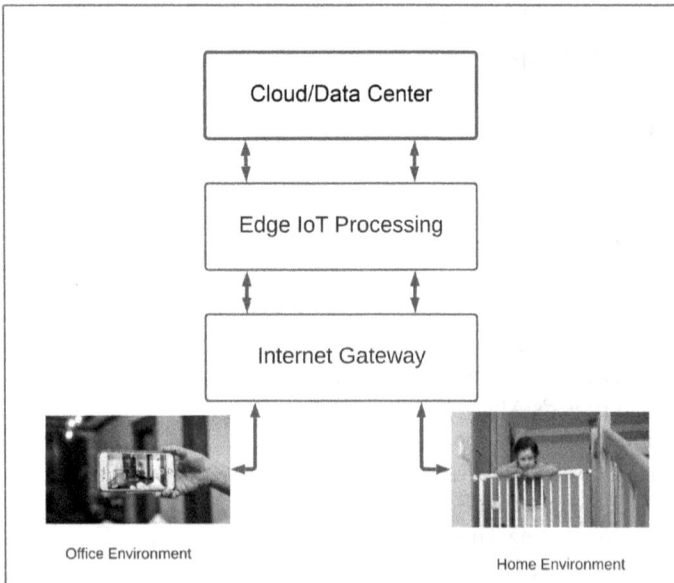

FIGURE 16.1   Machine interactions in IoT application.

guest or stranger has arrived at the doorstep; the parent wants to authenticate the person identity by CCTV camera at doorsteps and decide whether to allow the guest or restrict entry inside the home. All the household devices comprise sensors and actuator, which are connected in the first stage of IoT architecture. Internet gateway like Wi-Fi and wired LAN collect the information from sensors and compress it to the optimal size, further processing to edge-enabled system. In the third stage, simultaneously capturing and transferring data to the remote location takes place [4]. The last stage stores the data for deep analysis and performs predictive analysis. Latency or failure in interaction between the machines leads drastic effects, since the child is monitored by parent who relies on IoT machine and its features; therefore, computation speed of each device should be reliable.

All devices/machines are connected to the network to form an Internet of Devices or IoT for effective communication between them but lead to a decrease in computational speed in classical methods.

16.1.1 THE END OF MOORE'S LAW

As stated by Gordon Moore, the number of transistors on Integrated Circuits (ICs) will double every two years, with miniature IC's providing faster processing speed [5]. The validity of Moore's law is on the verge of extinct in modern era; therefore, faster device computability is not achievable by minimizing the ICs [6]. Further increasing the transistor count in integrated chip is not possible; therefore, computation capability of IoT devices will not increase, which leads to downfall of an IoT model. The computational mechanism of classical model is in sequential pattern in terms of computing; therefore, machine learning or prediction algorithm will take substantial amount of time for execution which leads to delay. A cutting-edge technology is required for improving a computation complexity so that the IoT model could efficiently be implemented and can be served in real-time use without any latency.

16.1.2 COMPUTATIONAL SPEEDUP THROUGH QUANTUM

The evolution of quantum has drastically changed the way of computing as compared with classical methods in computing. The classical computer will hold classical bits "0" and "1" and quantum will hold qubit for computing. Let us take a decimal number from $\{0,1,2,3\}$ that can be represented by two classical bits $\{00,01,10,11\}$, respectively. Conversion of binary number '00' to '11', it takes 2 sequential computation on first and second classical bit. In classical computing for "n" bit, 2^n possible combinations are present, but at a single point of time, only one bit can be changed. In quantum computing, a qubit can be present in pure state. By applying Hadamard gate, it moves to superposition state, where it can be either in state "0" or "1" simultaneously; therefore, the computational speedup is increased as compared to classical way [7]. For a two-qubit system, the qubit can be represented in superposition state; the four sequences $\{00,01,10,11\}$ can be represented parallel based on the probability as shown Figure 16.2.

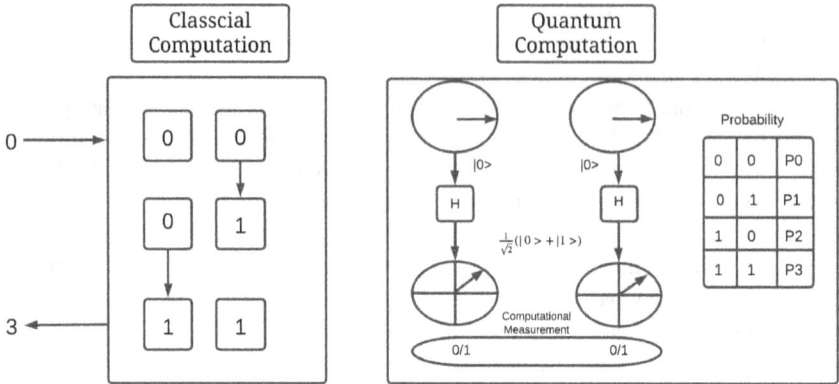

FIGURE 16.2   Classical vs. quantum computations.

16.1.3 PITFALL IN CURRENT SECURITY SYSTEMS

The current generation security entirely depends on public-key cryptography based on Rivest, Shamir, Adleman (RSA) algorithm, which performs the reverse operation of prime number multiplication. The RSA key sizes vary from 512 bits to 2048 bits; to crack the 2048 bits RSA algorithm by classical model, it takes nearly 300 trillion years to crack the encryption. Let us see the analogy of parent monitoring the child when a stranger came near the door; CCTV camera indicates the parent about the stranger with visual image. Now the parent wants to authenticate that person before releasing the door lock; if the unknown stranger uses the classical approach, it will take exponential time to crack the encryption. In the year 1994 [8], Peter Shor came up with an algorithm that uses classical periodic finding and quantum Fourier transform (QFT) for faster computation, then encryption can be cracked in polynomial time as shown in Figure 16.3.

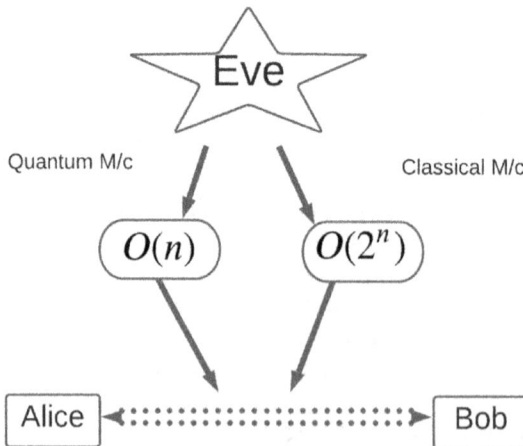

FIGURE 16.3   Quantum advantage in hacking.

Another classical method for secure communication is considered to be one-time pad (OTP), which is also known as Vernam Cipher [9], where the key is generated and communicated between two parties and will be considered to be safe under certain conditions.

1. Key length == message length.
2. Key can be used only one time; repeated usage leads to leaking of information.
3. Exchange of key between two parties takes places confidentially.

The current era uses Diffie-Hellman key exchange for secretly transmitting the key between parties considered being safe, due to the involvement of quantum hacking; therefore, new methods for key exchange protocols are needed with the help of quantum.

16.2 QUANTUM PARALLELISM

16.2.1 REVERSIBLE COMPUTING

The classical computing upon which the circuits are constructed is irreversible in implementation except for NOT operation. In classical computing, if we provide two inputs in AND gate and obtain a single output state, from output state, we cannot obtain which input states are used in processing the output. In classical physic, when an operation is performed on two inputs and obtains one output, there is a loss of heat energy while performing operation and have insufficient heat energy to do reverse operation. In quantum computing, the quantum circuits are reversible in an implementation having an ancillary bit, which hold the current computed value and the original input variable are retained from losses [10]. The quantum circuit is reversible while performing any unitary transformation on them, but once measured it cannot be reverted back.

Let us see with an example as shown in Figure 16.4; let's take an Exclusively-OR (XOR) gate and perform it in both classical and quantum circuit. The XOR gate

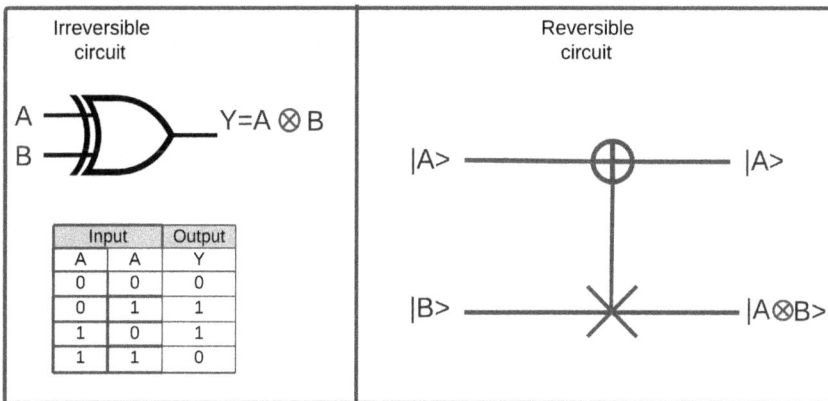

FIGURE 16.4 Reversible circuits.

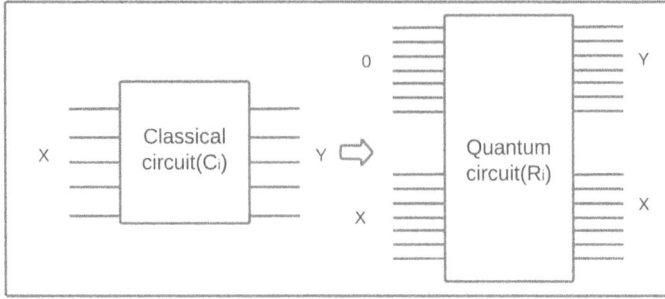

FIGURE 16.5   Generalizing classical circuits.

will obtain output bit "1" when both the input bits are different and output bit "0" when both bits are same. In classical circuit, with one output, two input bits cannot be generated, but in quantum circuit, we retain the original one input along with the XOR calculated value, helping to reverse the operation. For construction of efficient quantum circuits, many unitary operation gates are involved in circuits along with ancillary bits. The maximum number of ancillary bits involved during a construction of "n" bit input bits will be "n − 1" ancillary bits. The general construction of effectively converting a classical circuit to quantum circuits [11] is shown in Figure 16.5.

16.2.2   ORACLE OR BLACK BOX COMPUTATION

In classical computing, the complexity depends on the number of times the given statement is executed based on complexity notation like big (O), theta (θ), and omega (Ω). The complexity in quantum can depend on the concept of black box or oracle where the input and output are known, but the processing steps inside it are unknown. The quantum complexity can be defined by a number of queries called to an oracle, which solves the problem. Let us see with Deutsch's problem to determine whether a given function f(x) is constant or balanced. In classical computing to solve this problem [12], two steps are taken to prove the solution, but in quantum, it takes a single call to be executed inside oracle as shown in Figure 16.6. A function that takes

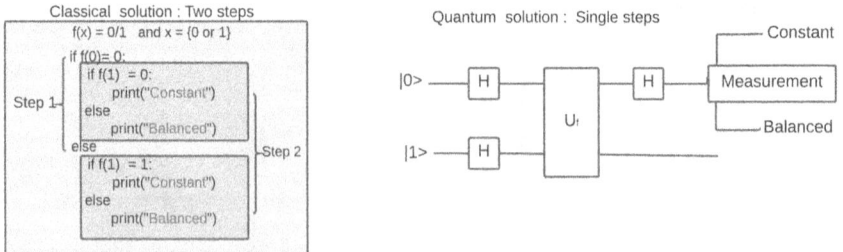

FIGURE 16.6   Deutsch algorithm.

any input and provides a solution as all "0"s or all "1"s is called a constant function and the function that provides half of the solution as "0" and the other half as "1" can be called a balanced function.

In quantum part, the inputs are taken to superposition state by applying Hadamard gate, then the oracle function applies the unitary function $U_f = |x>|y\oplus f(x)>$ will reflect the change in first qubit $|x>$. Once again applying the Hadamard gate to reverse from superposition state to the original state and measurement of solution will be constant if obtain $|0>$ as output or else $|1>$ will give balanced function. Let's see with an example in Figure 16.6, the input $|0>$ refers to $|x>$ and $|1>$ refers to $|y>$, which will be changed to $|+>$ and $|->$ after applying Hadamard gate where $|+> = (|0>+|1>)/\sqrt{2}$ and $|-> = (|0>-|1>)/\sqrt{2}$, respectively, are in a superposition state. During the oracle function, only a single query is required to perform the unitary function $U_f\left(\frac{1}{\sqrt{2}}\Sigma_{x=0}^1 |x>|->\right) = \frac{1}{\sqrt{2}}\Sigma_{x=0}^1 (-1)^{f(x)}|x>|->$. If $f(x)$ takes value 0, it means $(-1)^{f(x)}$ will give values 1; therefore, no change in output, and it remains $|+>|->$ as constant function otherwise the output will be $|->|->$ as an balanced function after applying Hadamard gate on $|x>$ qubits.

16.2.3 GROVER'S ALGORITHM

The best classical algorithm to search in unsorted element is linear search with complexity $O(N)$, but quantum Grover's algorithm [13, 14] improves the searching technique in $O(\sqrt{N})$. In classical part, searching items should compare with all items in list to obtain the solution found or not found. In Grover's search works, a number of times oracle (Grover's iteration) was called to obtain the presence of element in list or not.

Let us see with an example: given list = $\{0,\ldots\ldots, N-1\} \rightarrow \{0 \text{ (not found)}, 1 \text{ (found)}\}$ such that list(x) = 1 for exactly one x. The Grover's algorithms starts with superposition state. First, apply the uniform transformation on states, flip the sign of searching element (x*) and find the inversion mean (average) of all elements. The amplitude of the searched element will increase and the inversion means the average will be freezing after some iteration; therefore, the number of Grover's iteration leads to the complexity of Grover's search as shown in Figure 16.7.

16.3 QUANTUM PROPERTIES FOR COMMUNICATION

In earlier discussion, it is shown how the current era suffers security issue in sending secure information between two parties. Now let us see how the properties of quantum help us to provide unconditional security in message transfer among parties when an unknown adversary is present in that communication.

16.3.1 HEISENBERG'S PRINCIPLES

In quantum mechanics, Heisenberg's principles state that when an object is traveling in space, the velocity and position cannot be measured exactly at the same time; this approach is applied in transferring quantum information in communication

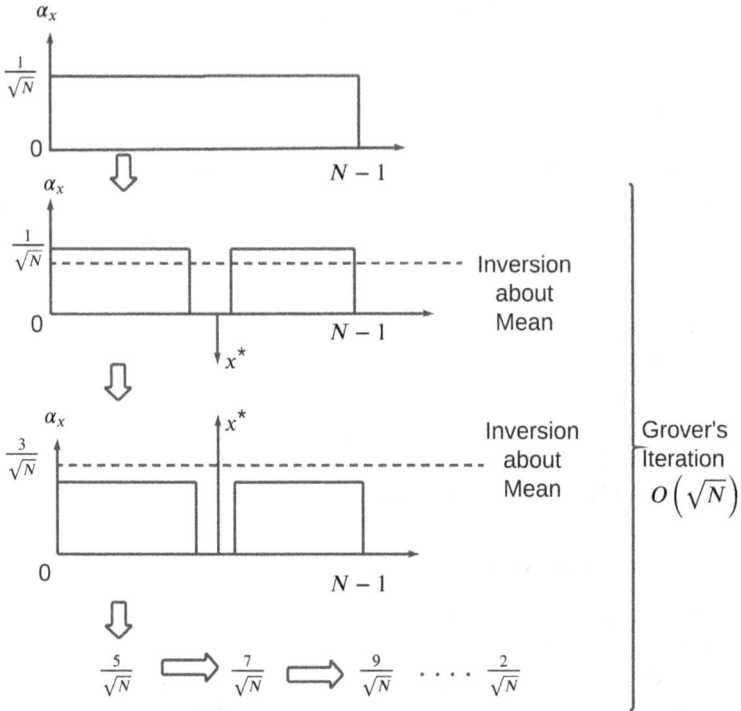

FIGURE 16.7   Grover's iteration.

channel. The Heisenberg uncertainty principle is given in an equation form, where h is Planck's constant (16.1).

$$\Delta v \Delta p \geq \frac{h}{4\pi} \text{ where h is Planck's constant } [15] \tag{16.1}$$

An unknown qubit is transferred between two parties when an adversary is trying to measure it; the qubit changes to a pure state and it is impossible to create the original unknown state sent by the sender. Once an adversary involved in communication, there will a mismatch between the sender and receiver qubit. The mismatch will lead to identification of involvement of unknown parties in communication lead to abort the current transmission by both message communicating parties.

Let us see with an example. Alice generates an unknown qubit state and sends to Bob. The unknown adversary Eavesdropping (Named as EVE) involved in communication tampers the quantum channels by measuring the qubit sent by Alice and prepares a new state and sends it to Bob for measurement. Once the communication in quantum channel is completed, both Alice and Bob will communicate to authenticate the public channel to discover the trustiness during communication as shown in Figure 16.8. When Alice and Bob discover that the sent bit is not similar to the receiver bit, they abort the current transmission.

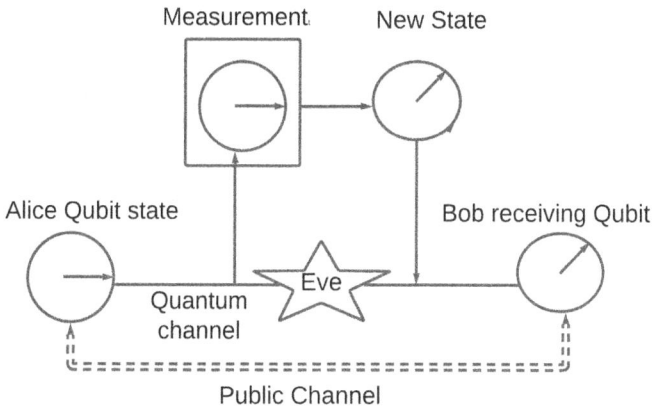

FIGURE 16.8 Adversary involvement.

16.3.2 ENTANGLEMENT PROPERTIES

Another important property of quantum is entanglement, when two qubits are maximally entangled or mixed; measurement of one qubit is correlated with another measurement. Let's see with an example, |0> will the first qubit and |1>will be the next qubit. Apply a Hadamard (H) and identity transformation on first $\frac{1}{\sqrt{2}}$ (|0>+|1>) and second qubit(I), respectively, which leads to unentangled state $\frac{1}{\sqrt{2}}$ (|01>+|11>). Now apply CNOT transformation on first as control qubit and another as target bit, which leads to an entangled state $\frac{1}{\sqrt{2}}$ (|01>+|11>) as shown in Figure 16.9.

There are four possible entangled state from {|0>|0>,|0>|1>,|1>|0>,|1>|1>} from a Bell state transformation [16] as shown in Table 16.1. The unknown qubit can be transferred with the help of two classical bits that will lead to teleportation [17] and transfer of two classical bits using an entangled state that will lead to superdense coding [18] in quantum communication.

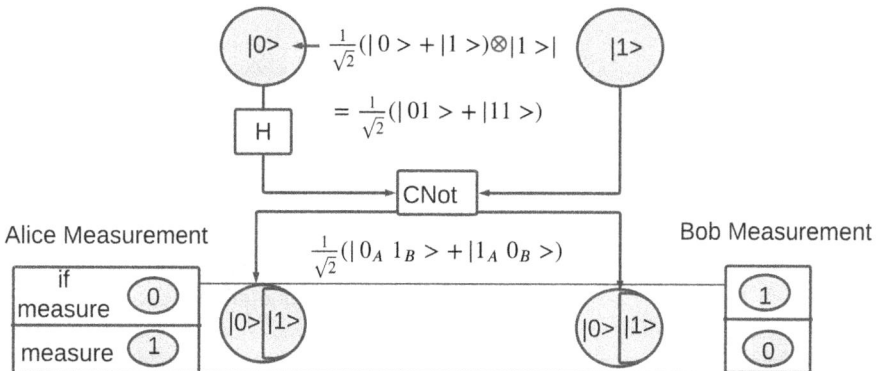

FIGURE 16.9 Entanglement principles.

TABLE 16.1

Bell States' Representation

$|00\rangle \rightarrow H(|0\rangle) \otimes |0\rangle \rightarrow CNOT(\frac{1}{\sqrt{2}} (|00\rangle + |10\rangle)) \rightarrow \frac{1}{\sqrt{2}} (|00\rangle + |11\rangle) \leftrightarrow |\varphi^+\rangle$

$|01\rangle \rightarrow H(|0\rangle) \otimes |1\rangle \rightarrow CNOT(\frac{1}{\sqrt{2}} (|01\rangle + |11\rangle)) \rightarrow \frac{1}{\sqrt{2}} (|01\rangle + |10\rangle) \leftrightarrow |\psi^+\rangle$

$|10\rangle \rightarrow H(|1\rangle) \otimes |0\rangle \rightarrow CNOT(\frac{1}{\sqrt{2}} (|00\rangle - |10\rangle)) \rightarrow \frac{1}{\sqrt{2}} (|00\rangle - |11\rangle) \leftrightarrow |\varphi^-\rangle$

$|11\rangle \rightarrow H(|1\rangle) \otimes |1\rangle \rightarrow CNOT(\frac{1}{\sqrt{2}} (|01\rangle - |11\rangle)) \rightarrow \frac{1}{\sqrt{2}} (|01\rangle - |10\rangle) \leftrightarrow |\psi^-\rangle$

16.3.3 NO CODING THEOREM

The major disadvantage of classical cryptography is the bit can be duplicated, which means ciphertext can be copied and sent to the receiver. Later the cipher can be decrypted to read the information hidden in that encrypted text. In quantum computing, no cloning theorem [19] states that it's impossible to clone or duplicate the unknown quantum states. Let us consider an unknown state $|c\rangle = \frac{1}{\sqrt{2}} (|a\rangle + |b\rangle)$; adversary wants to apply an auxiliary qubit ($|0\rangle$) to clone the unknown bit by applying a unitary transformation U.

$$U(|c\rangle |0\rangle) = \frac{1}{\sqrt{2}} ((U|a\rangle |0\rangle) + U(|b\rangle |0\rangle)) \qquad (16.2)$$

Suppose the unitary transformation on the pure state provides U|a>|0>=|a>|a> and U|b>|0>=|b>|b> in Equation (16.2), then the unknown state can be written as

$$U(|c\rangle |0\rangle) = \frac{1}{\sqrt{2}} ((U|a\rangle |a\rangle) + U(|b\rangle |b\rangle)) \qquad (16.3)$$

The actual cloning of U(|c>|0>)=|c>|c>=$\frac{1}{\sqrt{2}}$ (|a>+|b>)$\otimes \frac{1}{\sqrt{2}}$ (|a>+|b>)

$$= \frac{1}{2} (|a\rangle |a\rangle + |a\rangle |b\rangle + |b\rangle |a\rangle + |b\rangle |b\rangle) \qquad (16.4)$$

Equation (16.4) is not equal to Equation (16.3); therefore, no unitary operation can duplicate or clone the unknown quantum states.

16.4 QUANTUM SECURE DIRECT COMMUNICATION

The two-way protocol is a bidirectional protocol where both the parties can communicate the message securely without key. Quantum secure direct communication (QSDC) means it doesn't provide direct communication from Alice and Bob; it means Alice will initiate a communication, Bob will encode a message, and Alice will decode it in measurement. The two different approaches are explained in deterministic manner and secure a two-way communication; ping-pong protocol with entanglement [20] and without entanglement [21].

16.4.1 PING-PONG PROTOCOL (WITH ENTANGLEMENT)

The general procedure of ping-pong protocol has two steps: ping (Alice initiates communication process) and pong (Bob responds to Alice by either encoding or measuring). In ping methods, Alice generates the Einstein-Podolsky-Rosen (EPR) pair and keeps one bit with her and sends the other bit to Bob, and in pong method, Bob chooses either control mode or message mode based on some probability and performs needed operation based on selected mode. Let us see this with an example. Alice generates an EPR pair in $\psi+ = \frac{1}{\sqrt{2}}(0_H 1_T + 1_H 0_T)$ entangled state and keeps home qubit (H) in his lab and travel qubit (T), which he sends to Bob. Bob performs two operations with probability (c) in control mode and probability (c − 1) in message mode. In message mode, Bob will apply unitary transformation (I) on the travel bit to encode classical bit "0." Bob will perform transformation Z transformation to encode bit "1". During identity transformation, there are no changes in travel bit, but in Z transformation, the bit "1" changed to "−1." Alice will start measuring once the travel bit is received from Bob and maintains an entangled state, if Alice finds the same state which she prepared, Alice will come to know Bob bit is "0," for different state Bob bit "1."

The message mode is insecure because an involvement of adversary in communication cannot be detected and the entire message will be leaked to a third person. In control mode, instead of performing a gate operation, the measurement of travel bit takes place in computational basis, which provides bit 0 or 1 with equal probability. Once the measurement is done and the results will be transferred to Alice in public authenticated channels. If bits of Alice and Bob are the same, then there will be an involvement of adversary in communication; therefore, abort the current communication. If bits of Alice and Bob are different, then the message transferred consider being valid in peer to peer communication. The involvement of adversary cannot be detected in message mode; therefore requiring a control mode to check the presence of adversary in communication as shown in Figure 16.10. Enhancement

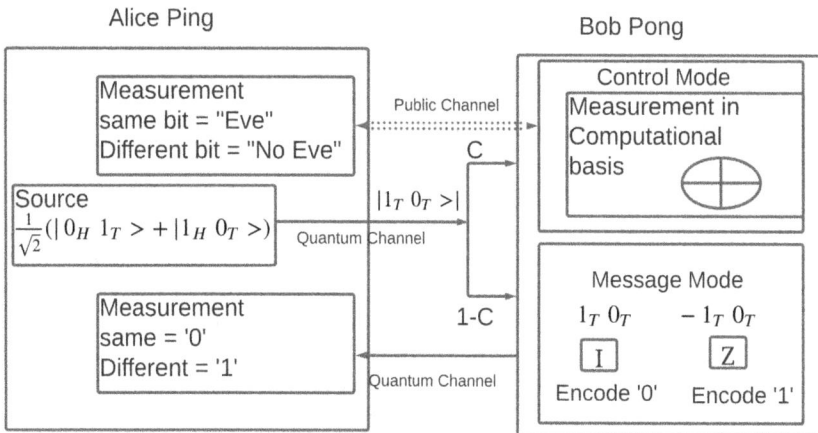

FIGURE 16.10 Adversary in communication.

Alice Ping   Bob Pong

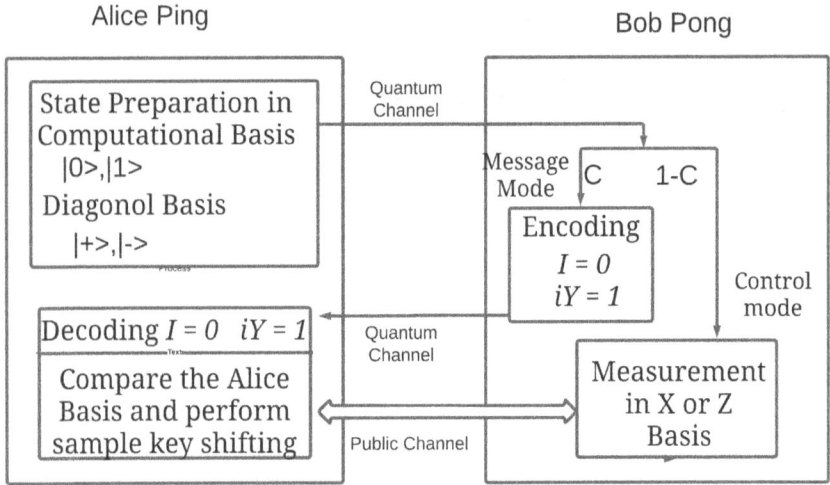

FIGURE 16.11   Unitary transformation.

and modification of ping-pong protocol using superdense coding mechanism by Bostrom-Felbinger [22] and high dimensional Greenberger-Horne-Zeilinger (GHZ) states used instead two qubit states along with entanglement swapping [23–25].

16.4.2   Ping-Pong Protocol (without Entanglement)

The orthogonal state is used in providing a two-way protocol communications instead of entanglement for secure direct communication. It was developed by Lucamarini and Stefano Mancini [21]. The general structure of the protocol as similar to ping-pong methods but instead of entangled state directly used unentangled states for a two-way communication as shown in Figure 16.11.

Let's see with an example, Alice generates the qubit in computational or diagonal basis and sends the qubit to Bob. Bob has to perform operation based on probability "c" for message mode or probability "c − 1" for control mode. In message mode, Bob encodes bit "0" by applying unitary transformation of identity and encodes bit "1" by applying the unitary transformation of X (swap) followed by Z gate as shown in Table 16.2

In control mode, Bob will conduct a measurement in computational or diagonal bases and compare the bases with Bob in classical mode to check the presence of an intruder in communication. Alice's operation depends on Bob's probability of choosing mode; if Bob chooses the message mode, Alice will decode the information sent by Bob. If Bob chooses control mode, Alice will have to compare the basis with Bob and discard the mismatch basis. In matched basis, some samples are taken to check the involvement of adversary in communication. Recent developments in QSDC without entanglement along with OTP [26], block transmission, and order rearrangement techniques use to carry single d-level systems as message carrier in communication process [27].Secret frequency-dependent phase modulation

TABLE 16.2
Bob Encoding Techniques using Identity and Y Gate

Encode Bit (0)		Encode Bit	
I|0>	I|0>	iY|0>	−|1>
I|1>	I|1>	iY|1>	|0>
I|+>	I|+>	iY|+>	|−>
I|−>	I|−>	iY|−>	−|+>

techniques are applied to achieve deterministic secure communication feasible in current environment [28]. Further advance techniques in QSDC lead to quantum dialogue or bidirectional QSDC where Alice and Bob use two QSDC protocol for authenticated communication [29, 30].

16.5 QUANTUM KEY DISTRIBUTION

The Vernam Cipher or OTP is the safest modern technique security protocol in the current era used in the treatment owing to the evolution of quantum computing. The difficulties lie in using OTP is key distribution between parties; since Diffie-Hellman considers being safe key exchange protocol in classical technology but lack is quantum technology. To upgrade the security features in modern times using modern technology is lacking; therefore, either quantum encryption or post-quantum technology (based on mathematical computation) should be used. The two key distribution protocols BB84 [31] and Ekret91 [32] are discussed below.

16.5.1 BB84 PROTOCOL

The first protocol for key distribution using quantum methods was developed by Charles Bennett and Gilles Brassard in the year 1984 known as BB84 protocol based on Heisenberg principles. The method is based on the polarization concept in quantum mechanics and how quantum behaves when it is transferred in two different basis. Let us see the working model of BB84 [31] protocol; Alice randomly generates "n" classical bit and performs polarization (computational or diagonal basis) using quantum random number generator (QRNG) [33]. QRNG provides truly randomness compared with classical used pseudorandom number generator (which follows some predictive pattern in generating randomness). After polarizing with random bit, the qubit will send to Bob through quantum channel and Bob will use QRNG for generating randomness in selecting basis. The selection of the same basis by Alice and Bob gives 100% probability to yield the same classical bit and for different basis 50% chance to yield the same basis. After the quantum communication, the classical communication takes place; where Alice and Bob will perform error correction to compare and hold the classical bit where selection of the same basis occurred. Key shifting steps are used to check the quantum bit error rate (QBER) [34] whether an intruder is present in communication while comparing some classical bit from both

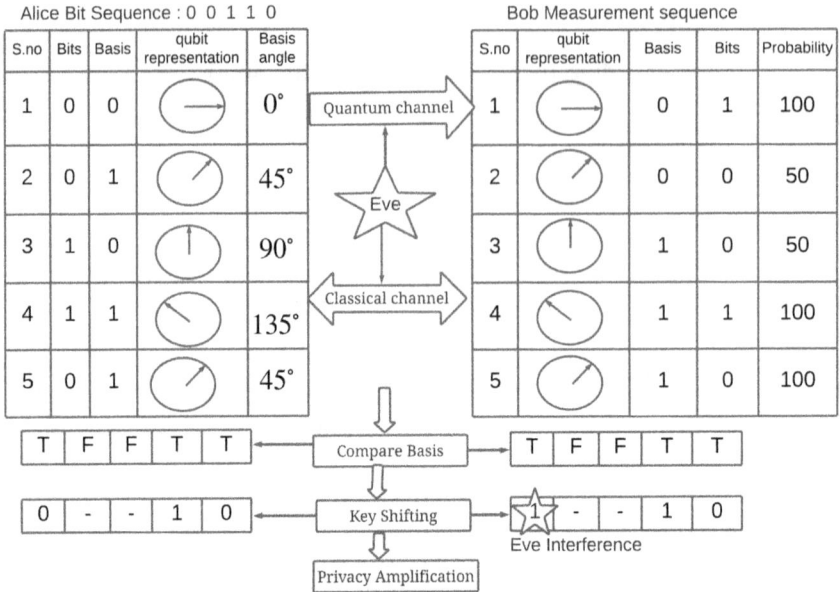

Alice Bit Sequence : 0 0 1 1 0   Bob Measurement sequence

S.no	Bits	Basis	qubit representation	Basis angle
1	0	0		0°
2	0	1		45°
3	1	0		90°
4	1	1		135°
5	0	1		45°

S.no	qubit representation	Basis	Bits	Probability
1		0	1	100
2		0	0	50
3		1	0	50
4		1	1	100
5		1	0	100

Quantum channel

Eve

Classical channel

| T | F | F | T | T |

Compare Basis

| T | F | F | T | T |

| 0 | - | - | 1 | 0 |

Key Shifting

| 1 | - | - | 1 | 0 |

Eve Interference

Privacy Amplification

FIGURE 16.12 BB84 model.

parties. The QBER can occur due to noise in communication channel or by presence of intruder. If the QBER is above the threshold value, then the current communication will be aborted and initiates the next iteration until the raw is generated. Once the raw key is generated, privacy amplifications are performed on raw key to reduce the key information obtained by Eve and makes a finite key that can be perfectly used to secure communication. Assume Alice generates n = 100, n will reduce to 1/2 while performing error correction and another ½ will reduce in key shifting process; therefore, only 1/4 of key is generated as raw key.

In Figure 16.12, Alice generates five classical bits, polarized with photons, and sends to Bob for measurement. In authenticated public channel, Bob sends the basis that he has used for measurement and Alice compares the basis, discards the second and third bits, the basis of which are not matched, and keeps first, fourth, and fifth bits that are matched for key shift process. In shifting process, Alice uses the first bit on the same basis as Bob as shown by circle but the classical bit value differs, then the presence of Eve can be identified. If substantial amounts of bit are found to different, then abort the protocol or else perform with privacy amplification. Instead of using orthogonal bases like computational and diagonal bases, single non-orthogonal bases were used for communication with less overhead [35]. Six state protocols [36, 37] were used with three orthogonal bases (X, Y, Z) for measurement result in one-third of randomness in selection.

The quantum key distribution (QKD) protocol is unconditional and secure in theoretical model but lacks implementation and suffers from attack by quantum itself. The generation of multiphoton leads to a photon splitting attack (or also known as storage attack) as one photon is transferred to Bob and the other taken by an adversary to seek the key in classical communication. Decoy state [38] protocol prevents the adversary not to differentiate the single- and multiphoton for generating keys. As another protocol to prevent the Photon Number splitting (PNS) attack by SARG04, [39] used the same technique as BB84, but instead of comparing the basis, i.e., one of four non-orthogonal states, the probability of detection reduced from 50% to 25%. Various other protocols have been used by the same technique in quantum communication part but attempted different procedure in classical to prevent the reduction of raw keys and generate finite key from communication [40–42].

16.5.2 EKRET91

Another important property of quantum mechanics in secure communication is entanglement. When two qubits are maximally entangled and each qubit is separated by a long distance, measurement of one qubit is directly correlated with the measurement of another qubit. Unlike in BB84 protocol, Alice generates a polarized photon and sends it to Bob, Eve has full access to the polarized photon, but in Ekret91 [32], only the part of entangled state is available for Eve to access information; therefore, it provides reliable security.

The working model of Ekret protocol is depicted in Figure 16.13. External sources generate an "n" EPR pair with a maximum entangled state and send individual pair of qubit to Alice and Bob. The measuring angle of Alice (a1 = 0, a2 = $\pi/8$, a3 = $\pi/4$) and Bob (b1 = 0, b2 = $\pi/8$, b3 = $-\pi/8$) for performing measurement obtains the results with −1 and +1 at both ends. Once quantum transmission has taken place, Alice and Bob publically announce the orientation and form two groups. 1. Same orientation is used for key generation and 2. Different orientation is used to identify the presence of adversary. In the first group, probability of obtaining the same orientation in both parties is 2/9 that is (a1, b1) and (a2, b2). The remaining unsuccessful probability is

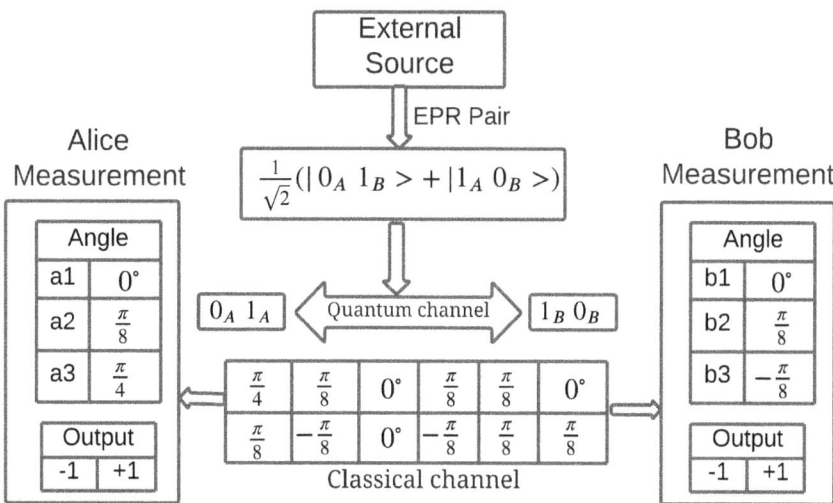

FIGURE 16.13 Ekret model.

used to identify the presence of Eve in communication by forming a correlation coefficient (16.5) in second group.

$$E(a_i, b_j) = P_{00}(a_i, b_j) - P_{01}(a_i, b_j) - P_{10}(a_i, b_j) + P_{11}(a_i, b_j) \qquad (16.5)$$

where P_{00} represents both Bob and Alice obtaining "0" bit similarly P_{01}, P_{10}, P_{11}. Equation (16.5) value is used in CHSH in equality to check the violation of Eve from Equation (16.6).

$$S = E(a_1, b_2) + E(a_1, b_3) - E(a_3, b_1) + E(a_3, b_3) \qquad (16.6)$$

The value of $S = -2\sqrt{2}$ should satisfy the Bell's inequality for secure communication, and if the value of S is in between $\{-2, 2\}$, it will predict that the qubits are not maximally entangled, disturbed, and presence of adversary in communication. Once the communication is free from disturbance, it will generate the key with the same orientation analyzer and use the key for encryption.

The practical implementation on entanglement-based protocol faces the problem for photon splitting attack; various protocols came up with a solution to avoid it. Initially, weak coherent pulses are used to generate photon with decoy states with probability "f" and without decoy states with "$(1 - f)/2$" probability along with timestamp; remove the encoded classical bit attached with decoy states in classical communication while obtaining raw keys [43]. Generate a weak coherent pulse with phase modulation along with time delay between two consecutive pulses in order to see the variation of receiver detector [44].

In recent times, various developments have been made in the field of QKD by considering the problem that persists in real-time usage. In every communication system, communicating channels are considered to be a threat and leak the information but what if a communicating device is at threats. Therefore, device-independent quantum key distribution (DIQKD) is implemented [45], a trusted party will generate a photon (either using Heisenberg principles or entanglement principles) and measurement will be done randomly to check the trustworthiness of source and generate a key. A scenario where the measuring side channels are being attacked by adversary, then measurement device-independent quantum key distribution (MIQKD) takes place by allowing a third party to perform the measurement [46]. In current scenario, everyone using quantum device is impossible; therefore, an SQKD (semi classical quantum key distribution) [47] is used where one user is fully quantum and the other user is partially using quantum capability. QSDC protocols using ping-pong methods are used to generate QKD key [48].

16.6   SECURITY ISSUES IN IoT

16.6.1   Impact of Security in IoT Application

The current era application works 24/7 for an entire year, managing and monitoring the work manually is difficult task, so required an automated process for doing it through IoT. Initial stage in the development of IoT makes things easier for usage but

Stealing customer detail Transactional frauds Financial loss Business	Stealing parent information Treats to patient Life Healthcare
Smart Homes Leakage of Private Information Monitoring of Daily Activites	Industries Misuse of protocols Stealing client bidding information Stealing product design and prototype

IoT Application

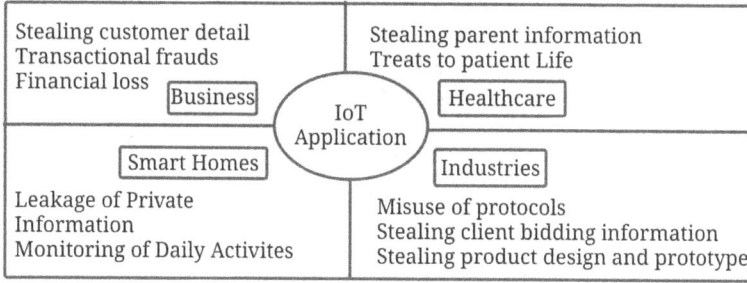

FIGURE 16.14   IoT application.

leads to access by unauthorized parties made big worries in modern times. Let us see some application [49] of IoT in real implementation and how the impact of security while functioning as shown in Figure 16.14.

16.6.2   SECURITY BREACHES IN IoT FRAMEWORK

The word internet integrates the entire communication system as a whole, where billions of devices are connected and communicate with each other without a human intervention. The increasing IoT devices lead to threats and security issues in communication [50]; therefore, various security factors should be taken care of before developing an IoT system in classical model.

1. User should create strong password and change frequently.
2. Frequent updation of IoT Software in IoT devices to prevent from new attacks.
3. Secure communication among IoT devices, gateway, edge node, edge server, and cloud.

The development of quantum technology had made the hacker community use it efficiently as compared with classical model due to Shor's algorithm. In the year 1994, Shor came up with an approach of finding the prime factor of a large number in polynomial time as classical model takes exponential time [51]. The current security entirely depends on RSA model where the complexity depends on the factoring a large number.

16.6.3   MOBILITY OF EDGE NODE IN HOME NETWORK OR TRANSIT NETWORK

All IoT devices communicate to the edge node (mobile phones) via gateway and edge node will provide instruction to devices for their functioning. The edge node will communicate important data to cloud for storage via edge server and perform further analysis of IoT devices functionality and reliability. In Figure 16.15, the hacker can access all the layers in IoT from devices to cloud network and if there is any leak or attack in any one layer leads to information gain.

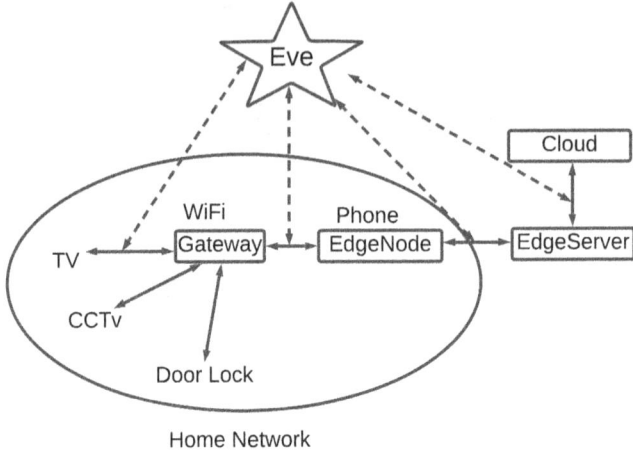

FIGURE 16.15 Edge node in home network.

Consider an IoT scenario where a parent wants to take care of child in real-time environment whether he/she is in home or office network. The edge device in home network is connected to the IoT devices through routers, since every device has a unique IP address. Edge node in home network considers being relatively secure compared to outside network [52], since the parent can physically monitor and control IoT device whenever threat occurs. The edge node can store the threats and attack information to cloud network for further identification of similar threats that occur.

When the edge node is outside home network, the IoT device messages have to travel a long way to hit the edge node. The security issues will increase because the data should transit via gateway and multiple edge servers; therefore, hackers have ample opportunities to hack the information. The parent uses the edge node application interface to check the data obtained from various IoT devices and act based on the current status as shown in Figure 16.16.

16.6.4   QKD-BASED IoT FRAMEWORK

If the edge node is in transit mode or office network, then the probability of adversary attack gets increased. The classical OTP is considered to be a safe protocol for secure communication only for single point of time but the problem arises in key distribution between IoT devices and edge node. QKD provides the secure way for generating and distributing the key between IoT node and edge node [53]. Whenever the edge node wants to communicate, it will inform the corresponding IoT devices through classical channels and IoT devices will generate an unknown qubit using QRNG sent through quantum channels. The qubit will relay it through multi-edge servers and finally finite secure key will be obtained after performing postprocessing and privacy amplification techniques in classical channels as shown in Figure 16.17.

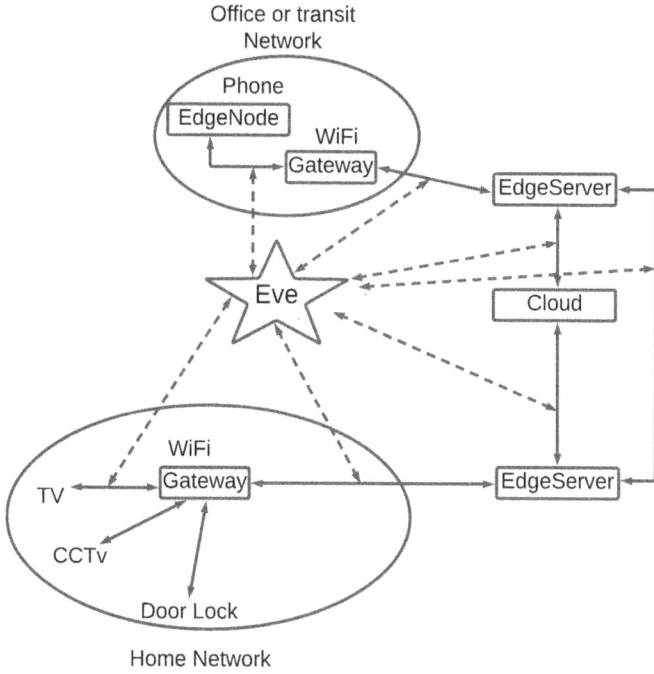

FIGURE 16.16   Edge node in office or transit network.

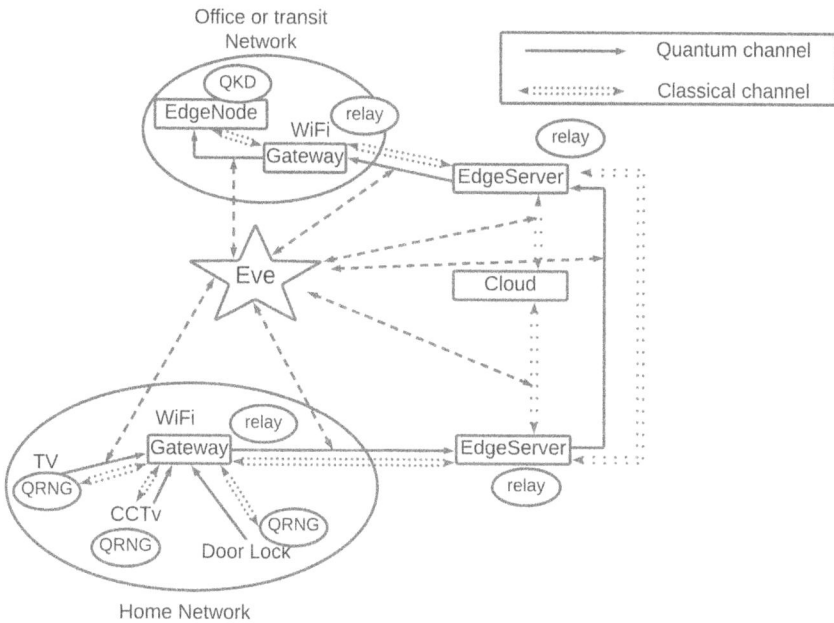

FIGURE 16.17   Quantum-enabled IoT framework.

Once the secure key shared between the node, then sender IoT devices will encrypt the sensor data, communicate in classical channels, and in receiver end, the decryption takes place to obtain the data. For every single communication, a secure key should be distributed.

16.7 CONCLUSION

IoT plays an important role in the development and monitoring of industries, smart home, healthcare, and business process through online channels as society demands in smart world. Rapid increase in IoT machine reduces the efficiency of interaction among machines due to classical computation and leads to insecurity in maintaining the privacy of the users. The development of sustainable cutting-edge technology known as quantum computing is boon as well as bane to the classical world. The quantum computing is boon as they speed up the computation and decrease the complexity of solving hard problem that takes exponential time to solve in classical world. The quantum computing deploys reversible computing to obtain the input from output results to prevent the loss of heat energy as in classical physics. Quantum oracle and QFT are used to increase the computation speed by exhibiting the qubit in superposition state to provide parallelism. The same quantum techniques are bane as the hacker can use speedup computation to crack the encryption algorithm in polynomial time and hack privacy information. Quantum hacking can be prevented by using quantum encryption as quantum behaves in deterministic nature and leads to identification of adversary in communication. Ping-pong and LM05 are the two-way communication protocols (QSDC) that provide an unconditional secure communication without generating a key but suffer from quantum attack. The quantum hacking can be prevented using QKD protocol using Heisenberg and entanglement principles of quantum mechanics as OTP in classical model is considered to be a safe method for communication for single point of usage. The IoT security issues can be overcome by incorporating quantum properties with IoT framework to guarantee secure transmission of information from end-to-end machine communication.

REFERENCES

1. Iera A. Atzori, G. Morabito, 'The Internet of Things: A survey'. *Computer Networks*. 2010; 54(15):2787–2805.
2. Mouzhi Ge, Hind Bangui, Barbora Buhnova, 'Big data for Internet of Things: A survey'. *Future Generation Computer Systems*. 2018; 4:053.
3. Sachin Kumar, Prayag Tiwari, Mikhail Zymbler, 'Internet of Things is a revolutionary approach for future technology enhancement: A review'. *Journal of Big Data*. 2019; 6(1):1–21.
4. Timothy Malche, Priti Maheshwary, 'Internet of Things (IoT) for building smart home system'. *International Conference on I-SMAC*; Palladam, India, Feb. 2017. IEEE: Oct 2017.
5. Gordon E. Moore, 'Cramming more components onto integrated circuits'. *Electronics*. 1965; 38(8):114–117.

6. H. Theis, S. Philip Wo, 'The end of Moore's Law: A new beginning for information technology'. *Computing in Science and Engineering.* 2017; 19(2):41–50.

7. Christopher Havenstein, Damarcus Thomas, Swami Chandrasekaran, 'Comparisons of performance between quantum and classical machine learning'. *SMU Data Science Review.* 2018; 1(4):11.

8. P.W. Shor, 'Polynomial-time algorithms for prime factorization and discrete logarithms on a quantum computer'. *SIAM Journal of Computing.* 1997; 26(5):1484–1509.

9. http://cryptomuseum.com/crypto/otp/index.html.

10. Marius Krumm, Markus P. Mueller, 'Quantum computation is the unique reversible circuit model for which bits are balls'. *npj Quantum Information.* 2019; 5(7): 1–8.

11. M.A. Nielsen, I.L. Chuang, *Quantum computation and quantum information.* Cambridge University Press, United States of America; 2012.

12. David Deutsch, 'Quantum computational networks'. *Proceedings of the Royal Society of London Series A, London,* Sep 1989; 425(1868):73–90.

13. Lov K. Grover, 'Quantum computers can search arbitrarily large databases by a single query'. *Physical Review Letters.* 1997; 79(23):4709–4712.

14. Lov K. Grover, 'A framework for fast quantum mechanical algorithms'. In *Proceedings of STOC.* 1998; pp. 53–62.

15. Donald C. Chang, 'Physical interpretation of the Planck's constant based on the Maxwell theory'. *Chinese Physics B.* 2017; 26(4):040301.

16. John S. Bell, 'On the Einstein-Podolsky-Rosen paradox'. *Physics.* 1964; 1:195–200.

17. Dirk Bouwmeester, Jian-Wei Pan, Klaus Mattle, Manfred Eibl, Harald Weinfurter, Anton Zeilinger, 'Experimental quantum teleportation'. *Nature.* 1997, 390:575.

18. Charles H. Bennett, Stephen J. Wiesner, 'Communication via one- and two-particle operators on Einstein-Podolsky-Rosen states'. *Physical Review Letters.* 1992; 69:2881–2884.

19. W.K. Wootters, W.H. Zurek, 'A single quantum cannot be clone'. *Nature.* 1982; 299(5886): 802–803.

20. K. Boström, Timo Felbinger, 'Deterministic secure direct communication using entanglement'. *Physical Review Letter.* 2002; 89(18):187902.

21. C. Qing-Yu, L. Bai-Wen, 'Deterministic secure communication without using Entanglement'. *China Physic Letters.* 2004; 21(4):601.

22. Q.Y. Cai, B.W. Li, 'Improving the capacity of the boström-felbinger protocol'. *Physical Review A.* 2004; 69(5):054301.

23. T. Gao, F.L. Yan, Z.X. Wang, 'Deterministic secure direct communication using GHZ states and swapping quantum entanglement'. *Journal of Physics A: Mathematical and General.* 2005; 38(25):5761.

24. Chuan Wang, Fu-Guo Deng, Yan-Song Li, Xiao-Shu Liu, Gui Lu Long, 'Quantum secure direct communication with high-dimension quantum superdense coding'. *Physical Review A.* 2005; 71(4):044305.

25. Jian Li, Zeshi Pan, Fengqi Sun, Yanhua Chen, 'Quantum secure direct communication based on dense coding and detecting eavesdropping with four-particle genuine entangled state'. *Entropy.* 2015; 17(10):6743–6752.

26. F.G. Deng, G.L. Long, 'Secure direct communication with a quantum one-time Pad'. *Physical Review A.* 2004; 69(5):052319.

27. Dong Jiang, Yuanyuan Chen, Xuemei Gu, Ling Xie, Lijun Chen, 'Deterministic secure quantum communication using a single d-level system'. *Scientific Reports.* 2017; 7:44934.

28. A.G.A.H. Guerra, F.F.S. Rios, R.V. Ramos, 'Quantum secure direct communication of digital and analog signals using continuum coherent states'. *Quantum Information Process.* 2016; 15(11):4747–4758.

29. B.A. Nguyen, 'Quantum dialogue'. *Physics Letters A.* 2004; 328(1):6–10.

30. H. Wang, Y.Q. Zhang, X.F. Liu, Y.P. Hu, 'Efficient quantum dialogue using entangled states and entanglement swapping without information leakage'. *Quantum Information Process.* 2016; 15(6):2593–2603.

31. C.H. Bennett, G. Brassard, 'Quantum cryptography: Public key distribution and coin tossing'. *Theoretical Computer Science.* 2014; 560(1):7–11.

32. A.K. Ekert, 'Quantum cryptography based on Bell's theorem'. *Physical Review Letter.* 1991; 67(6):661.

33. Mario Stipcevic, Cetin Kaya Koc, 'True random number generators'. *Open Problems in Mathematics and Computational Science.* 2014: 275–315.

34. Agoston Schranz, Eszter Udvary, 'Quantum bit error rate analysis of the polarization based BB84Protocol in the presence of channel errors'. *Proceedings of the 7th International Conference on Photonics, Optics and Laser Technology.* 2019; 1:181–189.

35. C.H. Bennett, 'Quantum cryptography using any two non orthogonal states'. *Physical Review Letters.* 1992; 68:3121–3124.

36. H.B. Pasquinucci, N. Gisin, 'Incoherent and coherent eavesdropping in the six-state protocol of quantum cryptography'. *Physical Review Letter A.* 1999; 59:4238–4248.

37. Dagmar Bru, 'Optimal eavesdropping in quantum cryptography with six states'. *Physical Review Letter.* 1998; 81(14):3018–3021.

38. Hoi-Kwong Lo, Xiongfeng Ma, Kai Chen, 'Decoy state quantum key distribution'. *Physical Review Letters.* 2005; 94:230504.

39. V. Scarani, A. Acin, G. Ribordy, N. Gisin, 'Quantum cryptography protocols robust against photon number splitting attacks for weak laser pulse implementations'. *Physical Review Letters.* 2004; 92(5):057901.

40. M.M. Khan, M. Murphy, A. Beige, 'High error-rate quantum key distribution for long distance communication'. *New Journal of Physics.* 2009; 11(6):63043.

41. E.H. Serna, Quantum Key Distribution from a Random Seed. Quantum Technology Lab, Colombia; 2013.

42. M. Kalra, Ramesh Chandra, 'Design a new protocol and compare with BB84 protocol for quantum key distribution'. *Advances in Intelligent Systems and Computing.* 2019; 2:969–978.

43. Damien Stucki, Nicolas Brunner, Nicolas Gisin, Valerio Scarani, Hugo Zbinden, 'Fast and simple one-way quantum key distribution'. *Applied Physics Letter.* 2005; 87:194108.

44. K. Inoue, E. Waks, Y. Yamamoto, 'Differential-phase-shift quantum key distribution using coherent light'. *Physical Review A.* 2003; 68:22317.

45. Stefano Pironio, Antonio Acin, Nicolas Brunner, Nicolas Gisin, Serge Massar, Valerio Scarani, 'Device-independent security of quantum cryptography against collective attacks'. *New Journal of Physics.* 2009; 11(4):25.

46. Feihu Xu, Marcos Curty, Bing Qi, Hoi-Kwong Lo, 'Measurement-device-independent quantum key distribution'. *IEEE Journal of Selected Topics in Quantum Electronics.* 2015; 21(3):238146.

47. Michel Boyer, Dan Kenigsberg, Tal Mor, 'Quantum key distribution with classical Bob'. *Physical Review Letters.* 2007; 99:140501.

48. Jian Li, Hengji Li, Chaoyang Li, Leilei Li, Yanyan Hou, Xiubo Chen et al., 'Quantum key distribution protocol based on the EPR pairs and its simulation'. *Mobile Networks and Application.* 2019; 295: 288–301.

49. Sajid Habib, Muhammad Ali, Saleem Ullah, 'Security issues in the Internet of Things (IoT): A comprehensive study'. *International Journal of Advanced Computer Science and Applications.* 2017; 8(6).

50. L. Tawalbeh, Fadi Muheidat, Mais Tawalbeh, Muhannad Quwaider, 'IoT privacy and security: Challenges and solutions'. *Applied Science.* 2020; 10(12):4102.

51. Enrique Mart n-Lopez, Anthony Laing, Thomas Lawson, Roberto Alvarez, Xiao-Qi Zhou, Jeremy L. O'Brieny, 'Experimental realization of Shor's quantum factoring algorithm using qubit recycling'. *Nature Photonics*. 2012; 6:773–776.
52. Sufyan Almajali, Haythem Bany Salameh, Moussa Ayyash, Hany Elgala, 'A framework for efficient and secured mobility of IoT devices in mobile edge computing'. *Third International Conference on Fog and Mobile Edge Computing (FMEC)*; Barcelona, Spain. IEEE: 2018.
53. S. Krithika, T. Kesavmurthy, 'Securing IOT network through quantum key distribution'. *International Journal of Innovative Technology and Exploring Engineering (IJITEE)*. 2019; 8(6S4):2278–3075.

Index

For Product Safety Concerns and Information please contact our EU
representative GPSR@taylorandfrancis.com
Taylor & Francis Verlag GmbH, Kaufingerstraße 24, 80331 München, Germany

www.ingramcontent.com/pod-product-compliance
Lightning Source LLC
Chambersburg PA
CBHW060330220326
41598CB00023B/2664